# Springer Series in Optical Sciences   Volume 35
Edited by Theodor Tamir

# Springer Series in Optical Sciences

Volumes 1–41 are listed on the back inside cover

# Keigo Iizuka

# Engineering Optics

Second Edition

With 385 Figures

Springer-Verlag Berlin Heidelberg New York
London Paris Tokyo

Professor KEIGO IIZUKA, Ph.D.

University of Toronto, Department of Electrical Engineering, Toronto, Canada M5S 1A4

Revised translation of the 2nd original Japanese edition:
Keigo Iizuka: *Hikarikogaku*
© by Keigo Iizuka 1983
Originally published by Kyoritsu Shuppan Co., Tokyo (1983)
English translation by Keigo Iizuka

ISBN 3-540-17131-2 2. Auflage Springer-Verlag Berlin Heidelberg New York
ISBN 0-387-17131-2 2nd edition Springer-Verlag New York Berlin Heidelberg

ISBN 3-540-11793-8 1. Auflage Springer-Verlag Berlin Heidelberg New York Tokyo
ISBN 0-387-11793-8 1st edition Springer-Verlag New York Berlin Heidelberg Tokyo

Library of Congress Cataloging-in-Publication Data. Iizuka, Keigo, 1931–. Engineering optics. (Springer series in optical sciences ; v. 35) Translation of: Hikari kógaku. Bibliography: p. Includes index. 1. Optics. I. Title. II. Series. TA1520.I3813 1987 621.36 86-27881

Typesetting, offsetprinting, and bookbinding: Graphischer Betrieb Konrad Triltsch, Würzburg
2153/3150-543210

# Preface to the Second Edition

The first edition of this textbook was published only last year, and now, the publisher has decided to issue a paperback edition. This is intended to make the text more affordable to everyone who would like to broaden their knowledge of modern problems in optics.

The aim of this book is to provide a basic understanding of the important features of the various topics treated. A detailed study of all the subjects comprising the field of engineering optics would fill several volumes. This book could perhaps be likened to a soup: it is easy to swallow, but sooner or later heartier sustenance is needed. It is my hope that this book will stimulate your appetite and prepare you for the banquet that could be yours.

I would like to take this opportunity to thank those readers, especially Mr. Branislav Petrovic, who sent me appreciative letters and helpful comments. These have encouraged me to introduce a few minor changes and improvements in this edition.

Toronto, September 1986 *Keigo Iizuka*

# Preface to the First Edition

"Which area do you think I should go into?" or "Which are the areas that have the brightest future?" are questions that are frequently asked by students trying to decide on a field of specialization. My advice has always been to pick any field that combines two or more disciplines such as Nuclear Physics, Biomedical Engineering, Optoelectronics, or even Engineering Optics. With the ever growing complexity of today's science and technology, many a problem can be tackled only with the cooperative effort of more than one discipline.

Engineering Optics deals with the engineering aspects of optics, and its main emphasis is on applying the knowledge of optics to the solution of engineering problems. This book is intended both for the physics student who wants to apply his knowledge of optics to engineering problems and for the engineering student who wants to acquire the basic principles of optics.

The material in the book was arranged in an order that would progressively increase the student's comprehension of the subject. Basic tools and concepts presented in the earlier chapters are then developed more fully and applied in the later chapters. In many instances, the arrangement of the material differs from the true chronological order.

The following is intended to provide an overview of the organization of the book. In this book, the theory of the Fourier transforms was used whenever possible because it provides a simple and clear explanation for many phenomena in optics. Complicated mathematics have been completely eliminated.

Chapter 1 gives a historical prospective of the field of optics in general. It is amazing that, even though light has always been a source of immense curiosity for ancient peoples, most principles of modern optics had to wait until the late eighteenth century to be conceived, and it was only during the mid-nineteenth century with Maxwell's equations that modern optics was fully brought to birth. The century following that event has been an exciting time of learning and a tremendous growth which we have been witnessing today.

Chapter 2 summarizes the mathematical functions which very often appear in optics and it is intended as a basis for the subsequent chapters.

Chapter 3 develops diffraction theory and proves that the far field diffraction pattern is simply the Fourier transform of the source (or aperture) function. This Fourier-transform relationship is the building block of the entire book (Fourier optics).

Chapter 4 tests the knowledge obtained in Chaps. 2 and 3. A series of practical examples and their solutions are collected in this chapter.

Chapter 5 develops geometrical optics which is the counterpart of Fourier optics appearing in Chap. 3. The power of geometrical optics is convincingly demonstrated when working with inhomogeneous transmission media because, for this type of media, other methods are more complicated. Various practical examples related to fiber optics are presented so that the basic knowledge necessary for fiber optical communication in Chap. 13 is developed.

Chapter 6 deals with the Fourier transformable and image formable properties of a lens using Fourier optics. These properties of a lens are abundantly used in optical signal processing appearing in Chap. 11.

Chapter 7 explains the principle of the Fast Fourier Transform (FFT). In order to construct a versatile system, the merits of both analog and digital processing have to be cleverly amalgamated. Only through this hybrid approach can systems such as computer holography, computer tomography or a hologram matrix radar become possible.

Chapter 8 covers both coherent and white light holography. The fabrication of holograms by computer is also included. While holography is popularly identified with its ability to create three-dimensional images, the usefulness of holography as a measurement technique deserves equal recognition. Thus, holography is used for measuring minute changes and vibrations, as a machining tool, and for profiling the shape of an object.

Descriptions of microwave holography are given in Chap. 12 as a separate chapter. Knowledge about the diffraction field and FFT which are found in Chaps. 3 and 6 are used as the basis for many of the discussions on holography.

Chapter 9 shows a pictorial cook book for fabricating a hologram. Experience in fabricating a hologram could be a memorable initiation for a student who wishes to be a pioneer in the field of engineering optics.

Chapter 10 introduces analysis in the spatial frequency domain. The treatment of optics can be classified into two broad categories: one is the space domain, which has been used up to this chapter, and the other is the spatial frequency domain, which is newly introduced here. These two domains are related by the Fourier-transform relationship. The existence of such dual domains connected by Fourier transforms is also found in electronics and quantum physics. Needless to say, the final results are the same regardless of the choice of the domain of analysis. Examples dealing with the lens in Chap. 6 are used to explain the principle.

Chapter 11 covers optical signal processing of various sorts. Knowledge of diffraction, lenses, FFT and holography, covered in Chaps. 3, 6, 7 and 8, respectively, is used extensively in this chapter. In addition to coherent and incoherent optical processing, Chap. 11 also includes a section on tomography.

Many examples are given in this chapter with the hope that they will stimulate the reader's imagination to develop new techniques.

Chapter 12 is a separate chapter on microwave holography. While Chap. 8 concerns itself primarily with light wave holography, Chap. 12 extends the principles of holography to the microwave region. It should be pointed out that many of the techniques mentioned here are also applicable to acoustic holography.

Chapter 13 describes fiber-optical communication systems which combine the technologies of optics and those of communications. The treatment of the optical fiber is based upon the geometrical-optics point of view presented in Chap. 5. Many of the components developed for fiber-optical communication systems find applications in other areas as well.

Chapter 14 provides the basics necessary to fully understand integrated optics. Many an integrated optics device uses the fact that an electro- or acousto-optic material changes its refractive index according to the external electric field or mechanical strain. The index of refraction of these materials, however, depends upon the direction of the polarization of the light (anisotropic) and the analysis for the anisotropic material is different from that of isotropic material. This chapter deals with the propagation of light in such media.

Chapter 15 deals with integrated optics, which is still such a relatively young field that almost anyone with desire and imagination can contribute.

Mrs. Mary Jean Giliberto played an integral role in proof reading and styling the English of the entire book. Only through her devoted painstaking contribution was publication of this book possible. I would like to express my sincere appreciation to her.

Mr. Takamitsu Aoki of Sony Corporation gave me his abundant cooperation. He checked all the formulas, and solved all problem sets. He was thus the man behind the scenes who played all of these important roles.

I am thankful to Dr. Junichi Nakayama of Kyoto Institute of Technology for helping me to improve various parts of Chap. 10. I am also grateful to Professor Stefan Zukotynski and Mr. Dehuan He of The University of Toronto for their assistance. The author also wishes to thank Professor T. Tamir of The Polytechnic Institute of New York, Brooklyn and Dr. H. K. V. Lotsch of Springer Verlag for critical reading and correcting the manuscript. Mr. R. Michels of Springer-Verlag deserves praise for his painstaking efforts to convert the manuscript into book form. Megumi Iizuka helped to compile the Subject Index.

Toronto, August 1985                                            *Keigo Iizuka*

# Contents

# 1. History of Optics

## 1.1 The Mysterious Rock Crystal Lens

It was as early as 4,000 B.C. that the Sumerians cultivated a high level civilization in a region of Mesopotamia which at present belongs to Iraq. The name Mesopotamia, meaning "between the rivers", [1.1] indicates that this area was of great fertility enjoying natural advantages from the Tigris and Euphrates rivers. They invented and developed the cuneiform script which is today considered to have been man's first usable writing.

Cuneiforms were written by pressing a stylus of bone or hard reed into a tablet of soft river mud. The tip was sharpened into a thin wedge and thus the cuneiform letters are made up of such wedge shaped strokes as shown in Fig. 1.1 [1.2]. These tablets were hardened by baking in the sun. Literally tens of thousands of the tablets were excavated in good condition and deciphered by curious archaeologists.

The inscriptions on the tablets have revealed education, religion, philosophy, romance, agriculture, legal procedure, pharmacology, taxation, and so on. It is astonishing to read all about them, and to discover that

**Fig. 1.1.** Cuneiform tablets excavated from Mesopotamia. The contents are devoted to the flood episode. (By courtesy of A. W. Sjörberg, University Museum, University of Pennsylvania)

they describe a highly sophisticated society more than five thousand years ago [1.3].

What is amazing about the tablets is that some of the cuneiform inscriptions are really tiny (less than a few millimeters in height) and cannot be easily read without some sort of a magnifying glass. Even more mysterious is how these inscriptions were made. A rock-crystal, which seemed to be cut and polished to the shape of a plano-convex lens, was excavated in the location of the tablets by an English archaeologist, Sir *Austen Layard*, in 1885 [1.4]. He could not agree with the opinion that the rock crystal was just another ornament. Judging from the contents of the tablet inscriptions, it is not hard to imagine that the Sumerians had already learned how to use the rock crystal as a magnifying glass. If indeed it were used as a lens, its focal length is about 11 cm and a magnification of two would be achieved with some distortion.

Another interesting theory is that these cuneiforms were fabricated by a group of nearsighted craftmen [1.5]. A myopic (nearsighted) eye can project a bigger image onto the retina than an emmetropic (normal) eye. If the distance $b$ between the lens and the retina is fixed, the magnification $m$ of the image projected onto the retina is

$$m = \frac{b}{f} - 1, \tag{1.1}$$

where $f$ is the focal length of the eye lens. The image is therefore larger for shorter values of $f$. In some cases, the magnifying power of the myopic eye becomes 1.7 times of that of the emmetropic eye, which is almost as big a magnification as the rock crystal "lens" could provide. However, it is still a mystery today how these tiny tablet inscriptions were fabricated and read.

Two other significant developments in optics during this period, besides the questionable rock crystal lens, were the use of a lamp and a hand-held mirror. Palaeolithic wall paintings are often found in caves of almost total darkness. Such lamps, as shown in Fig. 1.2, were used by the artists and have been excavated from the area. Animal fat and grease were used as

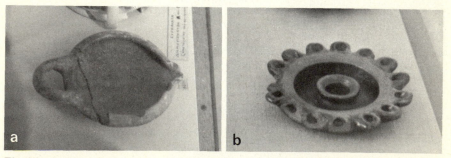

**Fig. 1.2 a, b.** One candle power at the beginning of Greek empire (**a**) evolved to 15 candle power (**b**) at the end of the empire

their fuel. The lamps were even equipped with wicks. The paintings themselves demonstrated a sophisticated knowledge about colour.

Other optical instruments of early discovery are metallic mirrors, which have been found inside an Egyptian mummy-case. By 2,000 B.C., the Egyptians had already mastered a technique of fabricating a metallic mirror. Except for the elaborate handle designs, the shape is similar to what might be found in the stores today.

## 1.2  Ideas Generated by Greek Philosophers

Greece started to shape up as a country around 750 B.C. as a collection of many small kingdoms. Greek colonies expanded all around the coast lines of the Mediterranean Sea and the Black Sea including Italy, Syria, Egypt, Persia and even northeast India [1.6].

A renowned Greek scientist, Thales (640−546 B.C.), was invited by Egyptian priests to measure the height of a pyramid [1.4]. For this big job, all that this brilliant man needed was a stick. He measured the height of the pyramid by using the proportion of the height of the stick to its shadow, as shown in Fig. 1.3. He also predicted the total solar eclipse that took place on May 28, 585 B.C.

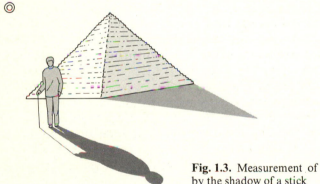

**Fig. 1.3.** Measurement of the height of a pyramid by the shadow of a stick

The greatest mathematician and physicist of all time, Euclid (315−250 B.C.), and his student Archimedes (287−212 B.C.) and many others were from the University of Alexandria in Egypt which was then a Greek colony. The "Elements of Geometry" written by Euclid has survived as a text book for at least twenty centuries. The progress of geometry has been inseparable from that of optics.

Democritus (460−370 B.C.), Plato (428−347 B.C.) and Euclid all shared similar ideas about vision. Their hypothesis was that the eyes emanate vision rays or eye rays and the returned rays create vision. Under this hypothesis, vision operated on a principle similar to that of modern day

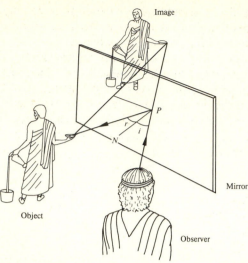

**Fig. 1.4.** Reversed direction of arrows as well as reversed designation of the incident and reflected angles

radar or sonar. As a result of the concept of the eye ray, arrows indicating the direction of a ray always pointed away from the eye. Not only the direction of the arrows, but also the designation of the incident and reflected angles were reversed from what these are today, as illustrated in Fig. 1.4. As a matter of fact, it took fourteen hundred years before the direction of the arrows was reversed by Alhazen (around 965–1039) of Arabia in 1026.

Democritus further elaborated this theory of vision. He maintained that extremely small particles chip off from the object and those chipped particles form a replica which is imprinted on the eyes by the moisture in the eyes. His proof was that, on the surface of the viewer's eye, one can see a small replica of the object going into the viewer's eyes as illustrated in Fig. 1.5. The basis of a corpuscular theory of light had already started percolating at this early age.

Aristotle (384–322 B.C.), who was one of Plato's pupils, showed dissatisfaction about the emission theory saying, "why then are we not able to seen in the dark?" [1.4, 7, 8].

**Fig. 1.5.** Democritus (460–370 BC) said "a replica of the object is incident upon the eyes and is imprinted by the moisture of the eye"

Hero(n) (50?) tried to explain the straight transmission of a ray by postulating that the ray always takes the shortest distance. An idea similar to Fermat's principle was already conceived in those early days.

The left and right reversal of the image of a vertical mirror such as shown in Fig. 1.4, or the upside-down image of a horizontal mirror, such as shown in Fig. 1.6, aroused the curiosity of the Greek philosophers. Plato attempted the explanation but without success. Hero even constructed a corner mirror which consisted of two mirrors perpendicular to each other. One can see the right side as right, and the left side as left, by looking into the corner mirror, as illustrated in Fig. 1.7.

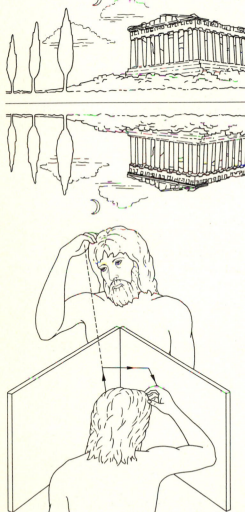

**Fig. 1.6.** Inversion of mirror images

**Fig. 1.7.** Corner mirror which forms "correct image" was invented by Hero (50?)

Around the second century A.D., more quantitative experiments had started. Claudius Ptolemy (100−160) [1.4, 7] made a series of experiments to study refraction using the boundary between air and water. He generated a table showing the relationship between the incident and refracted angles, and obtained an empirical formula of refraction

$$r = u\,i - k\,i^2, \tag{1.2}$$

where $u$ and $k$ are constants. This formula fitted the experiments quite nicely for small incidence angles. Again, man had to wait for another fifteen hundred years before Snell's law was formulated in 1620.

Ptolemy attempted to explain Ctesibiu's coin-in-a-cup experiment. Ctesibiu's experiment was first performed at the University of Alexandria around 50 B.C. As shown in Fig. 1.8, a coin placed at the point $c$ at the bottom of an empty opaque cup is not visible to an eye positioned at $e$, but the coin becomes visible as soon as the cup is filled with water. Ptolemy recognized that the light refracts at the interface between air and water at 0 and changes its course toward the coin and the coin becomes visible to the eye. Ptolemy even determined the apparent position of the coin at the intercept $c'$ of the extension of $\overline{e0}$ with the vertical line $hc$. The exact position is somewhat to the right of $c'$ [1.9].

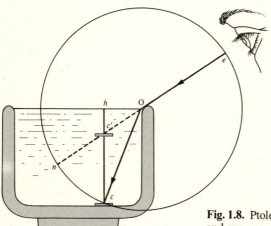

**Fig. 1.8.** Ptolemy's explanation of Ctesibiu's coin and cup experiment

After fierce fighting, the Romans invaded Greece. Greek gods had to bow their heads to Roman gods and eventually to the Christian god. The fall of the Roman Empire in the fifth century A.D. marked the beginning of the Dark Ages. Progress in the science of light was turned off. It was not until the Renaissance (14th−16th centuries) that science was roused from its slumber.

# 1.3  A Morning Star

A star scientist who twinkled against the darkness of the Middle Ages especially in the field of optics, was Abu Ali Al-Hasen ibn al-Hasan ibn Al-Haytham, in short Alhazen (around 965–1039) of the Arabian Empire [1.4, 7, 8, 10]. His contribution to the field of optics was very broad and prolific.

By measuring the time elapsed from the disappearance of the setting sun from the horizon to total darkness, i.e., by measuring the duration of the twilight which is caused by scattering from particles in the atmosphere, he determined the height of the atmosphere to be 80 km [1.4]. This value is shorter than 320 km which is now accepted as the height of the atmosphere, but nevertheless, his observation showed deep insight.

He investigated reflection from non-planar surfaces (concave, convex, spherical, cylindrical, etc.) and finally formulated Alhazen's law of reflection, i.e. the incident ray, the normal of the point of reflection, and the reflected ray, all three lie in the same plane. He also explained the mechanism of refraction taking place at the boundary of two different media by resolving the motion into two components: parallel and normal to the surface. While the component parallel to the surface is unaffected, that normal to the surface is impeded by the denser medium, and as a result, the resultant direction of the motion inside the second medium is different from that of the incident motion. Unfortunately, with this explanation, the refracted ray changes its direction away from the normal at the boundary when light is transmitted into a denser medium, contrary to reality.

Alhazen discarded the idea of the eye ray supported by the Greek philosophers. He could not agree with the idea of an eye ray searching for an object. He recognized that the degree of darkness and colour of an object changes in accordance with that of the illuminating light, and if eye rays were responsible for vision, then vision should not be influenced by external conditions. Alhazen finally reversed the direction of the arrows on the rays showing the geometry of vision.

Alhazen tried to understand the mechanism of vision from comprehensive anatomical, physical and mathematical viewpoints; in today's terms, his approach was interdisciplinary. From the fact that eyes start hurting if they are exposed to direct sunlight, and from the fact that the effect of the sun is still seen even after the eyes are shut, he concluded that the light ray must be causing some reaction right inside the human eye. Alhazen's ocular anatomy was quite accurate. Based upon his detailed anatomy, he constructed a physical model of an eye to actually identify what is going on inside the eye. In this model, several candles were placed in front of a wall with a small hole at the center. The upside down images of the candles were projected onto a screen placed on the other side of the hole. He verified that the position of the image was on a line diagonally across the hole by obstructing each candle one at a time, as shown in Fig. 1.9. He had to give

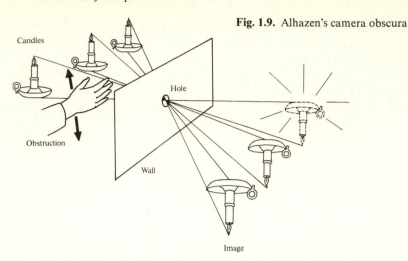

**Fig. 1.9.** Alhazen's camera obscura

an explanation why man can see images right side up, while those of his physical model were upside down. He was forced to say that the image was sensed by the first surface of the crystalline lens even though his anatomy diagram shows the retina and its nerve structure. This experiment is called camera obscura [1.8].

## 1.4 Renaissance

Renaissance means "rebirth" and it was during this period in history that many disciplines, including optics, experienced a revival. The Renaissance began in Italy in the fourteenth century and quickly spread to other countries.

A great Italian architect, sculptor and scientist, Leonardo da Vinci (1452–1519), followed up Alhazen's experiments and developed the pinhole camera. It was only natural that da Vinci, who was one of the most accomplished artists of all times, indulged in the study of colour. He made an analogy between acoustic and light waves. He explained the difference in colour of smoke rising from a cigar and smoke exhaled from the mouth. The former is lighter, and the frequency of oscillation is higher, so that its colour is blue; while the latter is moistened and heavier, and the frequency of oscillation is lower, and hence its colour is brown. This statement implies that he believed that light is a wave and that its colour is determined by the frequency.

Around da Vinci's time, the printing press was invented by Johannes Gutenberg and financed by Johann Fust both in Germany. This invention accelerated the spread of knowledge.

The astronomer Galileo Galilei (1565–1642), working on a rumor that a Dutchman had succeeded in constructing a tool for close viewing with significant magnification (the Dutchman could have been either Hans Lippershey [1.4] or De Waard [1.8] of Middleburg), managed to construct such a telescope, as shown in Fig. 1.10, for astronomical observation. Using the telescope of his own make, Galilei indulged in the observation of stars. During consecutive night observations from January 7 to January 15, 1610, he discovered that the movement of Jupiter relative to four adjacent small stars was contrary to the movement of the sky. He was puzzled but continued observation led him to conclude that it was not Jupiter that was moving in the opposite direction but rather it was the four small stars that were circling around Jupiter.

When Galilei first claimed the successful use of the telescope, academic circles were skeptical to use the telescope for astronomy. These circles believed that whatever was observed by using such a telescope would be distorted from reality because of its colour coma and aberration.

Galilei kept correspondence with Johann Kepler (1571–1630) of Germany who was rather skeptical about the telescope because he himself tried to build a telescope without much success. Galilei presented the telescope he built to Kepler. Kepler was amazed by the power of the telescope and praised Galilei without qualification in his Latin treatise published in September 1610.

Although Kepler is most well known for his laws of planetary motion, he was also interested in various other fields. Kepler turned his attention to

**Fig. 1.10.** Telescopes made by Galileo Galilei (1564–1642). (*a*) *Telescope of wood covered with paper,* 1.36 meters long; with a biconvex lens of 26 mm aperture and 1.33 m focal length; plano-concave eye-piece; magnification, 14 times. – (*b*) *Telescope of wood,* covered with leather with gold decorations: 0.92 m long; with a biconvex objective of 16 mm useful aperture and 96 m focal length; biconcave eye-piece (a later addition); magnification, 20 times

earlier results that Ptolemy and Alhazen worked out about refraction. During experiments, he discovered that there is an angle of total reflection. He also explored the theory of vision beyond Alhazen and identified the retina as the photo-sensitive surface. It thus took six centuries for the light to penetrate from the surface of the crystalline lens to the retina. Kepler was not worried about the upside down vision because he reached the conclusion that the inversion takes place physiologically.

René Descartes (1596 – 1650) proposed the idea of a luminiferous aether to describe the nature of light for the first time [1.11]. The aether is a very tenuous fluid-like medium which can pass through any transparent medium without being obstructed by that medium. The reason why it is not obstructed was explained by using the analogy of the so called Descartes' Vat, such as shown in Fig. 1.11. The liquid which is poured from the top can go through between the grapes and can come out of the vat at the bottom. Later, however, Descartes abandoned this idea and proposed a new postulate, namely, that light is not like fluid, but rather a stream of globules. The laws of reflection and refraction were then explained by using the kinematics of a globule.

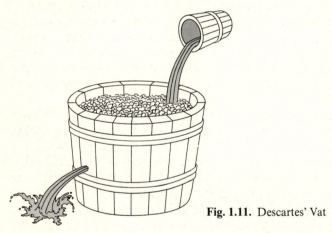

**Fig. 1.11.** Descartes' Vat

## 1.5 The Lengthy Path to Snell's Law

The one law which governs all refraction problems was finally formulated around 1620 by the Dutch professor Willebrod Snell (1591 – 1626) at Leyden. Starting with Ptolemy in ancient Greece, the problems of refraction had been relayed to Alhazen, and then to Kepler and finally to Snell's inspiration; this process covers fifteen hundred years of evolving ideas. Snell's law is indeed one of the most basic and important laws of optics to-date. The cosecant rather than sine of the angles was used when Snell introduced the law.

Descartes attempted to explain Snell's law by using the model that light is a stream of small globules. Like Alhazen, he decomposed the motion of a globule into tangential and normal components. However, Descartes, was different from Alhazen in treating the normal component of the motion. He realized that in order to be consistent with Snell's law, the normal component had to be larger so that the light would bend toward the normal. He therefore used the analogy that a ball rolling over a soft carpet is slower than the same ball rolled over a hard bare table, thus the motion in the optically dense medium is larger than that in a less-dense medium. Descartes used sine functions in his illustration and brought the results in today's form of Snell's law. He also proposed a hypothesis concerning the dispersion of white light into a spectrum of colours after refraction from glass. He postulated that the globules of light rotate due to impact at the interface, and consequently the light starts displaying colour. This hypothesis was, however, disproved by Robert Hooke (1635–1703) who arranged glass slabs such that the light refracts twice in opposite directions, in which case the colours should disappear according to Descartes' theory, but they did not.

Returning to the refraction problem, Pierre Fermat (1601–1665) of France did not agree with Descartes' explanation that the normal movement of the light globules in water is faster than that in air. Instead, Fermat focused his attention on Hero's idea of the shortest path length, which fitted very nicely with the observation of the path of reflected light. However, Hero's idea fails when one tries to use it to explain the path of refraction.

Fermat postulated the concept of the least time rather than the least length for the light to reach from one point in the first medium to another point in the second medium. He argued that, in the case of the boundary between air and water, the speed in water is slower than that in air, so that the light takes a path closer to the normal in the water so as to minimize the transit time. Fermat, being a great mathematician, could determine the exact path which makes the total time the least. It is noteworthy that, even though Descartes' globules approach made a wrong assumption of a faster speed of light in water than air, he reached a conclusion that is consistent with Snell's law.

## 1.6  A Time Bomb to Modern Optics

One of the most important books in the history of optics, "Physico – Mathesis de Lumine coloribus et iride, aliisque annexis libri", was written by a professor of mathematics at the University of Bologna, Italy, Father Francesco Maria Grimaldi (1618–1663) and was published in 1665 after his death. The book openly attacked most of the philosophers saying ". . . let us be honest, we do not really know anything about the nature of light and it is dishonest to use big words which are meaningless" [1.8]. Needless to say, his frank opinions did not go over well in academic circles, and he lost some friends. He developed experiments and an analysis based upon his own fresh viewpoint, at times totally disregarding earlier discoveries. His book planted many a seed which were later cultivated to full bloom by Huygens, Newton, Young and Fresnel. His major contributions were in the field of diffraction, interference, reflection and the colour of light.

Probably the most important contribution of Grimaldi was the discovery of the diffration of light. Even such minute obscure phenomena did not escape his eyes. His sharp observations can never be overpraised. Since he used white light as a source, he had to use light coming out of an extremely small hole in order to maintain a quasi-coherency across the source. This means that he had to make his observations with a very faint light source.

The small hole also helped to separate the direct propagation from that of the diffracted rays. Using this source, the shadow of an obstacle was projected onto a screen. He discovered that the light was bending into the region of the geometrical shadow, so that the edge of the shadow was always blurred. In addition to this blurriness of the edge of the shadow, he recognized a bluish colour in the inner side of the edge, and a reddish colour in the outer side of the edge. He named this mysterious phenomenon diffusion of light.

Could he ever have dreamed that this faint phenomenon would form the basis of optical computation or holography of today? He tried to explain the phenomenon by proposing an analogy with the spreading of fire by sparks. Wherever a spark is launched on the grass, a new circle of fire starts, and each point of fire again generates new sparks. Such sparks can even turn the corner of a building. Diffraction of light takes place in exactly the same manner. Even though he made this analogy, he was not satisfied with it because a secondary wave is generated wherever the light reaches, and soon all area will be filled with light. Despite this drawback of the analogy, Grimaldi's explanation has a lot in common with Huygens' principle.

Grimaldi shared Descartes' view that all materials have numerous tiny pores and the degree of transparency is determined by the density of the pores. A transparent material has a higher density and an opaque material has a lower density of pores. Based upon this hypothesis, refraction and reflection between air and water were explained by Grimaldi as follows. Since the density of pores is smaller in water than that in air, the light beam has to become wider in the water, as shown in Fig. 1.12, so that the flow of the light beam will be continuous. Thus the transmitted beam bends toward the normal at the boundary. Using the same hypothesis, he tried to explain reflection when a light beam is incident upon a glass slab held in air. When the light beam hits the surface of the glass whose density of pores is less than that of air, only a portion of the light beam can go into the glass and the rest has to reflect back into air. This was a convincing argument, but he

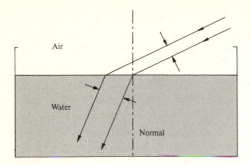

**Fig. 1.12.** Grimaldi's explanation of refraction

could not explain why the light transmitted into the glass reflects again when it goes out into air from glass.

His experiments were wide in scope and subtle in arrangement. He succeeded in detecting interference fringes even with such a quasi-coherent source as the pinhole source. He illuminated two closely spaced pinholes with a pinhole source and, in the pattern projected onto a screen, he discovered that some areas of the projected pattern were even darker than when one of the holes was plugged. His observation was basically the same as the observation of Thomas Young (1773−1829) in an experiment performed one and a half centuries later.

## 1.7 Newton's Rings and Newton's Corpuscular Theory

Sir Isaac Newton (1642−1726) was born a premature baby in a farmer's family in Lincolnshire, England. In his boyhood, he was so weak physically that he used to be bothered and beaten up by his school mates. It was a spirit of revenge that motivated little Isaac to beat these boys back in school marks.

When he entered the University of Cambridge in 1665, the Great Plague spread in London, and the University was closed. He went back to his home in Lincolnshire for eighteen months. It was during these months that Newton conceived many a great idea in mathematics, mechanics and optics, but it was very late in life at the age of 62 when he finally published "Opticks" which summarized the endeavours of his lifetime. Profound ideas expressed in an economy of words were the trademark of his polished writing style.

He was renowned for his exceptional skill and insight in experiments. Newton had been in favour of the corpuscular theory of light, but he conceded that the corpuscular theory alone does not solve all the problems in optics. His major objection to the undulatory theory or wave theory is that waves usually spread in every direction, like a water wave in a pond, and this seemed inconsistent with certain beam-like properties of light propagation over long distances. The light beam was considered as a jet stream of a mixture of corpuscles of various masses.

He explained that refraction is caused by the attractive force between the corpuscles and the refracting medium. For denser media, the attractive force to the corpuscles is stronger, when the ray enters into a denser medium from a less dense medium, the ray is bent toward the normal because of the increase in the velocity in that direction. A pitfall of this theory is that the resultant velocity inside a denser medium is faster than that in vacuum.

Quantitative studies on the dispersion phenomena were conducted using two prisms arranged as in Fig. 1.13. A sun beam, considered to be a heterogeneous mixture of corpuscles of different masses, was incident upon the first prism. The second prism was placed with its axis parallel to that of the first one. A mask $M_2$ with a small hole in the center was placed in between the two prisms. The spread of colour patterns from prism $P_1$ was first projected onto the mask $M_2$. The ray of one particular colour could be sampled by properly orienting the small hole of $M_2$. The sampled colour ray went through the second prism $P_2$ and reached the screen.

The difference in the amount of change in velocity between the light and heavy corpuscles was used to explain the dispersion phenomenon. While the output from the first prism was spread out, that of the second prism was not. This is because only corpuscles of the same mass had been sampled out by the hole $M_2$. Thus, the monochromatic light could be dispersed no more. Newton repeated a similar experiment with a few more prisms after prism $P_2$ to definitely confirm that the beam splits no further. The fact that monochromatic light cannot be further split is the fundamental principle of present-day spectroscopy. The angle of refraction from the second prism $P_2$ was measured with respect to the colour of light by successively moving the hole on $M_2$. He also examined the case when the second refraction is in a plane different from that of the first refraction; accordingly, he arranged the second prism in a cross position, as shown in Fig. 1.14. The first prism spread the spectrum vertically, and the second spread it side-ways at different angles according to the colour of light. A slant spectrum was observed on the screen.

Even the brilliant Newton, however, made a serious mistake in concluding that the degree of dispersion is proportional to the refractive index. It is most natural that he reached this conclusion because he maintained the theory that the dispersion is due to the differential attractive forces. He did not have a chance to try many media to find that the degree of dispersion varies even among media of the same refractive index. This is known as "Newton's error". He went as far as to say that one can never build a practical telescope using lenses because of the chromatic aberration. This error, however, led to a happy end. He wound up building an astronomical telescope with parabolic mirrors such as shown in Fig. 1.15. He also attempted to construct a microscope with reflectors but he did not succeed.

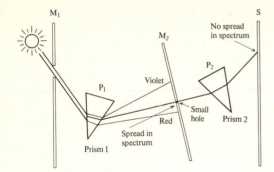

**Fig. 1.13.** Newton's two prism experiment

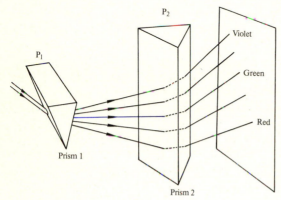

**Fig. 1.14.** Newton's cross prism experiment

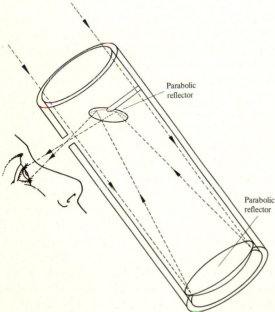

**Fig. 1.15.** Newton's reflector telescope

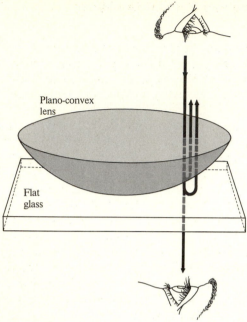

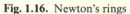

**Fig. 1.16.** Newton's rings

Next to be mentioned are the so-called Newton's rings. He placed a plano-convex lens over a sheet glass, as shown in Fig. 1.16, and found a beautiful series of concentric rings in colour. His usual careful in-depth observation led him to discover such fine details about the rings as: (i) The spacings between the rings are not uniform but become condensed as they go away from the center. To be more exact, the radius of the $n$th order ring grows with $\sqrt{n}$. (ii) The locations of the dark lines occur where the gap distance between the lens and plate is an integral multiple of a basic dimension. (iii) Within any given order of rings, the ratio of the gap distance where the red ring is seen and the gap distance where violet ring is seen is always 14 to 9. (iv) The rings are observable from the bottom as well as from the top and they are complimentary to each other.

For Newton, who had tried to maintain the corpuscular theory, it was most difficult and challenging to explain those rings. He introduced a "fit" theory as an explanation. Wherever reflection or refraction takes place, there is a short transient path. Because the gap is so small in the Newton ring case, before the first transient dies off the second starts; thus the transient is accentuated and the so called "fit" phenomenon takes place. The colourful rings are generated where the "fit" of easy reflection is located and the dark rings, where the "fit" of easy transmission is located. It is interesting to note that there is some similarity between the "fit" to explain the fringe pattern by Newton and the transition of an orbital electron to explain the emission of light by Niels Bohr (1885–1962).

# 1.8 Downfall of the Corpuscle and Rise of the Wave

Newton had acquired such an unbeatable reputation that any new theory that contradicted his corpuscular theory had difficulty gaining recognition. Christiaan Huygens (1629–1695) of The Hague, Holland, who was a good friend of Newton's, was the first to battle against the corpuscular theory.

Huygens postulated that light is the propagation of vibrations like those of a sound wave. He could not support the idea that matterlike corpuscles physically moved from point A to point B. Figure 1.17 illustrates Huygens' principle. Every point on the wave front becomes the origin of a secondary spherical wave which, in turn, becomes a new wave front. Light is incident from air to a denser medium having a boundary O–O'. A line A–O' is drawn perpendicular to the direction of incidence. Circles are drawn with their centers along O–O' and tangent to A–O'. The line A–O' would be the wave front if air were to replace the denser medium. Since the velocity in the denser medium is reduced by $1/n$, circles reduced by $1/n$ are drawn concentric with the first circles. The envelope A'–O' of the reduced circles is the wave front of the refracted wave. As a matter of fact, the envelope A''–O' is the wave front of the reflected wave.

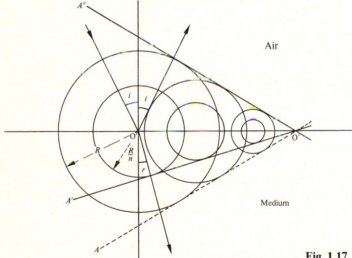

**Fig. 1.17.** Huygens' principle

Unlike Newton's or Descartes' theory, the velocity in the denser medium is smaller, and an improvement in the theory is seen. Huygens was also successful in treating the double refraction of Iceland spar (Calcite $CaCO_3$) by using an ellipsoidal wave front rather than a spherical wave front. Figure 1.17 could also be used for explaining total reflection.

A great scholar who succeeded Huygens and Newton was Thomas Young (1773–1829) of England. It is not surprising that Young, who was originally trained as a medical doctor, had a special interest in the study of vision. He introduced the concept of three basic colours and concluded that there are three separate sensors in the retina sensitive to each colour. This concept of basic colours is still supported to date.

Young's greatest contribution to optics was a successful explanation of interference phenomena. For the explanation of Newton's rings, he made an analogy with the sound wave. He considered that the path difference is the source of the dark and bright rings. Referring to Fig. 1.16, the path difference of a ray reflecting from the convex surface of the lens and that reflecting from the bottom flat plate is responsible for the generation of the ring pattern. He further explained that, if the path difference is a multiple of a certain length, the two rays strengthen each other but, if the path difference is the intermediate of these lengths, the two rays cancel each other. He also noted that this length varies with the colour of light.

Young conducted an experiment on interference that now bears his name. He arranged a pinhole such that a sunbeam passing through the pinhole illuminated two pinholes located close together. The interference fringe pattern from the two pinholes was projected onto a screen located several meters away. From measurements of the spacings between the maxima of the fringe pattern associated with each colour, the wavelengths of each colour light were determined. His measurement accuracy was very high and the wavelengths of 0.7 μm for red and 0.42 μm for violet were obtained.

## 1.9 Building Blocks of Modern Optics

Augustin Jean Fresnel (1788–1827) started out his career as a civil engineer and was engaged in the construction of roads in southern France while maintaining his interest in optics. Fresnel spoke openly against Napoleon, and as a consequence, he had to resign his government position. This freed him to devote himself entirely to optics and he asked Dominique François Jean Arago (1786–1853) at the Paris observatory for advice. It was suggested to Fresnel that he read such materials as written by Grimaldi, Newton and Young. Fresnel, however, could not follow Arago's advice because he could read neither English nor Latin [1.8].

Fresnel reexamined the distribution of light across an aperture. He measured the aperture distribution at various distances from the aperture to see how the positions of the maxima moved with an increase in the distance from the aperture. He discovered that the size of the pattern did not increase linearly with distance as Newton's corpuscular theory predicts. The positions of the maxima followed a hyperbola suggesting that the fringe pattern is due to the interference of two waves originating from the edge of the aperture and the incident light.

Fresnel earned Arago's confidence as the result of a skillful experiment on the fringe pattern made by a piece of thread. He noticed that, when light is incident onto a piece of thread, fringe patterns are observed on both sides of the thread, but as soon as the light on one of the sides is blocked by a sheet of paper, the fringes on the shadow side disappear. From this, he proved that the fringes are due to the interference of the two waves coming from both sides of the thread.

Arago worriedly reported to Fresnel that even a transparent sheet of glass that does not block the light creates the same result. With confidence, Fresnel replied that a thick glass destroys the homogeneity of the wave front and with a thin sheet of glass the fringes will come back even though they may be shifted and distorted. Arago verified this prediction by experiments and Fresnel thus made another point.

Fresnel participated in a scientific-paper contest sponsored by the French Academie des Sciences of which Arago was the chairman of the selection committee [1.8]. Fresnel realized that Huygens' principle could not explain the fact that, in the diffraction pattern, the added light can be even darker than the individual light. Essentially what Fresnel did was to develop a mathematical formulation of Huygens' principle but with a newly introduced concept of the phase of the contributing field element. With phase taken into consideration, the same field can contribute either constructively or destructively depending upon the distance. His theory agreed superbly with the experiments. His paper was nominated as the first prize in the contest.

Siméon Poisson (1781−1840), who was one of the examiners of the contest calculated the diffraction pattern of a disk using Fresnel's formula and reached the conclusion that the field at the center of the shadow is the brightest and he refuted Fresnel's theory. Arago performed the experiment by himself and verified that Fresnel's theory was indeed correct [1.12]. Again Fresnel was saved by Arago. Poisson himself acquired his share of fame in the event. This spot has become known as Poisson's spot. Thus, Fresnel established an unshakeable position in the field of optics.

Fresnel also challenged the experiment with double refraction of Iceland spar (Calcite $CaCO_3$). He discovered that the ordinary ray and the extra-ordinary ray can never form a fringe pattern. With his unparalleled ability of induction, this experimental observation led him to discard the idea of longitudinal vibration of the light wave. Because longitudinal vibration has only one direction of vibration, the effect of the direction of vibration cannot be used to explain the observation on the double refraction. He proposed that the vibration of the light is transverse, and that the directions of the vibration (or polarization) of the ordinary and extraordinary rays are perpendicular to each other. Two vibrations in perpendicular directions cannot cancel one another and thus a fringe pattern can never be formed by these two rays.

Fresnel kept in close contact with Young and the two advanced the wave theory to a great extent and together they laid a firm foundation for optics in the nineteenth century. Other noteworthy contributors to optics before the end of the nineteenth century were Armand H. Louis Fizeau (1819−1896) and Leon Foucault (1819−1868), both of whom were Arago's students and determined the speed of light by two different methods.

James Clerk Maxwell (1831−1879), who became the first director of the Cavendish Laboratory at Cambridge, opened up a new era in history by developing the electromagnetic theory of light. He established elegant mathematical expressions describing the relationship between the electric and magnetic fields and postulated that light is an electromagnetic wave which propagates in space by a repeated alternating conversion between the electric and magnetic energies. His postulate was experimentally verified by a German scientist, Heinrich Hertz (1857−1894), eight years after his death, and Maxwell's equations became the grand foundation of electromagnetic theory.

American scientists Albert Abraham Michelson (1852−1931) and Edward Williams Morley (1838−1923) collaborated and performed a sophisticated measurement of the speed of light to challenge the existence of the aether first proposed by Descartes. They believed that if there is an aether, there must also be an aether wind on the earth because the movement of the earth is as high as 30 km/s. If indeed there is such wind, it should influence the velocity of light. They put the so-called Michelson's interferometer in a big mercury tub so that it could be freely rotated, as shown in Fig. 1.18.

The interference between the two beams, $S-M_0-M_1-M_0-T$ and $S-M_0-M_2-M_0-T$, taking mutually perpendicular paths was observed with one of the beam axes parallel to the direction of the orbital motion of

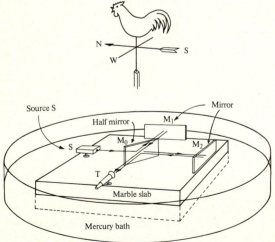

**Fig. 1.18.** Michelson-Morley experiment to examine the drag by aether wind

the earth. Then, the whole table was carefully rotated by 180° to observe any changes in the fringe pattern from before. The conclusion of the experiment was that there is no difference and thus there is no aether wind. This result eventually led to the theory of relativity by Einstein.


## 1.10 Quanta and Photons

In 1900, at the dawn of the twentieth century, Max Karl Ernst Ludwig Planck (1858–1947) made the first big step toward the concept of a quantum of energy. Radiation from a cavity made of a metal block with a hole was intensively investigated by such physicists as Stefan, Boltzmann, Wien and Jean. Radiant emission as a function of frequency was examined at various temperatures of the cavity. Planck introduced a bold postulate that the radiant emission is due to the contributions of tiny electromagnetic oscillators on the walls and these oscillators can take only discrete energy levels, and the radiation takes place whenever there is an oscillator that jumps from a higher level to a lower level. The discrete energy levels are integral multiples of $hv$, $h$ being Planck's constant and $v$ the light frequency. For this contribution, Planck was awarded the Nobel prize in 1918.

Philipp Eduard Anton von Lenard (1862–1947), a collaborator of Hertz, made experiments on the photoelectric effect by using a vacuum tube such as that shown in Fig. 1.19. When light is incident onto the cathode, electrons on the surface of the cathode are ejected from the surface with kinetic

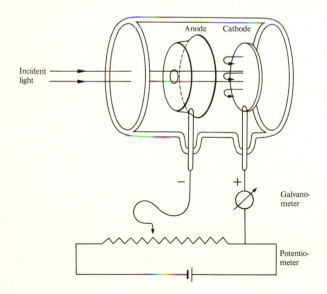

Fig. 1.19. Photoelectric effect experiment

energy $m v^2/2$ by absorbing energy from the light. When the electrons reach the anode, the current is registered by the galvanometer. The amount of kinetic energy can be determined from the negative potential $V_s$ needed to stop the anode current. When the condition

$$e V_s = \tfrac{1}{2} m v^2 \tag{1.3}$$

is satisfied, the emitted electrons will be sent back to the cathode before reaching the anode.

Lenard discovered that the stopping potential $V_s$ could not be changed no matter how the intensity of the incident light was raised. An increase in the intensity of the light resulted only in a proportionate increase in the anode current. Only a change in the frequency of the incident light could change the value of the stopping potential, as seen from the plot in Fig. 1.20a. Another observation was that there was practically no time delay between the illumination and the change in the anode current.

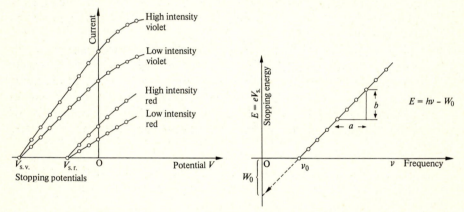

**Fig. 1.20 a, b.** Experimental results of the photoelectric effect. (**a**) Photoelectric current vs applied potentials with wavelength and intensity of the incident light as parameters. (**b**) Energy needed to stop photoelectric emission vs. frequency

In 1905, Albert Einstein (1879−1955) published a hypothesis to explain the rather mysterious photoelectric phenomena by the equation

$$E = h v - W_0. \tag{1.4}$$

Einstein supported Planck's idea of photo quanta. Radiation or absorption takes place only in quanta of $h v$. The amount of energy that can participate in the photoelectric phenomena is in quanta of $h v$. In (1.4), the energy of each photon is $h v$, the kinetic energy of an electron is $E$, and $W_0$ is the work function which is the energy needed to break an electron loose from the surface of the cathode.

Painstaking experimental results accumulated for ten years by Robert Andrews Millikan (1868–1953) validated Einstein's photoelectric formula (1.4). The plot of the stopping potential with respect to the frequency of the incident light, such as shown in Fig. 1.20b, matched (1.4) quite well. From such a plot, Planck's constant $h$ and the work function $W_0$ can be determined respectively from the intercept and the slope of the plot. The value of $h$ which Planck obtained earlier from the radiation experiment agreed amazingly well with Millikan's value. For this work, Millikan was awarded the Nobel prize in 1923.

Einstein's hypothesis denies that light interacts with the cathode as a wave by the following two arguments:

i)  If it interacts as a wave, the kinetic energy of the emitted electrons should have increased with the intensity of light, but it was observed that the stopping potential was independent of the intensity of light.
The cut-off condition was determined solely by wavelength. If it were a wave, the cut-off should be determined by the amplitude because the incident energy of the wave is determined by its amplitude and not by its wavelength.

ii) If light is a wave, there is a limit to how precisely the energy can be localized. It was calculated that it would take over 10 hours of accumulation of energy of the incident light "wave" to reach the kinetic energy of the emitted electron, because of the unavoidable spread of the incident light energy. But if light were a particle, the energy can be confined to the size of the particle, and almost instant response becomes possible.

Einstein added a second hypothesis that the photon, being a particle, should have a momentum $p$ in the direction of propagation where

$$p = \frac{h\nu}{c}. \tag{1.5}$$

Arthur Holly Compton (1892–1962) arranged an experimental set-up such as that shown in Fig. 1.21 to prove Einstein's hypothesis. He detected a small difference $\Delta\lambda$ in wavelengths between the incident and reflected x-rays upon reflection by a graphite crystal. He recognized that the difference in wavelength $\Delta\lambda$ (the wavelength of the x-ray was measured by Bragg reflection from a crystal) varies with the angle of incidence $\psi$ as

$$\Delta\lambda = 0.024(1 - \cos\psi). \tag{1.6}$$

In wave theory, the frequency of the reflected wave is the same as that of the incident wave so that this phenomena cannot be explained by the wave theory. Only after accepting Einstein's hypothesis that the x-ray is a particle just like the electron, could Compton interpret the difference in the wavelengths. From the conservation of momentum and energy among the

(a)

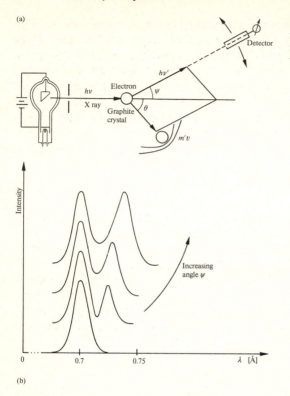

**Fig. 1.21 a, b.** Compton's experiment. (a) Experimental arrangement. (b) Change of wavelength with angle

(b)

colliding particles, he derived an expression that very closely agrees with the empirical formula (1.6).

## 1.11 Reconciliation Between Waves and Particles

Maxwell's equations, which worked so beautifully with wave phenomena like diffraction and refraction, were not capable of dealing with such phenomena as photoelectric emission or Compton's effect. Thus, Einstein's hypothesis of duality of light gained support.

Prince Louis-Victor de Broglie (1892−) extended this hypothesis even further by postulating that any particle, not just a photon, displays wave phenomena (de Broglie wave). This hypothesis was experimentally verified by Clinton Joseph Davisson (1881−1958) and Lester Germer (1896−) who succeeded in 1927 in detecting a Bragg type of reflection of electrons from a nickel crystal.

The idea of the de Broglie wave was further extended to a wave function by Erwin Schrödinger (1887−1961) in 1926. Schrödinger postulated that the de Broglie wave satisfies the wave equation just like the electromagnetic wave and he named its solution a wave function $\psi$. The wave function is a complex valued function but only $|\psi|^2$ has a physical meaning and represents the probability of a particle being at a given location. In other words, $|\psi|^2$ does not have energy but guides the movement of the particle which carries the energy. Interference of light takes place due to the relative phase of the $\psi$ functions. According to Richard P. Feynman (1918−) for example, the interference pattern between the light beams from the same source but taking two separate routes to the destination, like those in Young's experiment, is represented by

$$|\psi_1 + \psi_2|^2 \tag{1.7}$$

where $\psi_1$ and $\psi_2$ are wave functions associated with the two separate routes. The relative phase between $\psi_1$ and $\psi_2$ generates interference. It is important that both $\psi_1$ and $\psi_2$ satisfy Schrödinger's equation and their phase depends upon the path and the path has to conform with Fermat's principle.

It was only through inspiration and perspiration of such Nobel laureates as Paul Adrien Maurice Dirac (in 1933), Wolfgang Pauli (in 1945), Max Born (in 1954) and Sin Itiro Tomonaga, Julian Schwinger, Richard P. Feynman (all three in 1965), and others, that today's theories which reconcile waves and particles have been established.

## 1.12  Ever Growing Optics

Dennis Gabor (1900−1979) invented the hologram during his attempt to improve the resolution of an electron microscope image in 1948. The development of the technique of holography was accelerated by the invention of the laser in 1960. In 1962, E. N. Leith and J. Upatnieks developed a two-beam hologram which significantly improved the image quality and widened the scope of the engineering application of holography.

Nicolaas Bloembergen (1920−) introduced the concept of the three-level solid-state maser in 1956, and succeeded in building one by using chromium-doped potassium cobalt cyanide $K_3(Co_{1-x}Cr_x)(CN)_6$. Charles Townes (1915−), who was then at Columbia University, built a similar maser but with ruby. On the occasion when both of them were presented the 1959 Morris Leibmann award, Townes' wife received a medallion made from the ruby used for his maser. When Bloembergen's wife admired the medallion and hinted for a similar momento of his maser, he replied, "Well dear, my maser works with cyanide!" [1.13].

The maser principle was extended from microwaves to light, and the first laser was constructed by Theodore Maiman in 1960. The availability of a high-intensity light source opened up brand new fields of nonlinear optics, second-harmonic generation, and high-resolution nonlinear laser spectroscopy for which contribution Bloembergen was awarded the Nobel prize in 1981.

Charles Kuen Kao (1933–) predicted the possibility of using an optical fiber as a transmission medium in 1966, and the first relatively low-loss optical fiber of 20 dB/km was fabricated by Corning Glass in 1969. Integrated optics has been making rapid progress. This area covers such optical components as lenses, mirrors, wavelength division multiplexers, laser diode arrays, receiver arrays, spectrum analyzers and even a sophisticated signal processor, all of which can now be fabricated on glass, piezo-electric crystals, or on semi-conductor wafers smaller than a microscope deck glass.

Due to the extremely rapid development of present-day technology, its degree of complexity has been ever increasing. "Interdisciplinary" is the key word in today's technology. Optics has become exceptionally close to electronics. The rendez-vous of optics and electronics has given birth to many hybrid devices such as the solid-state laser, acoustooptic devices, magnetooptic devices, electrooptic devices, etc. All of these consist of interactions among photons, electrons and phonons in a host material. However, what has been discovered so far is only the tip of an iceberg floating in an ocean of human imagination. What a thrilling experience it will be to observe these developments!

# 2. Mathematics Used for Expressing Waves

Spherical waves, cylindrical waves and plane waves are the most basic wave forms. In general, more complicated wave forms can be expressed by the superposition of these three waves. In the present chapter, these waves are described phenomenologically while leaving the more rigorous expressions for later chapters. The symbols which are frequently used in the following chapters are summarized.

## 2.1 Spherical Waves

The wave front of a point source is a *spherical wave* [2.1]. Since any source can be considered to be made up of many point sources, the expression for the point source plays an important role in optics.

Let a point source be located at the origin and let it oscillate with respect to time as

$$e^{-j\omega t} .$$

The nature of the propagation of the light emanating from the point source is now examined. It takes $r/v$ seconds for the wave front emanating from the source to reach a sphere of radius $r$, $v$ being the speed of propagation. The phase at the sphere is the same as that of the source $r/v$ seconds ago. The field on the sphere is given by

$$
\begin{aligned}
E(t, r) &= E(r) \, e^{-j\omega(t-r/v)} \\
&= E(r) \, e^{-j\omega t + jkr} \qquad \text{where}
\end{aligned}
\qquad (2.1)
$$

$$k = \frac{\omega}{v} = \frac{2\pi}{\lambda} .$$

Next, the value of $E(r)$ is determined.

The point source is an omnidirectional emitter and the light energy is constantly flowing out without collecting anywhere. Hence the total energy $W_0$ passing through any arbitrary sphere centered at the source per unit time is constant regardless of its radius. The energy density of the electromagnetic wave is $\varepsilon |E|^2$, $\varepsilon$ being the dielectric constant of the medium.

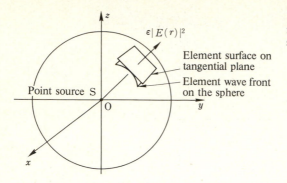

$z$

$\varepsilon |E(r)|^2$

Element surface on tangential plane

Element wave front on the sphere

Point source S

$O$

$y$

$x$

**Fig. 2.1.** Energy distribution emanating from a point source with a spherical wave front

This, however, is true only when the wave front is planar. If the sphere's radius is large compared to the dimensions of the selected surface area, the spherical surface can be approximated by a plane, as shown in Fig. 2.1. The contributions from the various elements of the surface area over the sphere are summed to give

$$4\pi r^2 \varepsilon |E(r)|^2 v = W_0, \qquad \text{with} \tag{2.2}$$

$$E(r) = \frac{E_0}{r}, \tag{2.3}$$

where $E_0$ is the amplitude of the field at unit distance away from the source. (It is difficult to define the field at $r=0$). Equation (2.3) is inserted into (2.1) to obtain the expression of the spherical wave

$$E(t,r) = \frac{E_0}{r} e^{-j(\omega t - kr)}. \tag{2.4}$$

In engineering optics, a planar surface such as a photographic plate or a screen is often used to make an observation. The question then arises as to what is the light distribution that appears on this planar surface when illuminated by a point source. Referring to Fig. 2.2, the field produced by the point source S located at $(x_0, y_0, 0)$ is observed at the point P $(x_i, y_i, z_i)$ on the screen. The distance from S to P is

$$r = \sqrt{z_i^2 + (x_i - x_0)^2 + (y_i - y_0)^2}$$

$$= z_i \sqrt{1 + \frac{(x_i - x_0)^2 + (y_i - y_0)^2}{z_i^2}}.$$

Applying the binomial expansion to $r$ gives

$$r = z_i + \frac{1}{2} \frac{(x_i - x_0)^2 + (y_i - y_0)^2}{z_i} + \dots \tag{2.5}$$

where the distance is assumed to be far enough to satisfy

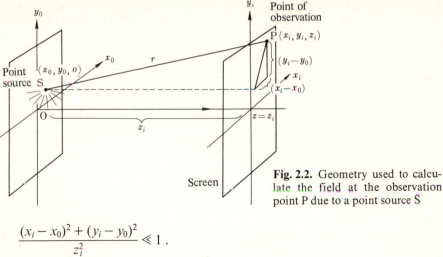

**Fig. 2.2.** Geometry used to calculate the field at the observation point P due to a point source S

$$\frac{(x_i - x_0)^2 + (y_i - y_0)^2}{z_i^2} \ll 1 \ .$$

Inserting (2.5) into (2.4) yields

$$E\,(t, x_i, y_i, z_i) = \frac{E_0}{z_i}\,\exp\left[j\,k\left(z_i + \frac{(x_i - x_0)^2 + (y_i - y_0)^2}{2\,z_i}\right) - j\,\omega\,t\right]. \quad (2.6)$$

Equation (2.6) is an expression for the field originating from a point source, as observed on a screen located at a distance $r$ from the source. The denominator $r$ was approximated by $z_i$.

## 2.2 Cylindrical Waves

The field from a line source will be considered. As shown in Fig. 2.3, an infinitely long line source is placed along the $y$ axis and the light wave

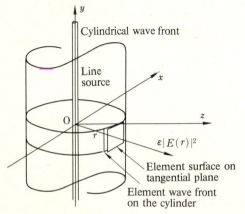

**Fig. 2.3.** Energy distribution emanating from a line source

expands cylindrically from it. The total energy passing through an arbitrary cylindrical surface per unit time is equal to the total energy $w_0 l$ radiated per unit time by the line source:

$$2\pi r l \varepsilon |E|^2 v = w_0 l, \tag{2.7}$$

where $l$ is the length of the source, and $w_0$ is the radiated light energy per unit time per unit length of the source. The spatial distribution of the *cylindrical wave* is therefore

$$E(r) = \frac{E_0}{\sqrt{r}}, \tag{2.8}$$

where $E_0$ is the amplitude of the field at unit distance away from the source. As in the case for the spherical wave, (2.8) is true only when the cylinder radius is large enough for the incremental area on the cylindrical surface to be considered planar. The expression for the line source, including the time dependence, is

$$E(t,r) = \frac{E_0}{\sqrt{r}} e^{-j(\omega t - kr)}. \tag{2.9}$$

*The amplitude of a spherical wave decays as $r^{-1}$ whereas that of a cylindrical wave decays as $r^{-1/2}$ with distance from the source.*

## 2.3 Plane Waves

This section is devoted to the plane wave. As shown in Fig. 2.4, a plane wave propagates at an angle $\theta$ with respect to the $x$ axis. The unit vector $\hat{k}$ denotes the direction of propagation. For simplicity, the $z$ dimension is

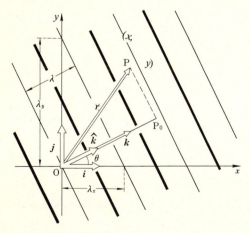

**Fig. 2.4.** Position vector and the wave vector of a plane wave propagating in the $\hat{k}$ direction

ignored. The point of observation, P, is specified by the coordinates $(x, y)$. The *position vector* $r$ joins P to the origin and is written as

$$r = \hat{i} x + \hat{j} y, \tag{2.10}$$

where $\hat{i}$ and $\hat{j}$ are unit vectors in the $x$ and $y$ directions.

Referring to Fig. 2.4, another point $P_0$ is determined by the intersection of the line through $\hat{k}$ and the line passing through P perpendicular to $\hat{k}$. P and $P_0$ have the same phase. If the phase at the origin is $\phi_0$, the phase at $P_0$ is

$$\phi = \frac{2\pi}{\lambda} \overline{OP_0} + \phi_0 \tag{2.11}$$

where $\lambda$ is the wavelength, and $\overline{OP_0}$ is the distance from O to $P_0$. The line segment $\overline{OP_0}$ is actually the projection of $r$ onto $\hat{k}$ and can be expressed as the scalar product of $r$ and $\hat{k}$. Thus,

$$\phi = \frac{2\pi}{\lambda} \hat{k} \cdot r + \phi_0 . \tag{2.12}$$

Defining a new vector as

$$k = \frac{2\pi}{\lambda} \hat{k} , \tag{2.13}$$

Eq. (2.12) becomes

$$\phi = k \cdot r + \phi_0 . \tag{2.14}$$

The vector $k$ is called the *wave vector* and can be decomposed into components as follows:

$$k = \hat{i} |k| \cos\theta + \hat{j} |k| \sin\theta . \tag{2.15}$$

From (2.13, 15), the wave vector is written as

$$k = \hat{i} \frac{2\pi}{\lambda/\cos\theta} + \hat{j} \frac{2\pi}{\lambda/\sin\theta} .$$

From Fig. 2.4, $\lambda_x = \lambda/\cos\theta$ and $\lambda_y = \lambda/\sin\theta$ are the wavelengths along the $x$ and $y$ axis respectively, so that

$$k = \hat{i} k_x + \hat{j} k_y = \hat{i} \frac{2\pi}{\lambda_x} + \hat{j} \frac{2\pi}{\lambda_y}$$

$$|k| = \frac{2\pi}{\lambda} = \sqrt{\left(\frac{2\pi}{\lambda_x}\right)^2 + \left(\frac{2\pi}{\lambda_y}\right)^2} . \tag{2.16}$$

Making use of the definitions (2.12), the plane wave in Fig. 2.4 can be expressed as

$$E(x, y) = E_0 \exp [j(k_x x + k_y y - \omega t + \phi_0)] ,$$

or upon generalizing to three dimensions

$$E(x, y, z) = E_0(x, y, z) \exp [j(k \cdot r - \omega t + \phi_0)] . \tag{2.17}$$

Here $E_0(x, y, z)$ is a vector whose magnitude gives the amplitude of the plane wave and whose direction determines the polarization. The mathematical analysis becomes quite complex when the direction of polarization is taken into consideration. A simpler approach, usually adopted for solving engineering problems, is to consider only one of the vector components. This is usually done by replacing the vector quantity $E(x, y, z)$ with a scalar quantity $E(x, y, z)$. Such a wave is called a "vector wave" or "scalar wave" depending on whether the vector or scalar treatment is applied.

Figure 2.4 is a time-frozen illustration of a plane wave. The sense of the propagation is determined by the combination of the signs of $k \cdot r$ and $\omega t$ in (2.17) for the plane wave. This will be explained more fully by looking at a particular example. Consider a plane wave, which is polarized in the $y$ direction and propagates in the $x$ direction, as given by

$$E = \text{Re} \{E_y e^{j(kx - \omega t)}\} = E_y \cos (kx - \omega t) . \tag{2.18}$$

The solid line in Fig. 2.5 shows the amplitude distribution at the instant when $t = 0$. The first peak (point Q) of the wave form $E$ corresponds to the case where the argument of the cosine term in (2.18) is zero. After a short time $(t = \Delta t)$ the amplitude distribution of (2.18) becomes similar to the dashed curve in Fig. 2.5. In the time $\Delta t$, the point Q has moved in the positive direction parallel to the $x$-axis. Since the point Q represents the amplitude of the cosine term when the argument is zero, the $x$ coordinate of Q moves from 0 to $\Delta x = (\omega/k) \Delta t$, as $t$ changes from 0 to $\Delta t$.

On the other hand, when $t$ changes from 0 to $\Delta t$ in an expression such as

$$E = E_y \cos (kx + \omega t) \tag{2.19}$$

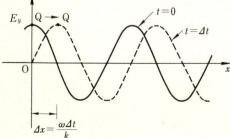

**Fig. 2.5.** Determination of the directional sense (positive or negative) of a wave

the position of Q moves from $x = 0$ to $\Delta x = -(\omega/k)\Delta t$ which indicates a backward moving wave. Similarly, $E = E_y \cos(-kx + \omega t)$ represents a forward moving wave and $E = E_y \cos(-kx - \omega t)$ denotes a backward moving wave. *In short, when the signs of $kx$ and $\omega t$ are different, the wave is propagating in the forward (positive) direction; when $kx$ and $\omega t$ have the same sign, the wave is propagating in the backward (negative) direction.*

As mentioned earlier, during the time interval $\Delta t$, the wave moves a distance $\Delta x = (\omega/k)\Delta t$ so that the propagation speed of the wave front (phase velocity) is

$$v = \frac{\Delta x}{\Delta t} = \frac{\omega}{k}. \tag{2.20}$$

**Exercise 2.1**  A plane wave is expressed as

$$E = (-2\hat{i} + 2\sqrt{3}\hat{j})\exp[j(\sqrt{3}x + y + 6 \times 10^8\,t)].$$

Find; 1) the direction of polarization, 2) the direction of propagation, 3) the phase velocity, 4) the amplitude, 5) the frequency, and 6) the wavelength.

*Solution*

1) $-\hat{i} + \sqrt{3}\hat{j}$

2) $k = \sqrt{3}\hat{i} + \hat{j}$. This wave is propagating in the $-k$ direction (210°)

3) $|k| = k = \sqrt{3+1} = 2$

$$v = \frac{\omega}{k} = 3 \times 10^8 \text{ m/s}$$

4) $|E| = \sqrt{2^2 + 2^2 \times 3} = 4$ v/m

5) $f = \frac{\omega}{2\pi} = \frac{3}{\pi} \times 10^8$ Hz

6) $\lambda = \frac{v}{f} = \pi$ m.

A summary of the results is shown in Fig. 2.6.

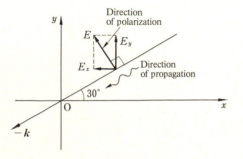

Fig. 2.6. Summary of the results of Exercise 2.1

It was shown that there are two ways of expressing waves propagating in the same direction. To avoid this ambiguity, we adopt the convention of always using a negative sign on the time function. Consequently, a plane wave propagating in the positive $x$ direction will be written as $\exp[j(kx - \omega t)]$ and not as $\exp[j(\omega t - kx)]$.

## 2.4 Interference of Two Waves

When considering the interference of two waves [2.1], the amplitudes and frequencies are assumed to be identical. In particular, consider the two waves

$$E_1 = E_0 \exp[j(\mathbf{k}_1 \cdot \mathbf{r} + \phi_1 - \omega t)]$$
$$E_2 = E_0 \exp[j(\mathbf{k}_2 \cdot \mathbf{r} + \phi_2 - \omega t)] . \tag{2.21}$$

First, as shown in Fig. 2.7, the vectors $\mathbf{k}_1$, $\mathbf{k}_2$ are decomposed into components in the same, and opposite directions. The line $\overline{PQ}$ connects the tips of $\mathbf{k}_1$, and $\mathbf{k}_2$. The vectors $\mathbf{k}_1$ and $\mathbf{k}_2$ are decomposed into $\mathbf{k}_1'$ and $\mathbf{k}_2'$, which are parallel to $\overline{OC}$, and $\mathbf{k}_1''$ and $\mathbf{k}_2''$, which are parallel to $\overline{CP}$ and $\overline{CQ}$, with

$$\mathbf{k}_1 = \mathbf{k}_1' + \mathbf{k}_1'' , \qquad \mathbf{k}_2 = \mathbf{k}_2' + \mathbf{k}_2'' ,$$
$$\mathbf{k}_1' = \mathbf{k}_2' = \mathbf{k}' , \qquad \mathbf{k}_1'' = -\mathbf{k}_2'' = \mathbf{k}'' . \tag{2.22}$$

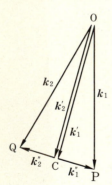

Fig. 2.7. Components of wave vectors $\mathbf{k}_1$ and $\mathbf{k}_2$

The resultant field $E = E_1 + E_2$ is obtained by inserting (2.22) into (2.21).

$$\underbrace{E = 2E_0 \cos(\mathbf{k}'' \cdot \mathbf{r} + \Delta\phi)}_{\text{amplitude}} \underbrace{e^{j(\mathbf{k}' \cdot \mathbf{r} + \phi - \omega t)}}_{\text{phase}} , \tag{2.23}$$

$$\phi = \frac{\phi_1 + \phi_2}{2} , \qquad \Delta\phi = \frac{\phi_1 - \phi_2}{2} . \tag{2.24}$$

The first factor in (2.23) has no dependence on time. Consequently, along the locus represented by

$$\boldsymbol{k}'' \cdot \boldsymbol{r} + \Delta\phi = (2\,n+1)\,\frac{\pi}{2}\,, \tag{2.25}$$

the field intensity is always zero and does not change with respect to time. These loci, shown in Fig. 2.8a, generate a stationary pattern in space known as an interference fringe. The presence of interference fringes is related only to the vectors $\boldsymbol{k}''$ and $-\boldsymbol{k}''$, which are components pointed in opposite directions, and is not dependent on $\boldsymbol{k}'$ at all. When the frequencies of the two waves are identical and when no components occur in opposite directions, no interference fringe pattern appears. The spacing between the adjacent null lines of the fringes in $\boldsymbol{k}''$ is obtained from (2.25) by subtracting the case for $n$ from that of $n+1$ and is $\pi/|\boldsymbol{k}''|$ and the period is twice this value. It is also interesting to note that the fringe pattern can be shifted by changing the value of $\Delta\phi$. The second factor in (2.23) is equivalent to the expression for a plane wave propagating in the $\boldsymbol{k}'$ direction.

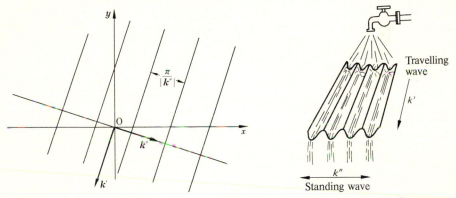

**Fig. 2.8a, b.** Interference pattern produced by the vector components $\boldsymbol{k}''$. (a) Geometry, (b) analogy to a corrugated sheet

To better understand the travelling-wave and standing-wave aspects of (2.23), an analogy can be drawn between the interference of the two waves and a corrugated plastic sheet. If one aligns $\boldsymbol{k}''$ in the direction cutting the ridges and grooves of the corrugation, the variation of the amplitude of the standing wave is represented by the corrugation, while the direction along the grooves aligns the direction $\boldsymbol{k}'$ of the propagation of the travelling wave (Fig. 2.8b).

## 2.5 Spatial Frequency

The word "frequency" is normally used to mean "temporal frequency", i.e., frequency usually denotes a rate of repetition of wave forms in unit time. In a similar manner, "spatial frequency" is defined as the rate of repetition of a particular pattern in unit distance.

The spatial frequency is indispensable in quantitatively describing the resolution power of lenses and films and in signal processing analysis. The units of spatial frequency are lines/mm, lines/cm, or lines/inch. In some cases, such as when measuring human-eye responses, the units of spatial frequency are given in terms of solid angles, e.g., lines/minute of arc or lines/degree of arc.

Since most wave form patterns in this text are treated as planar, the spatial frequencies of interest are those pertaining to the planar coordinates, usually $x$ and $y$. Fig. 2.9 illustrates the definition of these spatial frequencies, namely

$$\text{spatial frequency } f_x \text{ in } x = \frac{1}{\text{OA}} = \frac{1}{\lambda_x}$$

$$\text{spatial frequency } f_y \text{ in } y = \frac{1}{\text{OB}} = \frac{1}{\lambda_y}. \tag{2.26}$$

The spatial frequency $f = 1/\lambda$ in the $\overline{\text{OH}}$ direction is obtained from

$$\lambda = \lambda_x \cos \angle \text{HOA}$$
$$\lambda = \lambda_y \sin \angle \text{HOA}$$

whereby,

$$f = \frac{1}{\lambda} = \sqrt{\left(\frac{1}{\lambda_x}\right)^2 + \left(\frac{1}{\lambda_y}\right)^2} = \sqrt{(f_x)^2 + (f_y)^2} \ . \tag{2.27}$$

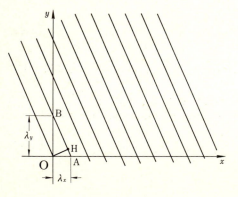

Fig. 2.9. Definition of spatial frequency

## 2.6 The Relationship Between Engineering Optics and Fourier Transforms

In this section, it will be shown that there is a Fourier transform relationship between the light distribution of a source and the light distribution appearing on a screen illuminated by the source [2.2]. Consider the configuration shown in Fig. 2.10. A one-dimensional slit of length $a$ is placed along the $x_0$ axis. The slit is illuminated from behind by a non-uniform light source. The amplitude distribution along the slit is $E(x_0)$.

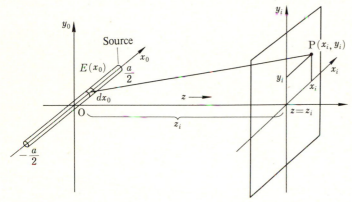

**Fig. 2.10.** Geometry used to calculate the field at the observation point P from a one dimensional slit

The slit can be approximated as a composite of point sources. From (2.6), the amplitude $E(x_i, 0)$ of the light observed along the $x_i$ axis is, to within a constant multiplying factor $K$, given by

$$E(x_i, 0) = \frac{K}{z_i} \exp\left[ j k \left( z_i + \frac{x_i^2}{2 z_i} \right) \right]$$

$$\times \int_{-a/2}^{a/2} E(x_0) \exp\left( -j 2\pi \frac{x_i}{\lambda z_i} x_0 \right) dx_0, \tag{2.28}$$

where $x_0^2/\lambda z_i \ll 1$ was assumed. It will be shown in the next chapter that $K = -j/\lambda$.

By carefully examining (2.28), one finds that this equation is in the same form as a Fourier transform, where the Fourier transform is defined as

$$G(f) = \mathscr{F}\{g(x_0)\} = \int_{-\infty}^{\infty} g(x_0) \exp(-j 2\pi f x_0) dx_0. \tag{2.29}$$

By letting

$$f = \frac{x_i}{\lambda z_i}, \qquad g(x_0) = E(x_0) \tag{2.30}$$

Eq. (2.28) can be rewritten with the help of (2.29) as

$$E(x_i, 0) = \frac{K}{z_i} \exp\left[jk\left(z_i + \frac{x_i^2}{2z_i}\right)\right] \mathscr{F}\{E(x_0)\}, \qquad f = x_i/\lambda z_i. \tag{2.31}$$

In short, *the amplitude distribution of light falling on a screen is provided by the Fourier transform of the source light distribution, where f in the transform is replaced by* $x_i/\lambda z_i$. In other words, by Fourier transforming the source function, the distribution of light illuminating a screen is obtained.

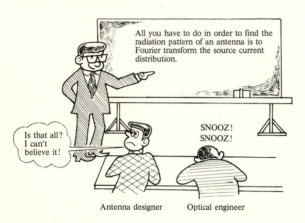

Antenna designer     Optical engineer

This fact forms the basis for solving problems of illumination and light scattering, and also plays a key role in optical signal processing. The branch of optics where the Fourier transform is frequently used is generally known as *Fourier optics*.

The usefulness of the Fourier transform is not limited to the optical region alone, but is applicable to lower frequency phenomena as well. For example, the radiation pattern of an antenna is determined by taking the Fourier transform of the current distribution on the antenna. Furthermore, the Fourier transform concept has been an important principle in radio astronomy. By knowing the distribution of the radiated field reaching the earth, the shape of the radiating source can be found. This is the reverse process of what was discussed in this chapter. The shape of the source is obtained by inverse Fourier transforming the observed radiated field. The difficulty with radio astronomy, however, is that the phase information about the radio field is not readily available to analyze.

Returning to (2.31), the problem of evaluating the Fourier transform in this equation is now considered. Most transforms are well tabulated in tables, but as an exercise the transform will be calculated for a simple example.

Let the distribution of the light along the slit be uniform and unity, and let the slit have unit length as well. The function describing the source is then

$$E(x_0) = \prod (x_0),$$ (2.32)

where

$$\prod(x) = \begin{cases} 1 & |x| \le 1/2, \\ 0 & \text{elsewhere.} \end{cases}$$ (2.33)

The value of the function is therefore unity in the range $|x| \le 1/2$ and zero elsewhere. This function, which has been given the symbol $\prod$, is known as either the *rectangle function* or the *gating function*. Its graph is shown in Fig. 2.11 a. The rectangle function in its more general form is written as

$$\prod\left(\frac{x-b}{a}\right)$$

and represents a function that takes on the value of unity in the interval of length $a$ centered at $x = b$, and is zero elsewhere. Equation (2.33) corresponds to the special case where $a = 1$ and $b = 0$.

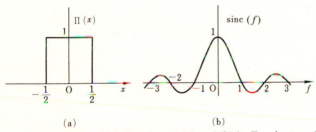

(a)                                   (b)

**Fig. 2.11.** (a) Rectangle function $\varPi(x)$ and (b) its Fourier transform

The Fourier transform of $\prod (x)$ is

$$\mathscr{F}\{\prod(x)\} = \int_{-\infty}^{\infty} \prod(x)\, e^{-j2\pi fx}\, dx = \int_{-1/2}^{1/2} e^{-j2\pi fx}\, dx$$

$$= \frac{e^{j\pi f} - e^{-j\pi f}}{j\,2\pi f} = \frac{\sin \pi f}{\pi f}.$$ (2.34)

Equation (2.34) is a sine curve that decreases as $1/\pi f$. This function is called the *sinc function*, so that

$$\mathscr{F}\{\prod(x)\} = \text{sinc}(f) = \frac{\sin \pi f}{\pi f}.$$ (2.35)

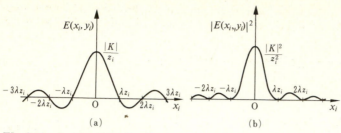

**Fig. 2.12 a, b.** The field distribution on an observation screen produced by a one dimensional slit. (**a**) Amplitude distribution, (**b**) intensity distribution

The graph of (2.35) is shown in Fig. 2.11 b. Inserting (2.35) into (2.31), the final expression for $E(x_i, y_i)$ is obtained as

$$E(x_i, y_i) = \frac{K}{z_i} \exp\left[j\,k\left(z_i + \frac{x_i^2 + y_i^2}{2z_i}\right)\right] \text{sinc}\left(\frac{x_i}{\lambda\,z_i}\right). \tag{2.36}$$

The amplitude distribution on the screen is shown in Fig. 2.12 a and the intensity distribution in Fig. 2.12 b.

## 2.7 Special Functions Used in Engineering Optics and Their Fourier Transforms

The gating function described in the previous section is one of several functions that appear often in engineering optics, and consequently, have been given special symbols to represent them [2.2, 3]. This section defines these functions and presents their Fourier transforms.

### 2.7.1 The Triangle Function

The *triangle function* is shaped like a triangle, as its name indicates. Figure 2.13 a illustrates the wave form. The function can be obtained by the convolution of $\prod(x)$ with itself

$$\Lambda(x) = \prod(x) * \prod(x). \tag{2.37}$$

Using the fact that

$$\mathscr{F}\{\prod(x) * \prod(x)\} = \mathscr{F}\{\prod(x)\}\mathscr{F}\{\prod(x)\},$$

the Fourier transform of (2.37) is found to be (Fig. 2.13 b)

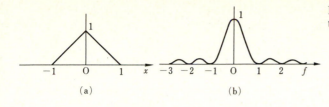

**Fig. 2.13.** (**a**) Triangle function $\Lambda(x)$ and (**b**) its Fourier transform $\mathrm{sinc}^2 f$

$$\mathscr{F}\{\Lambda(x)\} = \mathrm{sinc}^2 f. \tag{2.38}$$

### 2.7.2 The Sign Function

The function defined to be $+1$ in the positive region of $x$ and $-1$ in the negative region of $x$ is called the *sign function* (Fig. 2.14 a). It is written as sgn $x$:

$$\mathrm{sgn}\, x \begin{cases} -1 & x < 0, \\ 0 & x = 0, \\ +1 & x > 0. \end{cases} \tag{2.39}$$

The Fourier transform of the sign function is obtained as follows:

$$\mathscr{F}\{\mathrm{sgn}\, x\} = \int_{-\infty}^{0} -e^{-j2\pi fx}\, dx + \int_{0}^{\infty} e^{-j2\pi fx}\, dx. \tag{2.40}$$

The integral in (2.40) contains the improper integral

$$\left[ \frac{e^{-j2\pi fx}}{-j2\pi f} \right]_{0}^{\infty} = ?\,.$$

The exponential in the above expression is an oscillating function and such a function is not defined at $\infty$. However, the integral may still be evaluated by using appropriate limiting operations, as explained below. Adding '$\alpha x$' terms to the exponents in (2.40) and taking the limit as $\alpha$ approaches zero from the right gives

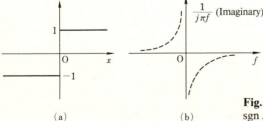

**Fig. 2.14 a, b.** Curves of (**a**) sign function sgn $x$ and (**b**) its Fourier transform $1/j\pi f$

$$\mathscr{F}\{\text{sgn } x\} = -\int_{-\infty}^{0} \lim_{\alpha \to 0+} e^{-j2\pi fx + \alpha x} \, dx + \int_{0}^{\infty} \lim_{\alpha \to 0+} e^{-j2\pi fx - \alpha x} \, dx.$$

Interchanging the operation of the limit with that of the integral results in

$$\mathscr{F}\{\text{sgn } x\} = \lim_{\alpha \to 0+} \left[ \frac{-e^{-j2\pi fx + \alpha x}}{-j2\pi f + \alpha} \right]_{-\infty}^{0} + \lim_{\alpha \to 0+} \left[ \frac{e^{-j2\pi fx - \alpha x}}{-j2\pi f - \alpha} \right]_{0}^{\infty}.$$

For positive $\alpha$ and negative $x$, the expression $\exp(\alpha x) \exp(-j2\pi fx)$ is the product of an exponentially decaying function and a bounded oscillating function. Thus, at $x = -\infty$, $\exp(\alpha x)$ is zero making $\exp(\alpha x) \exp(-j2\pi fx)$ zero. Likewise, for positive $\alpha$ and $x$, the expression $\exp(-\alpha x) \exp(-j2\pi fx)$ is zero at $x = \infty$. Equation (2.40) then reduces to

$$\lim_{\alpha \to 0+} \left[ -\frac{1}{-j2\pi f + \alpha} - 0 \right] + \lim_{\alpha \to 0+} \left[ 0 - \frac{1}{-j2\pi fx - \alpha} \right] = \frac{1}{j\pi f}.$$

In conclusion, the Fourier transform of the sign function is (Fig. 2.14 b)

$$\mathscr{F}\{\text{sgn } x\} = \frac{1}{j\pi f}. \tag{2.41}$$

### 2.7.3 The Step Function

The function whose value is zero when $x$ is negative and unity when $x$ is positive is defined to be the *step function*

$$H(x) = \begin{cases} 0 & \text{for} \quad x < 0, \\ \frac{1}{2} & \text{for} \quad x = 0, \\ 1 & \text{for} \quad x > 0. \end{cases} \tag{2.42}$$

The step function can be rewritten by using the sgn function as

$$H(x) = \tfrac{1}{2}(1 + \text{sgn } x).$$

The Fourier transform of $H(x)$ is obtained using (2.41).

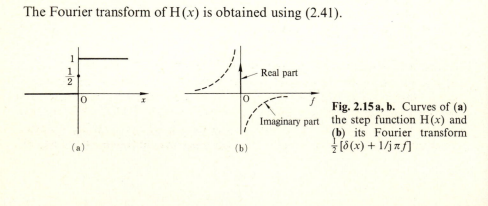

Fig. 2.15 a, b. Curves of (a) the step function $H(x)$ and (b) its Fourier transform $\frac{1}{2}[\delta(x) + 1/j\pi f]$

$$\mathscr{F}\{H(x)\} = \frac{1}{2}\left(\delta(f) + \frac{1}{j\pi f}\right), \qquad (2.43)$$

where the symbol $\delta$ represents the delta function to be explained in the next section. Figure 2.15 illustrates the step function and its Fourier transform.

### 2.7.4 The Delta Function

The *delta function* $\delta(x)$ possesses the following two properties [2.4]:

1) The value of the function is zero except at $x = 0$. The value at $x = 0$ is infinity.
2) The area enclosed by the curve and the $x$ axis is unity.

A simple description of the delta function can be made using an analogy. As shown by the picture in Fig. 2.16a, no matter how high a piece a dough is pulled up, the volume is constant (in the case of the delta function, area is constant), and the width decreases indefinitely with an increase in height.

The delta function $\delta(x)$ satisfies

$$\int_{-\varepsilon}^{\varepsilon} \delta(x)\, dx = 1. \qquad (2.44)$$

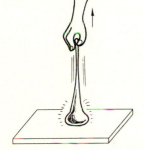
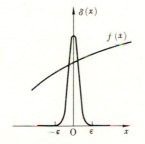

**Fig. 2.16.** (a) Analogy showing how the $\delta(x)$ function has constant area. (b) $\delta(x)$ and an arbitrary function $f(x)$

It is more common to use the delta function in an integral form than by itself, as follows.

$$\int_{-\varepsilon}^{\varepsilon} f(x)\,\delta(x)\, dx = f(0) \int_{-\varepsilon}^{\varepsilon} \delta(x)\, dx = f(0). \qquad (2.45)$$

Because the value of $\delta(x)$ is zero everywhere except in a small region of $|x| < \varepsilon$, the integration need only extend from $-\varepsilon$ to $+\varepsilon$. If the function $f(x)$ is slowly varying in the interval $|x| < \varepsilon$, then $f(x)$ can be considered constant over this interval, i.e., $f(x)$ is approximately $f(0)$ in the interval $|x| < \varepsilon$, as illustrated in Fig. 2.16b. Equation (2.45) was obtained under this assumption. Equation (2.45) thus expresses an important property of the delta function and is sometimes used as a definition of the delta function.

Another important property of the delta function is

$$\delta(a\,x) = \frac{1}{|a|}\,\delta(x). \tag{2.46}$$

Equation (2.46) is readily obtained from (2.45). Let $a$ be a positive number. Using $\delta(a\,x)$ in place of $\delta(x)$ in (2.45) and then making the substitution $a\,x = y$ gives

$$\int_{-\varepsilon}^{\varepsilon} f(x)\,\delta(a\,x)\,dx = \frac{1}{a}\int_{-a\varepsilon}^{a\varepsilon} f\left(\frac{y}{a}\right)\delta(y)\,dy = \frac{1}{a}f(0). \tag{2.47}$$

Comparing both sides of (2.45, 47), one concludes

$$\delta(a\,x) = \frac{1}{a}\,\delta(x). \tag{2.48}$$

Similarly, it can be shown that

$$\delta(-a\,x) = \frac{1}{a}\,\delta(x) \tag{2.49}$$

where $a$ again is a positive number. Equation (2.46) is a combination of (2.48, 49), and holds true for positive and negative $a$. The Fourier transform of the delta function is

$$\int_{-\infty}^{\infty} \delta(x)\,e^{-j2\pi f x}\,dx = e^{j2\pi f \cdot 0} = 1, \quad \text{and thus}$$

$$\mathscr{F}\{\delta(x)\} = 1. \tag{2.50}$$

### 2.7.5  The Comb Function

The *comb function*, shown in Fig. 2.17, is an infinite train of delta functions. The spacing between successive delta functions is unity. The comb function is given the symbol III and is expressed mathematically as

$$III(x) = \sum_{n=-\infty}^{n=\infty} \delta(x - n). \tag{2.51}$$

Since the symbol of III is a cyrillic character and is pronounced shah, the comb function is sometimes called the shah function.

The comb function with an interval of $a$ can be generated from (2.46, 51).

$$III\left(\frac{x}{a}\right) = \sum_{n=-\infty}^{\infty} \delta\left(\frac{x}{a} - n\right) = a \sum_{n=-\infty}^{\infty} \delta(x - a\,n). \tag{2.52}$$

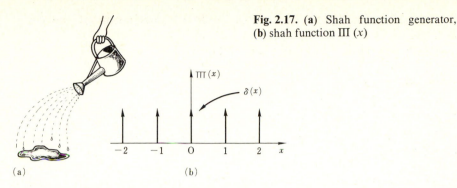

Fig. 2.17. (a) Shah function generator, (b) shah function III $(x)$

There is a scale factor $a$ appearing on the right hand side of (2.52) which can loosely be interpreted as an increase in the 'height' of each delta function with an increase in the interval $a$ thus keeping the total sum constant.

Equation (2.51) can be used to generate a periodic function from a function with an arbitrary shape. Since

$$g(x) * \delta(x - n) = g(x - n)$$

is true, a function generated by repeating $g(x)$ at an interval $a$ is

$$h(x) = \frac{1}{a} g(x) * \text{III}\left(\frac{x}{a}\right). \tag{2.53}$$

Figure 2.18 illustrates this operation.

Next, the Fourier transform of the comb function is obtained. The comb function III $(x)$ is a periodic function with an interval of unity and can be expanded into a Fourier series, as indicated in Fig. 2.19. The definition of the Fourier series is

$$f(x) = \sum_{n=-\infty}^{\infty} a_n \, e^{j 2 \pi n (x/T)} \qquad \text{where} \tag{2.54}$$

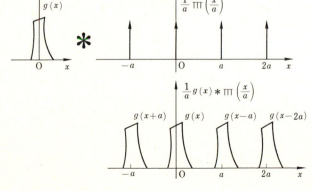

Fig. 2.18. Convolution of an arbitrary function $g(x)$ with the shah function resulting in a periodic function of $g(x)$

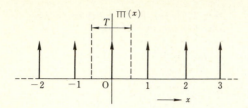

**Fig. 2.19.** Fourier expansion of the shah function

$$a_n = \frac{1}{T} \int_{-T/2}^{T/2} f(x)\, e^{-j2\pi n(x/T)}\, dx. \tag{2.55}$$

Inserting $f(x) = \text{III}(x)$ and $T = 1$ into (2.55), $a_n$ becomes unity, and the Fourier expansion of $\text{III}(x)$ is found to be

$$\text{III}(x) = \sum_{n=-\infty}^{\infty} e^{j2\pi nx}. \tag{2.56}$$

Equation (2.56) is an alternate way of expressing the comb function. The Fourier transform of (2.56) is

$$\mathscr{F}\{\text{III}(x)\} = \sum_{n=-\infty}^{\infty} \delta(f-n) = \text{III}(f), \tag{2.57}$$

where the relationship

$$\int_{-\infty}^{\infty} e^{j2\pi(f-n)}\, dx = \delta(f-n)$$

and then (2.46) were used.

The result is that the Fourier transform of a comb function is again a comb function. This is a very convenient property widely used in analysis.

## 2.8 Fourier Transform in Cylindrical Coordinates

In engineering optics, one deals with many phenomena having circular distributions. The cross-section of a laser beam or circular lenses are two such examples. In these cases, it is more convenient to perform the Fourier transform in cylindrical coordinates [2.5].

### 2.8.1 Hankel Transform

The two-dimensional Fourier transform in rectangular coordinates is

$$G(f_x, f_y) = \int_{-\infty}^{\infty} \int_{-\infty}^{\infty} g(x, y) \exp\left[-j\, 2\pi(f_x\, x + f_y\, y)\right] dx\, dy. \tag{2.58}$$

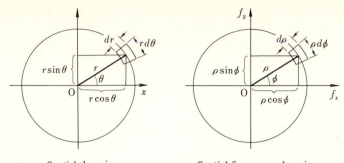

Spatial domain                    Spatial frequency domain

**Fig. 2.20 a, b.** Fourier transform in cylindrical coordinates. (**a**) Spatial domain, (**b**) spatial frequency domain

The expression for this equation in cylindrical coordinates is sought. Figure 2.20 shows the relationship between rectangular and cylindrical coordinates, namely,

$$x = r \cos \theta$$
$$y = r \sin \theta \tag{2.59}$$
$$dx\,dy = r\,dr\,d\theta$$

and in the Fourier transform plane

$$f_x = \varrho \cos \phi$$
$$f_y = \varrho \sin \phi \tag{2.60}$$
$$df_x\,df_y = \varrho\,d\varrho\,d\phi$$

Inserting (2.59, 60) into (2.58), the result is

$$G(\varrho, \phi) = \int\limits_0^\infty \int\limits_0^{2\pi} g(r, \theta)\, e^{-j2\pi\varrho r \cos(\theta - \phi)}\, r\,dr\,d\theta, \tag{2.61}$$

where $g(r \cos \theta, r \sin \theta)$ was written as $g(r, \theta)$. As long as $g(r, \theta)$ is a single valued function, $g(r, \theta)$ equals $g(r, \theta + 2n\,\pi)$. Therefore, $g(r, \theta)$ is a periodic function of $\theta$ with a period of $2\pi$ and can be expanded into a Fourier series. The coefficients of the Fourier series are a function of $r$ only, and referring to (2.54, 55),

$$g(r, \theta) = \sum_{n=-\infty}^{\infty} g_n(r)\, e^{jn\theta}, \quad \text{where} \tag{2.62}$$

$$g_n(r) = \frac{1}{2\pi} \int\limits_0^{2\pi} g(r, \theta)\, e^{-jn\theta}\, d\theta. \tag{2.63}$$

Equations (2.62, 63) are inserted into (2.61) to give

$$G(\varrho, \phi) = \sum_{n=-\infty}^{\infty} \int_0^{\infty} r\, g_n(r)\, dr \int_0^{2\pi} e^{-j2\pi\varrho r\cos(\theta-\phi)+jn\theta}\, d\theta . \tag{2.64}$$

Using the Bessel function expression in integral form

$$J_n(z) = \frac{1}{2\pi} \int_a^{2\pi+a} e^{j(n\beta - z\sin\beta)}\, d\beta , \tag{2.65}$$

Eq. (2.64) can be simplified. In order to apply (2.65), result (2.64) must be rewritten. Let $\cos(\theta - \phi)$ be replaced by $\sin\beta$, where [1]

$$\theta - \phi = \beta + \frac{3\pi}{2} . \tag{2.66}$$

Inserting (2.66) into the second integral in (2.64) gives

$$\int_0^{2\pi} e^{-j2\pi\varrho r\cos(\theta-\phi)+jn\theta}\, d\theta = \exp\left(j\, n\, \frac{3}{2}\, \pi + j\, n\, \phi\right) \int_{-\phi-3\pi/2}^{2\pi-\phi-3\pi/2} e^{j(n\beta - 2\pi\varrho r\sin\beta)}\, d\beta$$

$$= (-j)^n\, e^{jn\phi}\, 2\pi\, J_n(2\pi\varrho r) . \tag{2.67}$$

Inserting (2.67) into (2.64) gives

$$G(\varrho, \phi) = \sum_{n=-\infty}^{\infty} (-j)^n\, e^{jn\phi}\, 2\pi \int_0^{\infty} r\, g_n(r)\, J_n(2\pi\varrho r)\, dr, \quad \text{where} \tag{2.68}$$

$$g_n(r) = \frac{1}{2\pi} \int_0^{2\pi} g(r, \theta)\, e^{-jn\theta}\, d\theta . \tag{2.69}$$

Similarly, the inverse Fourier transform is

$$g(r, \theta) = \sum_{n=-\infty}^{\infty} (-j)^n\, e^{-jn\phi}\, 2\pi \int_0^{\infty} \varrho\, G_n(\varrho)\, J_n(2\pi r\varrho)\, d\varrho . \tag{2.70}$$

In the special case where no variation occurs in the $\theta$ direction, i.e., when $g(r, \theta)$ is cylindrically symmetric, the coefficients $g_n(r)$ can be written as

---

1 Another substitution

$$\theta - \phi = \frac{\pi}{2} - \beta$$

also seems to satisfy the condition. If this substitution is used, the exponent becomes $-j\, n\, \beta + j\, n\, (\phi + \pi/2)$ and the sign of $j\, n\, \beta$ becomes negative so that (2.65) cannot be immediately applied.

$$g_n(r) = \begin{cases} g_0(r) & \text{for} \quad n = 0, \\ 0 & \text{for} \quad n \neq 0, \end{cases} \tag{2.71}$$

and (2.68, 70) simplify to

$$G_0(\varrho) = 2\pi \int_0^\infty r\, g_0(r)\, J_0(2\pi \varrho r)\, dr, \tag{2.72}$$

$$g_0(r) = 2\pi \int_0^\infty \varrho\, G_0(\varrho)\, J_0(2\pi r \varrho)\, d\varrho. \tag{2.73}$$

Equations (2.68−73) are Fourier transforms in cylindrical coordinates. Essentially, they are no different than their rectangular counterparts. *Answers to a problem will be identical whether rectangular or cylindrical transforms are used.* However, depending on the symmetry of the problem, one type of transform may be easier to calculate than the other.

The integrals in (2.68, 70) are called the *Hankel transform of the nth order and its inverse transform,* respectively. The zero-th order of the Hankel transform (2.72, 73) is called the *Fourier Bessel transform* and its operation is symbolized by $\mathscr{B}\{\;\}$.

Now that the Fourier Bessel transform has been studied, another special function is introduced that can be dealt with very easily by using the Fourier Bessel transform. A function having unit value inside a circle of radius 1, and zero value outside the circle, as shown in Fig. 2.21 a, is called the *circle function.* The symbol 'circ' is used to express this function

$$\text{circ}(r) = \begin{cases} 1 & \text{for} \quad r \leq 1 \\ 0 & \text{for} \quad r > 1. \end{cases} \tag{2.74}$$

To obtain the Fourier transform of (2.74), result (2.72) can be used since $\text{circ}(r)$ has circular symmetry. Therefore, the Fourier Bessel transform of $\text{circ}(r)$ is

$$G(\varrho) = \mathscr{B}\{\text{circ}(r)\} = 2\pi \int_0^1 r\, J_0(2\pi \varrho r)\, dr. \tag{2.75}$$

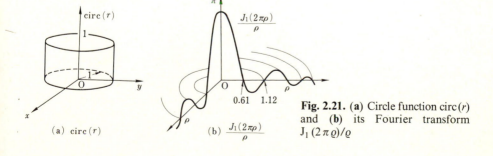

(a) $\text{circ}(r)$

(b) $\dfrac{J_1(2\pi\rho)}{\rho}$

**Fig. 2.21.** (a) Circle function $\text{circ}(r)$ and (b) its Fourier transform $J_1(2\pi\varrho)/\varrho$

Introducing a change of variables,

$$2\pi r \varrho = \tau,$$

the transform becomes

$$G(\varrho) = \frac{1}{2\pi \varrho^2} \int_0^{2\pi\varrho} \tau \, J_0(\tau) \, d\tau = \frac{J_1(2\pi\varrho)}{\varrho} \, . \tag{2.76}$$

where, the formula for the Bessel integral

$$\int_0^z z \, J_0(\alpha z) \, dz = \frac{z}{\alpha} J_1(\alpha z) \tag{2.77}$$

was used. Figure 2.21 b illustrates the shape of (2.76).

### 2.8.2 Examples Involving Hankel Transforms

A few examples of problems requiring the use of Hankel transforms will be presented.

**Exercise 2.2**   As shown in Fig. 2.22, a mirror is tilted at an angle $\gamma$ about the $x$-axis. Let the light reflected from the mirror be expressed by

$$g(x, y) = e^{j2ky\tan\gamma} \, \mathrm{circ}\left(\sqrt{x^2 + y^2}\right) . \tag{2.78}$$

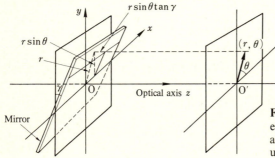

Fig. 2.22. Diagram showing geometry of Exercise 2.2 (mirror placed at an angle $\gamma$ to the plane perpendicular to the optical axis)

The projection of the mirror's light distribution on the $x\,y$ plane is a circle of unit radius.

a) Express $g(x, y)$ in cylindrical coordinates.
b) Expand the above result for $g(r, \theta)$ in a Fourier series with respect to $\theta$.
c) Find the Hankel transform of $g(r, \theta)$.

*Solution*   *(a)* From Fig. 2.22, $y$ can be expressed in terms of $r$ and $\theta$ as

$$y = r \sin \theta$$

so that

$$2k\, y \tan \gamma = k\, \beta\, r \sin \theta, \quad \text{where} \quad \beta = 2 \tan \gamma. \tag{2.79}$$

Inserting (2.79) into (2.78) gives

$$g(r, \theta) = e^{jk\beta r \sin\theta}\, \text{circ}(r). \tag{2.80}$$

(b) By using (2.63), one obtains

$$g_n(r) = \frac{\text{circ}(r)}{2\pi} \int\limits_0^{2\pi} e^{j(k\beta r \sin\theta - n\theta)}\, d\theta. \tag{2.81}$$

In order to use (2.65), the substitution $\theta = -\theta'$ must be made

$$g_n(r) = \frac{\text{circ}(r)}{2\pi} \int\limits_{-2\pi}^{0} e^{j(n\theta' - k\beta r \sin\theta')}\, d\theta'. \tag{2.82}$$

Setting $a = -2\pi$ in (2.65) results in

$$g_n(r) = \text{circ}(r)\, J_n(k\,\beta\, r). \tag{2.83}$$

Inserting this into (2.62) gives

$$g(r, \theta) = \sum_{n=-\infty}^{\infty} \text{circ}(r)\, J_n(k\,\beta\, r)\, e^{jn\theta}. \tag{2.84}$$

(c) Inserting the above result into (2.68) gives

$$G(\varrho, \phi) = 2\pi \sum_{n=-\infty}^{\infty} (-j)^n\, e^{jn\phi} \int\limits_0^1 r\, J_n(k\,\beta\, r)\, J_n(2\pi\varrho r)\, dr. \tag{2.85}$$

Applying the following Bessel integral formula,

$$\int\limits_0^1 J_n(\beta x)\, J_n(\alpha x)\, x\, dx$$
$$= \frac{1}{\alpha^2 - \beta^2}\, [\alpha\, J_n(\beta)\, J_{n+1}(\alpha) - \beta\, J_n(\alpha)\, J_{n+1}(\beta)], \tag{2.86}$$

Equation (2.85) becomes

$$G(\varrho, \phi) = \sum_{n=-\infty}^{\infty} 2\pi\, \frac{(-j)^n\, e^{jn\phi}}{(k\,\beta)^2 - (2\pi\varrho)^2}\, [k\,\beta\, J_n(2\pi\,\varrho)\, J_{n+1}(k\,\beta)$$
$$- 2\pi\,\varrho\, J_n(k\,\beta)\, J_{n+1}(2\pi\,\varrho)]. \tag{2.87}$$

This example was chosen from the mode analysis at a slant laser mirror.

**Exercise 2.3**   Express the plane wave $E = \exp(jky)$, propagating in the positive $y$ direction, in cylindrical coordinates and expand it into a Fourier series in $\theta$.

*Solution*   The results of Exercise 2.2 are used to obtain

$$E = e^{jky} = e^{jkr\sin\theta}. \tag{2.88}$$

Using (2.80, 84) with $kr$ instead of $k\beta r$ gives

$$e^{jkr\sin\theta} = \sum_{n=-\infty}^{\infty} J_n(kr)\, e^{jn\theta}. \tag{2.89}$$

A plane wave $\exp(-jky)$ propagating in the negative $y$ direction can be obtained in a similar way

$$e^{-jkr\sin\theta} = \sum_{n=-\infty}^{\infty} J_n(kr)\, e^{-jn\theta}. \tag{2.90}$$

Equations (2.89, 90) can be thought of as a group of waves which are propagating in the $\pm\,\theta$ direction and varying as $\exp(\pm\,j\,n\,\theta)$.

## 2.9  A Hand-Rotating Argument of the Fourier Transform

Consider the meaning of the Fourier transform

$$G(f) = \int_{-\infty}^{\infty} g(x)\, e^{-j2\pi fx}\, dx.$$

Suppose $g(x)$ contains many frequency components and is expressed as

$$g(x) = G_0 + G_1\, e^{j2\pi f_1 x} + G_2\, e^{j2\pi f_2 x} + G_3\, e^{j2\pi f_3 x} \dots .$$

The phasor of each term rotates with $x$. Now $g(x)$ is multiplied by the factor $\exp(-j2\pi fx)$. Only the term whose phasor rotates at the same speed as $f$, but in the opposite direction, stops rotation, whereas all other terms still rotate with $x$.

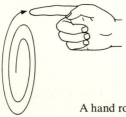

A hand rotating argument of the Fourier transform

When the integration with respect to $x$ is performed, all the terms whose phasor still rotates and whose real and imaginary parts oscillate with respect to the zero axis become zero, whereas the integration of the terms whose phasor has stopped rotating does not have a zero value.

Thus, the Fourier transform operation can be considered as a means of picking a term whose phasor rotates at a specific speed.

## Problems

**2.1**  A vector wave is expressed as

$$E = \left( -\frac{\hat{i}}{2} - \frac{\hat{j}}{2} + \hat{k} \right) e^{j2\pi[10(x+y+z)-3\times10^9 t]} .$$

By using MKS units, find:

1) the direction of polarization,      4) the amplitude,
2) the direction of propagation,      5) the frequency,
3) the velocity,                      6) the wavelength.

**2.2**  Two simultaneous measurements upon an unknown sinusoidal wave were made along the $x$ and $y$ axes. The phase distribution along the $x$ axis was $\phi_x = 10x$ radians and that along the $y$ axis was $\phi_y = \sqrt{3}\,10y$ radians. What is the direction of incidence of this unknown wave and what is its wavelength?

**2.3**  Prove the following relationships

a)  $\mathscr{F}\left\{ \dfrac{d}{dx} g(x) \right\} = j\,2\pi f\, G(f) ,$   where  $\mathscr{F}\{g(x)\} = G(f) .$

b)  $\dfrac{d}{dx} [f(x) * g(x)] = \left[ \dfrac{d}{dx} f(x) \right] * g(x) = f(x) * \left[ \dfrac{d}{dx} g(x) \right] .$

c)  $\mathrm{III}(-x) = \mathrm{III}(x) .$

d)  $\mathrm{III}(x - \tfrac{1}{2}) = \mathrm{III}(x + \tfrac{1}{2}) .$

e)  $\pi\,\delta(\sin\pi x) = \mathrm{III}(x) .$

f)  $\mathrm{III}\left( \dfrac{x}{2} \right) = \mathrm{III}(x) + \mathrm{III}(x)\, e^{j\pi x} .$

g)  $\mathrm{III}(x) = \lim\limits_{N \to \infty} \left| \dfrac{\sin N\pi x}{\sin \pi x} \right| .$

**2.4**  Find the formulae which describe the two transmittance distributions shown in Figure 2.23 and calculate their Fourier transforms.

(a)

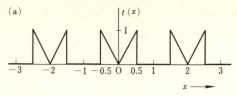

(b)

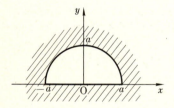

Fig. 2.23 a, b. Masks with transmission distributions $t(x)$

**2.5** Prove that the expression for the delta function in cylindrical coordinates is

$$\delta(\varrho) = 2\pi \int_0^\infty r\, J_0\,(2\pi\varrho r)\, dr\,.$$

**2.6** Find the Fourier transform $G(\varrho,\phi)$ of the transmittance function

$$g(r,\theta) = \text{circ}(r)\cos\theta\,.$$

**2.7** Let the transmittance function of a semi-circular window, as shown in Fig. 2.24, be $t(r,\theta)$. Find its Fourier transform $T(\varrho,\phi)$.

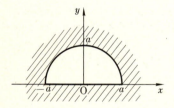

Fig. 2.24. Semi-circular mask (see Problem 2.7)

# 3. Basic Theory of Diffraction

In order to explain the diffraction phenomenon of light sneaking into shaded regions, C. Huygens (1678) claimed, "In the process of propagation of the wave, new wave fronts are emanated from every point of the old wave front". This principle, however, had a serious drawback in that the wave front, as drawn, generated an unwanted wave front in the backward direction in addition to the wave front in the forward direction. This draw-back was overcome by A. J. Fresnel (1818) and later by G. Kirchhoff (1882) by taking the periodic nature of the wave into consideration.

## 3.1 Kirchhoff's Integral Theorem

Let the light wave be expressed by $E(x, y, z)$. As an electromagnetic wave, $E$ satisfies the wave equation

$$\nabla^2 E + k^2 E = 0 . \tag{3.1}$$

A simplified theory [3.1−5] will be adopted here which focuses on only one of the vector components. This component will be represented by the scalar quantity $v$.

In a region where there is no source, $v$ satisfies the wave equation

$$\nabla^2 v + k^2 v = 0 . \tag{3.2}$$

A spherical wave in free space can be expressed in terms of the solution of (3.2) in spherical coordinates. Since the only variation of $v$ is in the $r$ direction, $\nabla^2 v$ becomes

$$\nabla^2 v = \frac{1}{r} \frac{d^2 (rv)}{dr^2} ,$$

so that (3.2) reduces to

$$\frac{d^2 (rv)}{dr^2} + k^2 (rv) = 0 . \tag{3.3}$$

The general solution of (3.3) is

$$r v = A\,e^{jkr} + B\,e^{-jkr}, \quad \text{or} \tag{3.4}$$

$$v = A\frac{e^{jkr}}{r} + B\frac{e^{-jkr}}{r}. \tag{3.5}$$

Since this book uses the $\exp(-j\omega t)$ convention, the first term of (3.5) represents a wave diverging from the origin and the second, a wave converging toward the origin.

Next, an approximate expression is derived for the field when the radiation medium is not free space. According to Green's theorem the following relationship holds for given functions $v$ and $u$ inside a volume $V$ enclosed completely by the surface $S$, i.e.

$$I = \int_V (v\nabla^2 u - u\nabla^2 v)\,dV = \int_S (v\nabla u - u\nabla v)\cdot\hat{\boldsymbol{n}}\,dS. \tag{3.6}$$

where $\hat{\boldsymbol{n}}$ is the outward normal to the surface $S$. Green's theorem is used for converting a volume integral into a surface integral. The only condition imposed on (3.6) is that the volume $V$, as shown in Fig. 3.1, be formed in such a way as to exclude the points $P_1$, $P_2$, $P_3$, ..., $P_n$ where the second derivatives of $v$ and $u$ are discontinuous. Other than this condition, there are no restrictions on the choice of the functions $v$ and $u$.

Since the choice of $v$ and $u$ is arbitrary so long as the above-mentioned condition is satisfied, $v$ will be taken as

$$v = e^{jkr}/r, \tag{3.7}$$

which is the solution of (3.2), and $u$ as the solution of

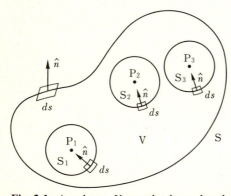

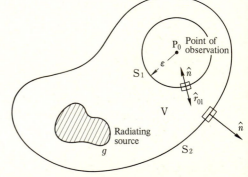

**Fig. 3.1.** A volume $V$ completely enclosed by a surface $S$. Note that the volume should not include any singularities such as $P_1, P_2, P_3, \ldots P_n$

**Fig. 3.2.** A volume $V$ containing a source $g$. Note that the observation point is excluded from the vloume but the source $g$ is not

$$\nabla^2 u + k^2 u = -g \, .$$ (3.8)

Equation (3.8) is the wave equation inside a domain with a radiating source like the one in Fig. 3.2. The integral $I$ was obtained by inserting (3.2, 8) into the left-hand side of (3.6) so that

$$I = - \int_V g \frac{e^{jkr}}{r} \, dV \, .$$ (3.9)

Since the first and second derivatives of $v$ are both discontinuous at the origin $r = 0$, the volume must not include this point. The origin is taken at the observation point $P_0$. A small sphere centered at $P_0$ with radius $\varepsilon$ is excluded from the volume $V$. The removal at this sphere creates a new surface $S_1$ and the total surface of the volume becomes $S_1 + S_2$, where $S_2$ is the external surface as shown in Fig. 3.2. Inserting (3.7, 9) into (3.6) gives

$$\int_V g \frac{e^{jkr}}{r} \, dV + \int_{S_1} \left[ \frac{e^{jkr}}{r} \nabla u - u \nabla \left( \frac{e^{jkr}}{r} \right) \right] \cdot \hat{n} \, dS$$

$$+ \int_{S_2} \left[ \frac{e^{jkr}}{r} \nabla u - u \nabla \left( \frac{e^{jkr}}{r} \right) \right] \cdot \hat{n} \, dS = 0 \, .$$ (3.10)

As the first step toward evaluating (3.10), the integration over the surface $S_1$ will be examined. Expressing the integrand in spherical coordinates gives

$$\left( \nabla \frac{e^{jkr_{01}}}{r_{01}} \right) \cdot \hat{n} = \left[ \hat{r}_{01} \frac{d}{dr_{01}} \left( \frac{e^{jkr_{01}}}{r_{01}} \right) \right] \cdot \hat{n} = \left( j k - \frac{1}{r_{01}} \right) \frac{e^{jkr_{01}}}{r_{01}} \hat{r}_{01} \cdot \hat{n}$$ (3.11)

where $\hat{r}_{01}$ is a unit vector pointing radially outward from the center $P_0$ of the sphere, to the surface $S_1$ of integration. Since the unit vectors $\hat{r}_{01}$ and $\hat{n}$ are parallel but in the opposite directions on the spherical surface with radius $\varepsilon$, the value of $\hat{r}_{01} \cdot \hat{n}$ is $-1$. Hence the value of (3.11) on the surface of the sphere $S_1$ is

$$- \left( j k - \frac{1}{\varepsilon} \right) \frac{e^{jk\varepsilon}}{\varepsilon} \, .$$ (3.12)

Inserting this result into the second term on the left-hand side of (3.10) and taking the limit as $\varepsilon \to 0$, and assuming that the integrand on the surface of this small sphere is constant, yields

$$\lim_{\varepsilon \to 0} 4 \pi \varepsilon^2 \left[ \frac{e^{jk\varepsilon}}{\varepsilon} (\nabla u) \cdot \hat{n} + \left( j k - \frac{1}{\varepsilon} \right) \frac{e^{jk\varepsilon}}{\varepsilon} u (\varepsilon) \right] = - 4 \pi u (0) \, ,$$ (3.13)

where $u(0)$ is the value of $u$ at the observation point $P_0$. Inserting (3.13) into (3.10) and replacing $u(0)$ by $u_p$, produces the result

$$u_{\mathrm{p}} = \frac{1}{4\pi} \int_V g \frac{e^{jkr_{01}}}{r_{01}} dV + \frac{1}{4\pi} \int_{S_2} \left[ \frac{e^{jkr_{01}}}{r_{01}} \nabla u - u \nabla \left( \frac{e^{jkr_{01}}}{r_{01}} \right) \right] \cdot \hat{n} \, dS .$$  (3.14)

This formula gives the field at the point of observation.

The amplitude of the illumination is, of course, influenced by the shape of the source, or the location of the observation, but should not be influenced by the choice of the volume of integration. A domain $V'$ such as shown in Fig. 3.3, which excludes the light source, is equally applicable and the result for $u_{\mathrm{p}}$ is identical to that obtained using domain $V$. A rigorous proof of this statement is left as an exercise, but a general outline on how to proceed is given here. In the case of $V'$, the value of

$$\frac{1}{4\pi} \int g \frac{e^{jkr_{01}}}{r_{01}} dV$$

is zero, but new surfaces $S_3$, $S_4$ and $S_5$ are created in the process of digging a hole to exclude the source, as illustrated in Fig. 3.3. The surface integral over $S_3$ makes up for the loss of

$$\frac{1}{4\pi} \int g \frac{e^{jkr_{01}}}{r_{01}} dV$$

and the value of $u_{\mathrm{p}}$ does not change. The surface integral over $S_4 + S_5$ is zero because the integral over $S_4$ exactly cancels that over $S_5$ due to the fact that $\hat{n}_4 = -\hat{n}_5$. The value of $u_{\mathrm{p}}$ in this case is the surface integral over $S = S_2 + S_3$, excluding a small spherical surface at the point of observation $P_0$.

$$u_{\mathrm{p}} = \frac{1}{4\pi} \int_S \left[ \frac{e^{jkr_{01}}}{r_{01}} \nabla u - u \nabla \left( \frac{e^{jkr_{01}}}{r_{01}} \right) \right] \cdot \hat{n} \, dS .$$  (3.15)

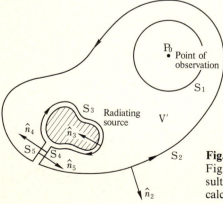

**Fig. 3.3.** A domain $V'$ which is similar to $V$ in Fig. 3.2 but with the source excluded. The resulting $u_{\mathrm{p}}$ using this geometry is identical to that calculated using the geometry in Fig. 3.2

Either (3.14) or (3.15) is called the *integral theorem* of *Kirchhoff*. Using this theorem, the amplitude of the light at an arbitrary observation point can be obtained by knowing the field distribution of light on the surface enclosing the observation point. The next section rewrites the Kirchhoff formula in a form better suited for solving diffraction problems, and we derive a result which is known as the *Fresnel-Kirchhoff diffraction formula*.

## 3.2 Fresnel-Kirchhoff Diffraction Formula

The Kirchhoff diffraction theorem will be used to find the diffraction pattern of an aperture when illuminated by a point source and projected onto a screen [3.2, 3]. Consider a domain of integration enclosed by a masking screen $S_c$, a surface $S_A$ bridging the aperture, and a semi-sphere $S_R$ centered at the observation point $P_0$ with radius $R$, as shown in Fig. 3.4. The integral on the surface of the masking screen $S_c$ is zero since the light amplitude is zero here. The integral of (3.15) becomes

$$u_p = \frac{1}{4\pi} \int_{S_A+S_R} \left[ \frac{e^{jkr_{01}}}{r_{01}} \nabla u - u \nabla \left( \frac{e^{jkr_{01}}}{r_{01}} \right) \right] \cdot \hat{n} \, dS . \tag{3.16}$$

It can also be shown that, under certain conditions, the integral of (3.16) over $S_R$ is zero. The integral being examined is

$$\int_{S_R} \left[ \frac{e^{jkR}}{R} \nabla u - u \left( j\,k - \frac{1}{R} \right) \frac{e^{jkR}}{R} \hat{r}_{03} \right] \cdot \hat{n} \, dS ,$$

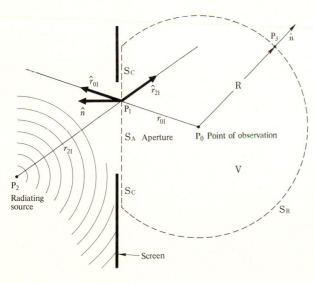

Fig. 3.4. The shape of the domain of integration conveniently chosen for calculating the diffraction pattern of an aperture using the Kirchhoff diffraction theorem

where $\hat{r}_{03}$ is a unit vector pointing from $P_0$ to point $P_3$ on $S_R$. When $R$ is very large, the integral over $S_R$ can be approximated as

$$\int_0^{4\pi} \frac{e^{jkR}}{R} [(\nabla u) \cdot \hat{n} - j k u] R^2 \, d\Omega \, , \tag{3.17}$$

where $\Omega$ is the solid angle from $P_0$ to $S_R$. Since the directions of $\hat{r}_{03}$ and $\hat{n}$ are identical, $\hat{r}_{03} \cdot \hat{n}$ is unity. It is not evident immediately that the value of (3.17) is zero, but if the condition

$$\lim_{R \to \infty} R [(\nabla u) \cdot \hat{n} - j k u] = 0 \tag{3.18}$$

is satisfied, the integral indeed vanishes. This condition is called the *Sommerfeld radiation condition*.

The boundary condition at an infinite distance is rather delicate. The condition that $u = 0$ at $R = a$ with $a \to \infty$ is not enough because there are two cases which satisfy this condition. Case (i) corresponds to a single smoothly decaying outgoing wave with no reflected wave (incoming wave). This outgoing wave dies down with an increase in distance. Case (ii) can occur when the field is set equal to zero by placing a totally reflecting screen at $R = a$ such that the resultant field of the outgoing and reflected waves satisfies the condition of $u = 0$ at $R = a$. One should note that in case (ii), both $u_1 = R^{-1} \exp(j k R)$ and $u_2 = R^{-1} \exp(-j k R)$ exist and both $u_1$ and $u_2$ approach zero as $R$ is increased so that the simple boundary condition of $u = 0$ at $R = \infty$ is applicable.

At present, the field at an infinite domain without reflection, i.e. case (i) is under consideration. Equation (3.18) is useful to guarantee case (i) since only the outgoing wave can satisfy (3.18). The first factor of (3.18) specifies the conservation of energy and the second factor specifies the direction of propagation. The solution that makes the second factor zero is $A \exp(j k R)$, which is the outgoing wave. [When j in (3.18) is replaced by $-$j, the corresponding solution becomes $B \exp(-j k R)$ and such a condition is called the *Sommerfeld absorption condition*].

If the expression for $u$ is simple, it is easy to discern the outgoing wave from the incoming wave, but quite often the expression is so complicated that the radiation condition has to be utilized.

After having established that the integral over $S_R$ is zero, Fig. 3.4 will be used to find the value of the integral over $S_A$. Taking $u$ as the amplitude of a spherical wave emanating from the source $P_2$, the radius $r_{21}$ is the distance from $P_2$ to a point $P_1$ on the aperture. The distribution on the aperture is therefore

$$u = A \frac{e^{jkr_{21}}}{r_{21}} ,$$

where $A$ is a constant representing the strength of the source.

The direction of the vector $\nabla u$ is such that the change in $u$ with respect to a change in location is maximum, so that $\nabla u$ is in the same direction as $\hat{\boldsymbol{r}}_{21}$. Hence, $\nabla u$ becomes

$$\nabla u = A\,\hat{\boldsymbol{r}}_{21}\,\frac{d}{dr_{21}}\left(\frac{e^{jkr_{21}}}{r_{21}}\right) = \hat{\boldsymbol{r}}_{21}\,A\left(jk - \frac{1}{r_{21}}\right)\frac{e^{jkr_{21}}}{r_{21}} \qquad \text{and} \qquad (3.19)$$

$$\nabla\left(\frac{e^{jkr_{01}}}{r_{01}}\right) = \hat{\boldsymbol{r}}_{01}\left(jk - \frac{1}{r_{01}}\right)\frac{e^{jkr_{01}}}{r_{01}}, \qquad (3.20)$$

where $r_{01}$ is the distance from the observation point $P_0$ to the point $P_1$ on the aperture. When $r_{01}$ and $r_{21}$ are much longer than a wavelength, the second term inside the brackets on the right hand-side of (3.19, 20) can be considered very small and therefore ignored when compared to the first term inside the brackets.

Inserting all the above results into (3.16) one obtains

$$u_{\mathrm{p}} = \frac{jA}{2\lambda}\int_{S_A}\frac{\exp\left[j\,k\,(r_{01}+r_{21})\right]}{r_{01}\,r_{21}}\,(\hat{\boldsymbol{r}}_{21}\cdot\hat{\boldsymbol{n}} - \hat{\boldsymbol{r}}_{01}\cdot\hat{\boldsymbol{n}})\,dS . \qquad (3.21)$$

Letting the angle between unit vectors $\hat{\boldsymbol{r}}_{21}$ and $\hat{\boldsymbol{n}}$ be $(\hat{\boldsymbol{r}}_{21}, \hat{\boldsymbol{n}})$ and letting the angle between $\hat{\boldsymbol{r}}_{01}$ and $\hat{\boldsymbol{n}}$ be $(\hat{\boldsymbol{r}}_{01}, \boldsymbol{n})$, we have

$$u_{\mathrm{p}} = \frac{A}{j\,2\lambda}\int_{S_A}\frac{\exp\left[j\,k\,(r_{01}+r_{21})\right]}{r_{01}\,r_{21}}\,[\cos(\hat{\boldsymbol{r}}_{01}, \hat{\boldsymbol{n}}) - \cos(\hat{\boldsymbol{r}}_{21}, \hat{\boldsymbol{n}})]\,dS . \qquad (3.22)$$

Equation (3.22) is called the *Fresnel-Kirchhoff diffraction formula*.

Among the factors in the integrand of (3.22) is the obliquity factor

$$[\cos(\hat{\boldsymbol{r}}_{01}, \hat{\boldsymbol{n}}) - \cos(\hat{\boldsymbol{r}}_{21}, \hat{\boldsymbol{n}})], \qquad (3.23)$$

which relates to the incident and transmitting angles. For the special case in which the light source is centrally located with respect to the aperture, the obliquity factor becomes $(1 + \cos\chi)$, $\chi$ being the angle between $\hat{\boldsymbol{r}}_{01}$ and the normal to the sphere $S_A$ centered at the light source $P_2$, as shown in Fig. 3.5. One then obtains

$$u_{\mathrm{p}} = \frac{A}{j\,2\lambda}\int_{S_A}\frac{\exp\left[j\,k\,(r_{01}+r_{21})\right]}{r_{01}\,r_{21}}\,(1 + \cos\chi)\,dS . \qquad (3.24)$$

When both $r_{01}$ and $r_{21}$ are nearly perpendicular to the mask screen, the obliquity factor becomes approximately 2 and (3.22, 24) become

$$u_{\mathrm{p}} = \frac{1}{j\lambda}\int_{S_A} u_{S_A}\cdot\frac{e^{jkr_{01}}}{r_{01}}\,dS \qquad \text{where} \qquad u_{S_A} = A\,\frac{e^{jkr_{21}}}{r_{21}}. \qquad (3.25)$$

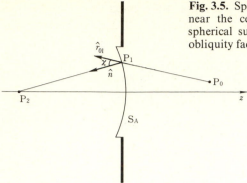

**Fig. 3.5.** Special case of the source $P_2$ being placed near the center of the aperture, where $S_A$ is a spherical surface centered around the source. The obliquity factor is $(1 + \cos \chi)$

Hence, if the amplitude distribution $u_{S_A}$ of the light across the aperture is known, the field $u_p$ at the point of observation can be obtained.

Equation (3.25) can be interpreted as a mathematical formulation of *Huygens' principle*. The integral can be thought of as a summation of contributions from innumerable small spherical sources of amplitude $(u_{S_A} dS)$ lined up along the aperture $S_A$. According to (3.24), the obliquity factor is zero when the point $P_0$ of observation is brought to the left of the aperture and $\chi = 180°$, which means there is no wave going back to the source. This successfully solved the difficulty associated with Huygens' principle!

## 3.3 Fresnel-Kirchhoff's Approximate Formula

Despite the seeming simplicity and power of (3.25), there are only a limited number of cases for which the integration can be expressed in a closed form. Even when one can perform the integral, it may be too complicated to be practical. In this section a further approximation is imposed on (3.25).

In developing the theory of the Fresnel-Kirchhoff diffraction formula, it was convenient to use $P_0$ to denote the point of observation, since the integrations were performed with $P_0$ as the origin. However, in most engineering problems, the origin is taken as either the location of the source or the location of the source aperture. In order to accommodate the usual engineering convention, a re-labelling in adopted, as shown in Fig. 3.6. Here, the origin $P_0$ is located in the plane of the aperture, and the coordinates of any given point in this plane are $(x_0, y_0, 0)$. The observation point $P$ is located in the plane of observation, and the coordinates of a point in this plane are $(x_i, y_i, z_i)$. The distance $r_{01}$ appearing in (3.25) will simply be referred to as $r$.

Using the rectangular coordinate system shown in Fig. 3.6, the diffraction pattern $u(x_i, y_i)$ of the input $g(x_0, y_0)$ is considered. Expressing (3.25) in rectangular coordinates, one obtains

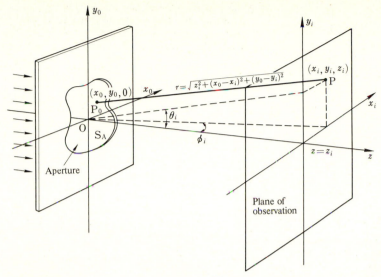

**Fig. 3.6.** Geometry for calculating the diffraction pattern $\{u(x_i, y_i, z_i)\}$ on the screen produced by the source $\{g(x_0, y_0, 0)\}$

$$u(x_i, y_i) = \frac{1}{j\lambda} \int\!\!\int_{-\infty}^{\infty} g(x_0, y_0) \frac{e^{jkr}}{r} dx_0\, dy_0 \quad \text{where}$$

$$r = \sqrt{z_i^2 + (x_0 - x_i)^2 + (y_0 - y_i)^2}\,.$$

(3.26)

Equation (3.26) has a squared term inside the square root, which makes integration difficult.

In a region where $(x_0 - x)^2 + (y_0 - y_i)^2$ is much smaller than $z_i^2$, the binomial expansion is applicable and

$$r = \sqrt{z_i^2 + (x_0 - x_i)^2 + (y_0 - y_i)^2} = z_i \sqrt{1 + \frac{(x_0 - x_i)^2 + (y_0 - y_i)^2}{z_i^2}}$$

$$\cong z_i + \frac{(x_0 - x_i)^2 + (y_0 - y_i)^2}{2z_i} - \frac{[(x_0 - x_i)^2 + (y_0 - y_i)^2]^2}{8z_i^3} \cdots.$$

(3.27)

Thus,

$$r = \underbrace{\underbrace{z_i + \frac{(x_0 - x_i)^2 + (y_0 + y_i)^2}{2z_i}}_{\text{Fraunhofer approximation}} = z_i + \frac{x_i^2 + y_i^2}{2z_i} - \frac{x_i x_0 + y_i y_0}{z_i} + \frac{x_0^2 + y_0^2}{2z_i} - \frac{[(x_0 - x_i)^2 + (y_0 - y_i)^2]^2}{8z_i^3}}_{\text{Fresnel approximation}}.$$

(3.28)

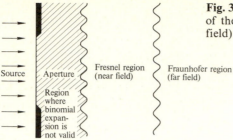

Source    Aperture    Fresnel region
                        (near field)

          Region
          where
          binomial
          expan-
          sion is
          not valid

                                    Fraunhofer region
                                    (far field)

**Fig. 3.7.** Diagram showing the relative positions of the Fresnel (near field) and Fraunhofer (far field) regions

The region where only the first 3 terms are included is called the farfield region or the *Fraunhofer region.* The region where the first 4 terms are included is called the near field region or the *Fresnel region.* It should be remembered that in the region very close to the aperture neither of these approximations are valid because use of the binominal expansion is no longer justified.

It is the value of the 4th term of (3.28) that determines whether the Fresnel or Fraunhofer approximation should be used. Generally speaking, it is determined according to whether the value of $k \, (x_0^2 + y_0^2)/2 z_i$ is larger than $\pi/2$ (Fresnel) or smaller than $\pi/2$ (Fraunhofer). For instance, in the case of a square aperture with side dimension $D$, the value of $k \, (x_0^2 + y_0^2)/2 z_i$ becomes $\pi/2$ radian when $z_i = D^2/\lambda$. This value of $z_i$ is often used as a criterion for distinguishing the Fresnel region from the Fraunhofer region. It should be realized that the distance $z_i$ required to reach the Fraunhofer region is usually very long. For example, taking $D = 6$ mm, and $\lambda = 0.6 \times 10^{-3}$ mm, the value of $z_i = D^2/\lambda$ is 60 m. It, however, is possible to observe the Fraunhofer pattern within the Fresnel region by use of a converging lens which has the effect of cancelling the 4th term of (3.28). The thickness of a converging lens decreases, as the rim is approached, and the phase of its transmission coefficient is of the form

$$- k \frac{x_0^2 + y_0^2}{2f}. \tag{3.29}$$

The phase of the transmitted light, therefore, is the sum of the phases in (3.28, 29). When $z_i = f$, the 4th term of (3.28) is cancelled by (3.29) and the far-field pattern can be observed in the near field. This is the basic principle behind performing a Fourier transform by using a lens, and will be discussed in more detail in Chap. 6.

## 3.4 Approximation in the Fraunhofer Region

As mentioned earlier, the Fraunhofer approximation makes use of only the first three terms of (3.28). When these three terms are inserted into (3.26), the field distribution on the observation screen is

$$u\,(x_i, y_i) = \frac{1}{j\,\lambda\,z_i}\,\exp\left[j\,k\left(z_i + \frac{x_i^2 + y_i^2}{2\,z_i}\right)\right]$$

$$\times \int\!\!\!\int_{-\infty}^{\infty} g\,(x_0, y_0)\,\exp\left[-j\,2\,\pi\left(\frac{x_i\,x_0}{\lambda\,z_i} + \frac{y_i\,y_0}{\lambda\,z_i}\right)\right] dx_0\,dy_0 ,\qquad (3.30)$$

where the approximation $r \cong z_i$ was used in the denominator of (3.30). Since (3.30) is of the form of a two-dimensional Fourier transform, the diffraction pattern can be expressed in Fourier transform notation as follows

$$G\,(f_x, f_y) = \mathscr{F}\,\{g\,(x_0, y_0)\}\,,$$

$$u\,(x_i, y_i) = \frac{1}{j\,\lambda\,z_i}\,\exp\left[j\,k\left(z_i + \frac{x_i^2 + y_i^2}{2\,z_i}\right)\right] [G\,(f_x, f_y)] \qquad (3.31)$$

with $f_x = x_i/\lambda\,z_i$ and $f_y = y_i/\lambda\,z_i$. Equation (3.31) is the Fraunhofer approximation to the Fresnel-Kirchhoff diffraction formula in rectangular coordinates.

An alternative expression for the diffraction pattern involves the use of the angles $\phi_i$ and $\theta_i$ shown in Fig. 3.6, where $\phi_i$ is the angle between the $z$-axis and a line connecting the origin to the point $(x_i, 0, z_i)$, and $\theta_i$ is the angle between the $z$-axis and a line connecting the origin to the point $(0, y_i, z_i)$. Since $\sin\phi_i \doteq x_i/z_i$ and $\sin\theta_i \doteq y_i/z_i$, the values of $f_x$, $f_y$ in (3.31) can be rewritten as

$$f_x = \frac{\sin\phi_i}{\lambda},\qquad f_y = \frac{\sin\theta_i}{\lambda}. \qquad (3.32)$$

## 3.5 Calculation of the Fresnel Approximation

The Fresnel approximation is obtained by inserting the first four terms of (3.28) into (3.26). Two types of expressions for the Fresnel approximation can be obtained depending on whether or not $(x_0 - x_i)^2 + (y_0 - y_i)^2$ is expanded; one is in the form of a convolution and the other is in the form of a Fourier transform.

If $(x_0 - x_i)^2 + (y_0 - y_i)^2$ is inserted into (3.26) without expansion,

$$u\,(x_i, y_i) = \frac{1}{j\,\lambda\,z_i}\,e^{jkz_i}\int\!\!\int g\,(x_0, y_0)\,\exp\left[j\,k\,\frac{(x_0 - x_i)^2 + (y_0 - y_i)^2}{2\,z_i}\right] dx_0\,dy_0$$
$$(3.33)$$

is obtained. By recognizing that (3.33) is a convolution, the diffraction pattern can be written using the convolution symbol as follows

$$u(x_i, y_i) = g(x_i, y_i) * f_{z_i}(x_i, y_i), \tag{3.34}$$

where

$$f_{z_i}(x_i, y_i) = \frac{1}{j\lambda z_i} \exp\left[j k \left(z_i + \frac{x_i^2 + y_i^2}{2z_i}\right)\right].$$

Here $f_{z_i}(x_i, y_i)$ is of the same form as the approximation obtained by binomially expanding $r$ in the expression $\exp(j k r)/r$ for a point source located at the origin. For this reason, $f_{z_i}(x_i, y_i)$ is called the point-source transfer function.

If $(x_0 - x_i)^2 + (y_0 - y_i)^2$ of (3.28) is expanded and used in (3.26), one gets

$$u(x_i, y_i) = \frac{1}{j\lambda z_i} \exp\left[j k \left(z_i + \frac{x_i^2 + y_i^2}{2z_i}\right)\right] \tag{3.35}$$

$$\times \iint_{-\infty}^{\infty} g(x_0, y_0) \exp\left[j k \frac{x_0^2 + y_0^2}{2z_i} - j 2\pi \left(\frac{x_0 x_i}{\lambda z_i} + \frac{y_0 y_i}{\lambda z_i}\right)\right] dx_0 \, dy_0.$$

Since (3.35) resembles the Fourier-transform formula, it can be rewritten as

$$u(x_i, y_i) = \frac{1}{j\lambda z_i} \exp\left[j k \left(z_i + \frac{x_i^2 + y_i^2}{2z_i}\right)\right] \mathscr{F} \left\{g(x_0, y_0) \exp\left[\frac{j k (x_0^2 + y_0^2)}{2z_i}\right]\right\}$$

$$= \exp\left[\frac{j k (x_i^2 + y_i^2)}{2z_i}\right] \mathscr{F} \{g(x_0, y_0) f_{z_i}(x_0, y_0)\} \tag{3.36}$$

with $f_x = x_i/\lambda z_i$ and $f_z = y_i/\lambda z_i$.

*One should get the same answer regardless of whether (3.34) or (3.36) is used, but there is often a difference in the ease of computation.* Generally speaking, it is more convenient to use (3.34) when $u(x_i, y_i)$ has to be further Fourier transformed because the Fourier transform of the factor

$$\exp\left[\frac{j k (x_i^2 + y_i^2)}{2z_i}\right] \tag{3.37}$$

appearing in (3.34) is already known. The Fourier transform of (3.37) is quite easily obtained by using the following relationship for the Fourier transform with respect to $x_0$ alone.

$$\mathscr{F} \left\{\exp\left(j \pi \frac{x_0^2}{\lambda z_i}\right)\right\} = \sqrt{z_i \lambda} \; e^{j\pi/4} e^{-j\pi\lambda z_i f_x^2}. \tag{3.38}$$

Since an analogous relation holds for $y_0$, the Fourier transform of the point-source transfer function is

$$\mathscr{F} \{f_{z_i}(x_0, y_0)\} = e^{j k z_i} \exp\left[-j \pi \lambda z_i (f_x^2 + f_y^2)\right]. \tag{3.39}$$

**Exercise 3.1**    As shown in Fig. 3.8, a point source is located at the origin. Find its Fresnel diffraction pattern on the screen, and compare with its Fraunhofer diffraction pattern.

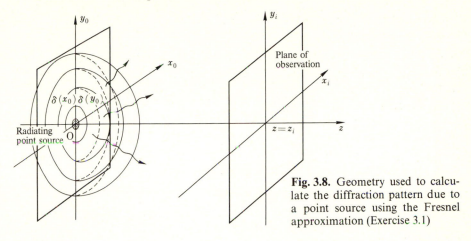

**Fig. 3.8.** Geometry used to calculate the diffraction pattern due to a point source using the Fresnel approximation (Exercise 3.1)

*Solution*    The source function $g(x_0, y_0)$ is

$$g(x_0, y_0) = \delta(x_0)\,\delta(y_0) \, . \tag{3.40}$$

Both (3.34) and (3.36) will be used in the Fresnel region to demonstrate that they give the same results. First, applying (3.34),

$$u(x_i, y_i) = [\delta(x_i)\,\delta(y_i)] * \frac{1}{j\,\lambda\,z_i}\,\exp\left[j\,k\left(z_i + \frac{x_i^2 + y_i^2}{2\,z_i}\right)\right]$$

$$= \frac{1}{j\,\lambda\,z_i}\,\exp\left[j\,k\left(z_i + \frac{x_i^2 + y_i^2}{2\,z_i}\right)\right] \tag{3.41}$$

is obtained. Next using (3.36), one finds

$$u(x_i, y_i) = \frac{1}{j\,\lambda\,z_i}\,\exp\left[j\,k\left(z_i + \frac{x_i^2 + y_i^2}{2\,z_i}\right)\right]$$

$$\times\,\mathscr{F}\left\{\delta(x_0)\,\delta(y_0)\,\exp\left(j\,k\,\frac{x_0^2 + y_0^2}{2\,z_i}\right)\right\} \tag{3.42}$$

with $f_x = x_i/\lambda\,z_i$ and $f_y = y_i/\lambda\,z_i$. Since the Fourier transform appearing in (3.42) is unity, the diffraction pattern simplifies to

$$u(x_i, y_i) = \frac{1}{j\,\lambda\,z_i}\,\exp\left[j\,k\left(z_i + \frac{x_i^2 + y_i^2}{2\,z_i}\right)\right], \tag{3.43}$$

which is exactly the same as (3.41).

When the screen is in the Fraunhofer region, (3.31) is used

$$u\,(x_i, y_i) = \frac{1}{j\,\lambda\,z_i}\,\exp\left[j\,k\left(z_i + \frac{x_i^2 + y_i^2}{2\,z_i}\right)\right]\mathscr{F}\{\delta(x_0)\,\delta(y_0)\}_{f_x = x_i/\lambda\,z_i,\,f_y = y_i/\lambda\,z_i}$$

$$= \frac{1}{j\,\lambda\,z_i}\,\exp\left[j\,k\left(z_i + \frac{x_i^2 + y_i^2}{2\,z_i}\right)\right]. \tag{3.44}$$

The results of the Fresnel and Fraunhofer approximations are identical. This, however, is true only with very special cases. The result is identical to the point-source transfer function, as expected.

## 3.6 One-Dimensional Diffraction Formula

As shown in Fig. 3.9, when the light source or aperture varies only in the $x_0$ direction, the aperture function can be written

$$g\,(x_0, y_0) = g\,(x_0)\,, \tag{3.45}$$

and hence the Fresnel-Kirchhoff approximation, (3.26), becomes

$$u\,(x_i, y_i) = \frac{1}{j\,\lambda} \int\limits_{-\infty}^{\infty} g\,(x_0) \left(\int\limits_{-\infty}^{\infty} \frac{e^{jkr}}{r}\,dy_0\right) dx_0 \tag{3.46}$$

where

$$r = \sqrt{z_i^2 + (x_0 - x_i)^2 + (y_0 - y_i)^2}\,.$$

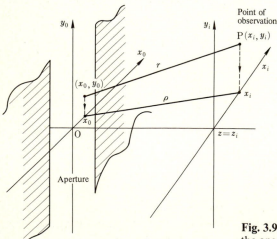

**Fig. 3.9.** Geometry used for calculating the one-dimensional diffraction pattern

Of special interest is the integration over $y_0$ in (3.46) which will be referred to as $I$, where

$$I = \int_{-\infty}^{\infty} \frac{e^{jkr}}{r} \, dy_0 . \tag{3.47}$$

As shown in Fig. 3.9, the projection of $r$ onto the plane $y_0 = 0$ is denoted by $\varrho$, where

$$\varrho = \sqrt{z_i^2 + (x_0 - x_i)^2} . \tag{3.48}$$

In terms of $\varrho$, $r$ is rewritten as

$$r = \sqrt{\varrho^2 + (y_0 - y_i)^2} . \tag{3.49}$$

A change of variable is introduced as follows

$$y_0 - y_i = \varrho \sinh t , \tag{3.50}$$

$$\frac{dy_0}{dt} = \varrho \cosh t . \tag{3.51}$$

Combining (3.49, 50), $r$ is expressed as a function of $t$

$$r = \varrho \cosh t . \tag{3.52}$$

Equations (3.51, 52) are inserted into (3.47) to obtain

$$I = \int_{-\infty}^{\infty} e^{jk\varrho\cosh t} \, dt = j \, \pi \, H_0^{(1)} (k \, \varrho) \tag{3.53}$$

where $H_0^{(1)} (k \, \varrho)$ is the zeroth-order Hankel function of the first kind. The formula

$$H_0^{(1)} (a) = \frac{1}{j \, \pi} \int_{-\infty}^{\infty} e^{ja\cosh t} \, dt \tag{3.54}$$

was used. Inserting (3.53) into (3.46) gives

$$u(x_i) = \frac{\pi}{\lambda} \int_{-\infty}^{\infty} g(x_0) \, H_0^{(1)} (k \, \varrho) \, dx_0 , \quad \text{where} \tag{3.55}$$

$$\varrho = \sqrt{z_i^2 + (x_0 - x_i)^2} . \tag{3.56}$$

This is the *one-dimensional Fresnel Kirchhoff formula*.

Under certain conditions, the integration in (3.55) can be simplified. When $k \, \varrho \gg 1$ the Hankel function can be approximated as

$$H_0^{(1)}(k\varrho) \cong \frac{1}{\pi} \sqrt{\frac{\lambda}{\varrho}}\ e^{j(k\varrho - \pi/4)}\ ,$$

(3.57)

and when

$$z_i^2 \gg (x_0 - x_i)^2\ ,$$

(3.58)

$\varrho$ can be written as

$$\varrho = z_i + \frac{(x_0 - x_i)^2}{2 z_i}\ .$$

(3.59)

Inserting (3.57, 59) into (3.55) gives

$$u(x_i) = \frac{\exp[j(k z_i - \pi/4)]}{\sqrt{z_i \lambda}} \int_{-\infty}^{\infty} g(x_0) \exp\left[\frac{j k (x_0 - x_i)^2}{2 z_i}\right] dx_0\ .$$

(3.60)

Equation (3.60) is the formula for the one-dimensional Fresnel approximation. When $\lambda z_i \gg x_0^2$, the formula for the one-dimensional Fraunhofer approximation is obtained.

$$u(x_i) = \frac{1}{\sqrt{z_i \lambda}} \exp\left[j\left(k z_i + k\frac{x_i^2}{2 z_i} - \frac{\pi}{4}\right)\right] [\mathscr{F}\{g(x_0)\}]_{f = x_i/\lambda z_i}\ .$$

(3.61)

## 3.7 The Fresnel Integral

In this section, the Fresnel diffraction pattern of a square aperture such as shown in Fig. 3.10 will be calculated using (3.36) [3.2, 5]. Let the size of the aperture be $2a$ by $2a$, and the amplitude of the incident wave be unity. The aperture is expressed in terms of the rectangular function as

$$\Pi\left(\frac{x_0}{2a}\right)\Pi\left(\frac{y_0}{2a}\right)\ .$$

(3.62)

By definition, the aperture function is zero outside the limits $(-a, a)$ and is unity inside. Equation (3.62) is inserted into (3.36) to yield

$$u(x_i, y_i) = \frac{1}{j \lambda z_i} \exp\left[j k \left(z_i + \frac{x_i^2 + y_i^2}{2 z_i}\right)\right]$$

$$\times \int_{-a}^{a}\int \exp\left[j\frac{\pi}{\lambda z_i}(x_0^2 + y_0^2) - j 2\pi(f_x x_0 + f_y y_0)\right] dx_0\, dy_0$$

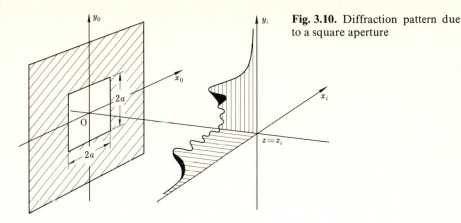

**Fig. 3.10.** Diffraction pattern due to a square aperture

$$= \frac{1}{j\,\lambda\,z_i} \exp\left[j\,k\left(z_i + \frac{x_i^2 + y_i^2}{2\,z_i}\right) - j\,\pi\,\lambda\,z_i(f_x^2 + f_y^2)\right]$$

$$\times \int_{-a}^{a} \int \exp\left\{j\,\frac{\pi}{\lambda\,z_i}[(x_0 - f_x\,\lambda\,z_i)^2 + (y_0 - f_y\,\lambda\,z_i)^2]\right\} dx_0\,dy_0. \qquad (3.63)$$

Changing variables

$$\frac{2}{\lambda\,z_i}(x_0 - f_x\,\lambda\,z_i)^2 = \xi^2, \quad \frac{2}{\lambda\,z_i}(y_0 - f_y\,\lambda\,z_i)^2 = \eta^2,$$

Eq. (3.63) becomes

$$u(x_i, y_i) = \frac{1}{2\,j}\, e^{jkz_i} \int_{-\sqrt{2/\lambda z_i}\,(a+f_x\lambda z_i)}^{\sqrt{2/\lambda z_i}\,(a-f_x\lambda z_i)} e^{j\pi\xi^2/2}\,d\xi \int_{-\sqrt{2/\lambda z_i}\,(a+f_y\lambda z_i)}^{\sqrt{2/\lambda z_i}\,(a-f_y\lambda z_i)} e^{j\pi\eta^2/2}\,d\eta \qquad (3.64)$$

with $f_x = x_i/\lambda\,z_i$ and $f_y = y_i/\lambda\,z_i$. If the limits of integration in (3.64) were $-\infty, +\infty$, the value of the integral would be

$$(\sqrt{2}\,e^{j\pi/4})^2$$

However, with finite limits, the computation is more complicated.

To assist with the integration in (3.64) a new function $F(\alpha)$, known as the *Fresnel integral*, is introduced, namely

$$F(\alpha) = \int_0^{\alpha} e^{j\pi x^2/2}\,dx. \qquad (3.65)$$

Before using the Fresnel integral, some of its properties will be investigated. The Fresnel integral can be separated into real and imaginary parts as

$$C(\alpha) = \int_0^{\alpha} \cos\left(\frac{\pi}{2}\,x^2\right) dx, \qquad (3.66)$$

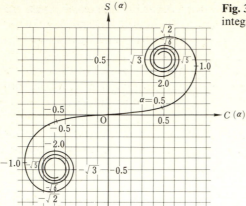

**Fig. 3.11.** Cornu's spiral (plot of the Fresnel integral in the complex plane)

$$S(\alpha) = \int_0^{\alpha} \sin\left(\frac{\pi}{2} x^2\right) dx. \tag{3.67}$$

The former is called the *Fresnel cosine integral* and the latter is the *Fresnel sine integral*. Equation (3.65) is now rewritten as

$$F(\alpha) = C(\alpha) + j\, S(\alpha). \tag{3.68}$$

Figure 3.11 is a plot of (3.68) by taking $\alpha$ as a parameter. As $\alpha$ is increased from zero to plus infinity the curve starts from the origin and curls up around the point (0.5, 0.5). As $\alpha$ is decreased from zero to minus infinity it curls up around the point $(-0.5, -0.5)$. The curve $F(\alpha)$ is called *Cornu's spiral.* The points of convergence of the spiral are

$$C(\infty) = S(\infty) = -C(-\infty) = -S(-\infty). \tag{3.69}$$

In the $C(\alpha) - S(\alpha)$ complex plane, the value of

$$\int_{\alpha_1}^{\alpha_2} e^{j\pi x^2/2}\, dx = [C(\alpha_2) + j\, S(\alpha_2)] - [C(\alpha_1) + j\, S(\alpha_1)]$$

can be expressed as the length of a vector whose end points are the points $\alpha = \alpha_1$ and $\alpha = \alpha_2$.

Returning to the problem of the Fresnel diffraction pattern of a square aperture, (3.64) with the insertion of the Fresnel integral $F(\alpha)$ and $f_x = x_i/\lambda z_i, f_y = y_i/\lambda z_i$ becomes

$$u(x_i, y_i) = \frac{1}{2j}\, e^{jkz_i}\left[ F\left(\sqrt{\frac{2}{\lambda z_i}}\, (a - x_i)\right) - F\left(-\sqrt{\frac{2}{\lambda z_i}}\, (a + x_i)\right)\right]$$

$$\times \left[ F\left(\sqrt{\frac{2}{\lambda z_i}}\, (a - y_i)\right) - F\left(-\sqrt{\frac{2}{\lambda z_i}}\, (a + y_i)\right)\right]. \tag{3.70}$$

Equation (3.70) can be calculated using mathematical tables of $F(\alpha)$, but a qualitative result may be obtained using Cornu's spiral. Since the factors involving $x_i$ in (3.70) are essentially the same as those involving $y_i$, only the $x_i$ terms will be considered. The value inside the brackets of (3.70) is the difference between the two values of the Fresnel integral. This can be expressed in the $C(\alpha) - S(\alpha)$ complex plane as a line starting from the point represented by the second term and ending at the point represented by the first term.

First, the value at $x_i = 0$ is obtained. At this point, the values of the first and second terms are situated diametrically across the origin in the complex plane and is represented by vector $A$ in Fig. 3.12.

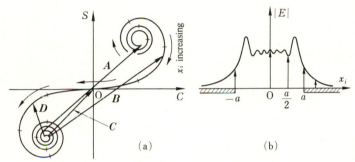

(a)                              (b)

**Fig. 3.12 a, b.** Relationship between the phasor on Cornu's spiral and the magnitude of the diffracted field. (**a**) Phasors representing the magnitude of the light. (**b**) The magnitude of the field vs position (the Fresnel diffraction pattern)

Next, the value at $x_i = a/2$, is obtained. The first term moves toward the origin on the Cornu's spiral and the second term moves away from the origin. It can be expressed by vector $B$.

Next, the value at $x_i = a$ is obtained. The first term is now at the origin and the second term moves further away from the origin. It can be represented by vector $C$.

Finally, the value at $x_i > a$, corresponding to a point inside the geometrical-optics shadow region, is considered. The first term enters the third quadrant and the second term moves even further away from the origin. It is represented by vector $D$. The length of $D$ monotonically decreases with an increase in $x_i$ and approaches zero.

In summary as the point $x_i$ moves away from the center of the aperture, the ends of the vector move in the direction of the arrows in Fig. 3.12 a. The results of plotting the length of the vectors versus $x_i$ gives the Fresnel diffraction pattern shown in Fig. 3.12 b.

## Problems

**3.1**  Show that, in the process of obtaining the Kirchhoff integral $u_p$, the same results are obtained by taking the domain including the source, as shown in Fig. 3.2, and by taking the domain excluding the source, as shown in Fig. 3.3. Show that the result is solely dependent on the relative positions of the point of observation and the source and is independent of the shape of the domain.

**3.2**  *Babinet's principle* states that the sum of the diffraction pattern $u_1$ of a mask with an aperture, as shown in Fig. 3.13a, and the diffraction pattern $u_2$ of its compliment, as shown in Fig. 3.13b, is the same as the field distribution without anything. If either one of the patterns is known, the other can be determined immediately by this principle. Prove Babinet's principle using the Fresnel-Kirchhoff diffraction formula.

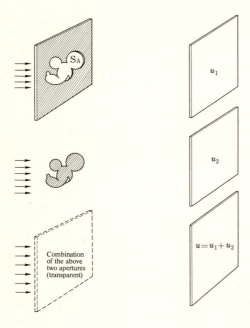

**Fig. 3.13.** Illustration of Babinet's principle

**3.3**  When there is no variation in the $y$ direction so that the input function can be expressed as

$$g(x_0, y_0) = g(x_0),$$

prove that (3.61) can be derived by applying the Fraunhofer approximation in the $x_0$ direction and the Fresnel approximation in the $y_0$ direction to (3.33).

**3.4**  Prove that the angle between the tangent of Cornu's spiral taken at $\alpha$ and the $C$ axis is $\pi \alpha^2/2$. Using this result, obtain the values of $\alpha$ that give the extrema of $C(\alpha)$ and $S(\alpha)$.

**3.5**  Prove that Cornu's spiral has point symmetry with respect to the origin.

# 4. Practical Examples of Diffraction Theory

The best way to understand the principles associated with a given discipline is through examples which apply these principles. This chapter devotes itself to illustrating methods for solving specific problems, and also serves as a collection of practical examples involving diffraction phenomena.

## 4.1 Diffraction Problems in a Rectangular Coordinate System

**Exercise 4.1**  Obtain the diffraction pattern of a rectangular aperture with dimensions $a \times b$ when projected onto the screen located at $z = z_i$ in the Fraunhofer region, as shown in Fig. 4.1.

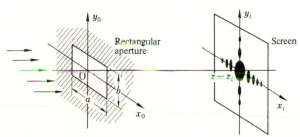

**Fig. 4.1.** Geometry of diffraction from a rectangular aperture

*Solution*  The input function $g(x_0, y_0)$ is given by

$$g(x_0, y_0) = \Pi\left(\frac{x_0}{a}\right) \Pi\left(\frac{y_0}{b}\right). \tag{4.1}$$

From (2.35) and (3.31), the diffraction pattern is

$$u(x_i, y_i) = \frac{a\,b}{j\,\lambda\,z_i} \exp\left[j\,k\left(z_i + \frac{x_i^2 + y_i^2}{2\,z_i}\right)\right] \operatorname{sinc}\left(a\,\frac{x_i}{\lambda\,z_i}\right) \operatorname{sinc}\left(b\,\frac{y_i}{\lambda\,z_i}\right). \tag{4.2}$$

If the first factor of (4.2) is suppressed, the distributions of the amplitude and intensity along the $x_i$ axis are represented by the curves in Fig. 4.2a and b. Similar curves can be drawn along the $y_i$ axis, while keeping in mind that the width of the main beam of the diffraction pattern is inversely

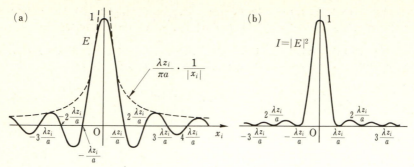

(a)    (b)

Fig. 4.2 a, b. Diffraction pattern of a rectangular aperture. (a) Amplitude. (b) Intensity

proportional to the dimension of the side of the aperture. The narrower beam is in the direction of the wider side of the rectangular aperture. A rule of thumb is that the beam width of an antenna with an aperture of $60\,\lambda$ is $1°$.

**Exercise 4.2**  Consider a light source with dimensions $2a \times 2a$. An obstructing square with dimensions $a \times a$ is placed with its center at $(\xi, \eta)$ inside the light source, as shown in Fig. 4.3 a. Compare the Fraunhofer

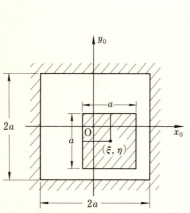

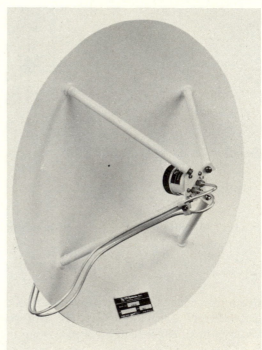

Fig. 4.3 a, b. Square light source obstructed by a square. (a) Geometry. (b) Antenna with center feed (EM Systems, Inc.)

diffraction patterns with and without the obstructing square. Obstruction caused by an antenna feed shown in Fig. 4.3b is a practical example.

*Solution*  The input function $g(x_0, y_0)$ representing the light source is obtained by subtracting the smaller square from the larger square.

$$g(x_0, y_0) = \Pi\left(\frac{x_0}{2a}\right) \Pi\left(\frac{y_0}{2a}\right) - \Pi\left(\frac{x_0 - \xi}{a}\right) \Pi\left(\frac{y_0 - \eta}{a}\right). \tag{4.3}$$

The Fourier transform $G(f_x, f_y)$ of (4.3) is

$$\begin{aligned} G(f_x, f_y) &= 4a^2 \operatorname{sinc}(2 a f_x) \operatorname{sinc}(2 a f_y) \\ &\quad - a^2 \operatorname{sinc}(a f_x) \operatorname{sinc}(a f_y) e^{-j2\pi(f_x\xi + f_y\eta)}. \end{aligned} \tag{4.4}$$

Using (3.31), the final result is

$$u(x_i, y_i) = \frac{1}{j\lambda z_i} \exp\left[jk\left(z_i + \frac{x_i^2 + y_i^2}{2 z_i}\right)\right] \{G(f_x, f_y)\} \tag{4.5}$$

with $f_x = x_i/\lambda z_i$ and $f_y = y_i/\lambda z_i$. The curve determined by (4.5) when $\xi = \eta = 0$ is drawn with respect to $f_x$ in a solid line in Fig. 4.4. The curve in the absence of the obstructing square is drawn by a dotted line. It can be seen from the figure that the existence of the mask does not cause too much change in the diffraction pattern.

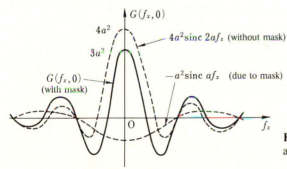

Fig. 4.4. Diffraction patterns with and without the obstructing square

**Exercise 4.3**  Find the far-field diffraction pattern of a triangular aperture. The edges of the triangle are expressed by $x_0 = a$, $y_0 = x_0$, $y_0 = -x_0$ (Fig. 4.5). The observation screen is placed at $z = z_i$.

*Solution*  It should be noticed that the limits of the integration with respect to $y_0$ are not constant but vary with $x_0$. Hence, integration with respect to $y_0$ in a thin strip between $x_0$ and $x_0 + dx_0$ ought to be performed first, followed by integration of $x_0$ from 0 to $a$. Integrating with respect to $y_0$ gives

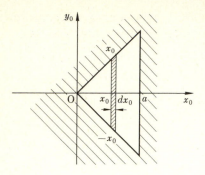

**Fig. 4.5.** Aperture in a triangular shape

$$G(f_x,f_y) = \int\limits_0^a dx_0 \int\limits_{-x_0}^{x_0} e^{-j2\pi(f_x x_0+f_y y_0)}\,dy_0$$

$$= \int\limits_0^a e^{-j2\pi f_x x_0}\left(\frac{e^{-j2\pi f_y x_0} - e^{+j2\pi f_y x_0}}{-j2\pi f_y}\right)dx_0$$

$$= \frac{j}{2\pi f_y}\int\limits_0^a (e^{-j2\pi(f_x+f_y) x_0} - e^{-j2\pi(f_x-f_y) x_0})\,dx_0. \tag{4.6}$$

Integrating with respect to $x_0$ and rearranging terms yields the result

$$G(f_x,f_y) = \frac{ja}{2\pi f_y}\{e^{-j\pi(f_x+f_y)a}\operatorname{sinc}[(f_x+f_y)a] - e^{-j\pi(f_x-f_y)a}\operatorname{sinc}[(f_x-f_y)a]\}. \tag{4.7}$$

Since (4.7) for the far-field diffraction pattern of a triangular aperture is somewhat more complex than previous examples using square apertures, a closer examination of the final result is worth making. First, the distribution along the $f_x = x_i/\lambda z_i$ axis is considered. If the value of $f_y = 0$ is directly inserted into (4.7), $G(f_x, 0)$ takes on the indefinite 0/0 form. It is therefore necessary to evaluate the limit of $G(f_x,f_y)$ as $f_y$ approaches zero. By defining $\Delta$ as

$$\Delta = \pi a f_y \ll \pi a f_x,$$

Eq. (4.7) can be approximated as

$$G(f_x, \Delta) = a^2\left(\frac{\sin\Delta}{\Delta}\right)e^{-j\pi f_x a}\operatorname{sinc}(f_x a). \tag{4.8}$$

Taking the limit as $\Delta \to 0$ gives

$$G(f_x) = \lim_{\Delta\to 0} G(f_x, \Delta) = a^2\,e^{-j\pi f_x a}\operatorname{sinc}(f_x a). \tag{4.9}$$

The curve determined by (4.9) is shown in Fig. 4.6. It is interesting to note that even though the phase distribution is antisymmetric with respect to the

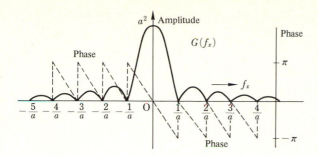

**Fig. 4.6.** Amplitude and phase distribution along $f_x$ for the diffraction pattern of a triangular aperture

$f_y$ axis, the amplitude and intensity distributions are both symmetric. It is quite surprising to see such symmetry exist considering that the shape of the aperture is far from symmetric.

The expression for the distribution along the $f_y$ axis is obtained from (4.7) and is

$$G(f_y) = G(0, f_y) = a^2 [\text{sinc}(f_y a)]^2. \tag{4.10}$$

The decay of $G(f_y)$ with respect to $f_y$ is much faster than that of $G(f_x)$ with respect to $f_x$.

Next, the distribution of the intensity along a line 45° with respect to the $f_x$ axis, namely, along the line $f_y = f_x$, is examined. In order to make the interpretation of the expression simpler, the coordinates will be rotated by 45 degrees. The variables in (4.7) are converted according to

$$\sqrt{2}\, X = f_y + f_x, \quad \sqrt{2}\, Y = f_y - f_x \tag{4.11}$$

to obtain

$$G(X, Y) = \frac{j\,a}{\sqrt{2}\,\pi(X + Y)}$$
$$\times [e^{-j\sqrt{2}\,\pi a X} \text{sinc}(\sqrt{2}\, a X) - e^{j\sqrt{2}\,\pi a Y} \text{sinc}(\sqrt{2}\, a Y)]. \tag{4.12}$$

The distribution along the $X$ axis is

$$G(X, 0) = \frac{j\,a}{\sqrt{2}\,\pi X} [e^{-j\sqrt{2}\,\pi a X} \text{sinc}(\sqrt{2}\, a X) - 1]. \tag{4.13}$$

The expression for $G(0, Y)$ is exactly the same as $G(X, 0)$ with $Y$ substituted for $X$. Fig. 4.7a shows a photograph of the diffraction pattern of the aperture. The photograph clearly shows that the dominant features of the diffraction pattern are aligned along directions which are perpendicular to the aperture edges. The longest trail of diffracted light lies along the direction perpendicular to the longest aperture edge. It is generally true that the diffraction is predominantly determined by the shape of the edges. A rule of thumb is that the trails of diffracted light are in the directions

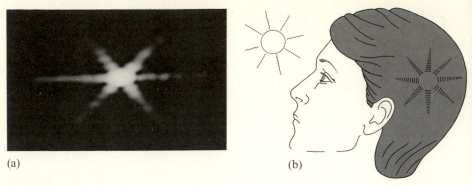

(a)     (b)

**Fig. 4.7 a, b.** Aperture diffraction. (**a**) Photograph of the diffraction pattern of a right-angled triangular aperture. (**b**) Diffraction due to iris

perpendicular to the tangent of the aperture edges. This rule of thumb is handy when an analytical approach is difficult.

When gazing directly at the sun, bright star like streaks are seen. These are considered due to the diffraction from the edges of the iris.

## 4.2 Edge Diffraction

As demonstrated by the above example, the edge plays an important role in the diffraction pattern. This section is devoted to a detailed study of the diffraction pattern of a semi-infinite plane aperture having one infinitely long straight edge. This aperture is ideally suited to studying the effect of the edge alone.

**Exercise 4.4**   Find the diffraction pattern of the aperture which is open in the entire right half plane, as shown in Fig. 4.8.

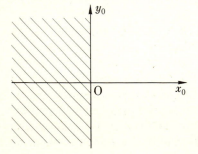

**Fig. 4.8.** Semi-infinite aperture

*Solution*  The step function provides a good description of such an aperture. The source function $g(x_0, y_0)$ is expressed by

$$g(x_0, y_0) = H(x_0). \tag{4.14}$$

The Fourier transform of (4.14) with respect to $x_0$ was given by (2.43) and that with respect to $y_0$ is $\delta(f_y)$. Using (3.31), the diffraction pattern is given by

$$u(x_i, y_i) = \delta(f_y) \frac{1}{j \lambda z_i} \exp\left(j k z_i + j k \frac{x_i^2}{2 z_i}\right)\left[\frac{1}{j 2 \pi f_x} + \frac{1}{2} \delta(f_x)\right] \tag{4.15}$$

with $f_x = x_i / \lambda z_i$ and $f_y = y_i / \lambda z_i$. The curve of (4.15) excluding the first factor is shown in Fig. 4.9, which shows that light is diffracted not only into the region $f_x > 0$ (where the direct ray is incident) but also into the region $f_x < 0$ (where the direct ray is blocked by the aperture).

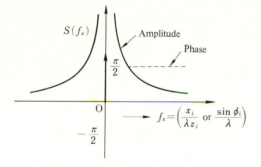

Fig. 4.9. Diffraction pattern of the semi-infinite aperture with respect to $f_x$

The diffraction pattern of the semi-infinite aperture is now compared with that of an aperture of infinite dimension. The value of the diffraction pattern of the infinite aperture, which corresponds to the bracketed portion of (4.15), is $\delta(f_x)\,\delta(f_y)$.

In the case of the semi-infinite aperture, obviously one half of the incident light energy is transmitted through the aperture. Examining the bracket of (4.15) reveals that one quarter of the incident light energy, or one half of the energy transmitted through the aperture, is not diffracted and goes straight through while the rest of the energy is diffracted equally on each side of the edge.

It is a somewhat puzzling fact that the diffraction of an infinite aperture is a point expressed by $\delta(f_x)\,\delta(f_y)$ and that of a semi-infinite aperture is of delta function form only in the $f_y$ direction. Does this really mean that the far-field diffraction pattern of the infinite aperture (no edge) becomes a point? The answer is "yes" and "no". According to (3.28), in order for the region of observation to be considered far field, the distance to the plane of observation has to be much larger than the dimension of the source. Due to the fact that the dimensions of the infinite aperture are indeed infinite, it is theoretically impossible to find the region of observation that satisfies the far-field condition of (3.28) and therefore no legitimate far-field region exists.

There is yet another more enlightening interpretation. In reality, it is impossible to make a source of truly infinite size. A beam that travels straight should have the same cross-section no matter how far the plane of the observation is moved back. However, due to diffraction, the beam deviates from the center line, and this deviation distance increases continuously as the screen is moved indefinitely further away. If the finite cross-section of the incident beam is compared with the deviation distance, there must be a distant place where the cross-section of light can be considered as a point (delta function) compared with the increasingly large deviation of the diffracted beam away from the center line.

Another characteristic of the far-field pattern of the edge is that its amplitude decreases monotonically with $f_x$ and there is no interference phenomenon. This is because the non-diffracted beam contributes to the $\delta$ function at the origin, but does not contribute in any other region.

Now that the diffraction patterns of the "no edge" and "one edge" aperture have been explored, the next logical step is to look at the diffraction produced by an aperture containing two edges. The aperture considered is a slit constructed by two infinitely long edges.

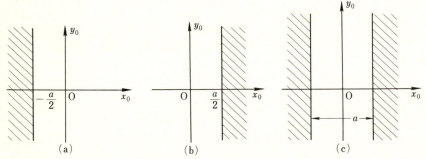

**Fig. 4.10a−c.** Composition of a slit. (**a**) Half plane shifted by $a/2$ to the left. (**b**) Half plane shifted by $a/2$ to the right. (**c**) Slit made up of two half planes

The semi-infinite plane displaced by $a/2$ to the left, as shown in Fig. 4.10a and that displaced the same amount to the right, as shown in Fig. 4.10b, are combined to make up the slit shown in Fig. 4.10c. The Fourier transform of the functions shown in Fig. 4.10a and b can be obtained by applying the shift theorem to the result of Exercise 4.4. The source function of Fig. 4.10a is expressed by

$$H_a(x_0) = \frac{1}{2}\left[\mathrm{sgn}\left(x_0 + \frac{a}{2}\right) + 1\right] \tag{4.16}$$

and its Fourier transform is

$$\mathcal{H}_a(f_x) = \frac{1}{2}\left[\frac{1}{j\pi f_x}e^{j\pi a f_x} + \delta(f_x)\right]. \tag{4.17}$$

Similarly, the source function of Fig. 4.10 b is expressed by

$$H_b(x_0) = \frac{1}{2}\left[-\operatorname{sgn}\left(x_0 - \frac{a}{2}\right) + 1\right] \tag{4.18}$$

and its Fourier transform is

$$\mathcal{H}_b(f_x) = \frac{1}{2}\left[-\frac{1}{j\pi f_x}e^{-j\pi a f_x} + \delta(f_x)\right]. \tag{4.19}$$

The source function displayed in Fig. 4.10 c has been already obtained as

$$H_c(x_0) = \Pi\left(\frac{x_0}{a}\right) \tag{4.20}$$

and its Fourier transform is

$$\mathcal{H}_c(f_x) = a\operatorname{sinc}(a f_x).$$

By comparing these three figures (Figs. 4.10 a−c) it is easy to see that proper manipulation of (a) and (b) generates (c). The mathematical expression for (c) is obtained with the aid of Fig. 4.11, i.e.

$$H_c(x_0) = H\left(x_0 + \frac{a}{2}\right) - H\left(x_0 - \frac{a}{2}\right)$$

$$= \frac{1}{2}\left[\operatorname{sgn}\left(x_0 + \frac{a}{2}\right) + 1\right] - \frac{1}{2}\left[\operatorname{sgn}\left(x_0 - \frac{a}{2}\right) + 1\right].$$

Its Fourier transform is

$$H_c(f_x) = \left[\frac{e^{j\pi f_x a}}{j 2\pi f_x} - \frac{e^{-j\pi f_x a}}{j 2\pi f_x}\right] = a\frac{\sin \pi f_x a}{\pi f_x a} = a\operatorname{sinc}(f_x a) \tag{4.21}$$

which is the same result as above.

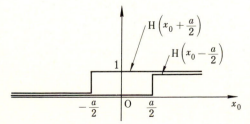

Fig. 4.11. Expression of a slit using step functions

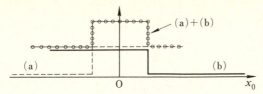

**Fig. 4.12.** Mere addition of two half plane functions. Note that it does not represent a slit. The law of superposition is applied to the field but not to the boundary conditions

*Caution*   It should be noted that, as shown in Fig. 4.12, the expression for (c) is not the sum of the expressions for (a) and (b), but rather the difference of the two. The law of superposition is not applicable here. *Only the values of field can be superposed but not the geometry, medium or boundary conditions.* Cases where superposition may be applied occur in calculating the field due to multiple sources. In these cases, only one source is assumed at a time and the field due to each such individual source is calculated separately inside the same geometry. The fields due to all of the individual sources are then superposed to obtain the final result.

Another point to be aware of is the existence of multiple reflections between the two edges of a slit. The wave diffracted by the left edge illuminates the right edge whose diffraction field again illuminates the left edge. Due to this mutual coupling between the edges, the distribution of the actual field in the close vicinity of the slit is very complicated. It should be noted that *Kirchhoff's integral approximation uses the original field which was present before the edges were introduced to this location to calculate the integral, thus excludes the effect of this mutual coupling between the edges.*

## 4.3  Diffraction from a Periodic Array of Slits

By using the comb function, periodic functions of various shapes can be easily generated [4.1 − 3].

**Exercise 4.5**   Consider a mask which is made up of periodic slits, as shown in Fig. 4.13. The width of each slit is $a$ and its period is $b$. The total length of the array is $c$. Find the Fraunhofer diffraction pattern of the mask.

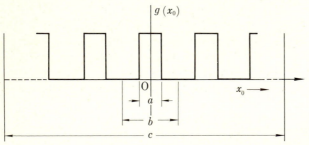

**Fig. 4.13.** One-dimensional periodic source

*Solution*   The input source function $g(x_0)$ is

$$g(x_0) = \frac{1}{b} \left[ \Pi\left(\frac{x_0}{a}\right) * \text{III}\left(\frac{x_0}{b}\right) \right] \cdot \Pi\left(\frac{x_0}{c}\right). \tag{4.22}$$

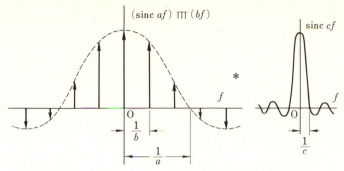

**Fig. 4.14.** Convolution of $(\text{sinc}\,(af)) * \text{III}\,(bf)$ with $\text{sinc}\,(cf)$

The factor inside the bracket is the expression for a periodic slit function of infinite length. The last factor truncates this function to a length $c$. The Fourier transform of (4.22) is

$$G(f) = a\,c\,[(\text{sinc}\,(af) \cdot \text{III}\,(bf)] * \text{sinc}\,(cf). \tag{4.23}$$

A separate plot for each factor of (4.23) is shown in Fig. 4.14. The factor in the bracket represents a chain of delta functions whose amplitude is modulated by $\text{sinc}\,(af)$, as shown by the left graph. The period of the delta functions is inversely related to the period of the slits and is given by $1/b$. In other words, the delta function period $1/b$ becomes shorter as the period of the slits becomes longer. It is also noticed that the zero-crossing point of $\text{sinc}\,(af)$ is inversely proportional to the width of the slit. The curve on the right of Fig. 4.14 is $\text{sinc}\,(cf)$ and its zero-crossing point is likewise inversely proportional to the total length of the array $c$. By convolving the left with the right graph, the final result shown in Fig. 4.15 is obtained.

It is interesting to note that, as far as the source dimensions are concerned, $c$ is the largest, $b$ is the next largest and $a$ is the smallest, but as far as the dimensions in the Fourier transform are concerned, the smallest is the width $1/c$ of $\text{sinc}\,(cf)$, the next is the period $1/b$ of the comb function $\text{III}\,(bf)$, and the largest is the width $1/a$ of $\text{sinc}\,(af)$. The Fourier widths are therefore in reverse order of the geometrical widths.

As seen from Fig. 4.15, when $c$ is large, the order of the calculation of (4.23) can be altered as

$$G(f) = a\,c\,\underbrace{\text{sinc}\,(af)}_{\substack{\text{element}\\\text{factor}}}\,\underbrace{[\text{III}\,(bf) * \text{sinc}\,(cf)]}_{\substack{\text{array}\\\text{factor}}}. \tag{4.24}$$

Equation (4.24) is now separated into two factors, one of which is associated with the characteristics of the slit element itself, and the other is

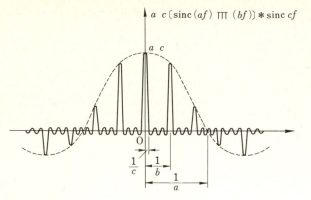

**Fig. 4.15.** Graph of $ac[\text{sinc}(af)\,\text{III}(bf)]*\text{sinc}(cf)$

associated with the arrangement of the slit elements (period and the overall length of the array).

*Because the element factor and the array factor can be separated, the procedure for designing an antenna array can be greatly simplified.* Changing the shape of the slit results in a change only in the shape of the dotted line in Fig. 4.15 and all other details such as the period and the width of each peak will not be affected. Therefore, the choice of the element antenna and that of the period can be treated separately. Generalizing, it is possible to express the radiation pattern of an antenna array as

$$(\text{pattern}) = (\text{element factor}) \times (\text{array factor}). \tag{4.25}$$

**Exercise 4.6** Consider an $N$ element linear array, as shown in Fig. 4.16. The width of each element antenna is $a$. Find the spacing $b$ between the elements such that the height of the first side lobe in the radiation pattern equals $2/\pi$ times that of the main lobe. Assume that the number of antenna elements $N$ is very large. The radiation pattern considered is in the $xz$ plane.

*Solution* Exercise 4.5 dealt with an array having an element width of $a$, a spacing between elements of $b$, and a total length of $c$. The results of Exercise 4.5 are shown in Fig. 4.15 and can be applied to the current

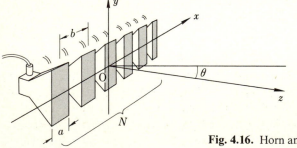

**Fig. 4.16.** Horn antenna array

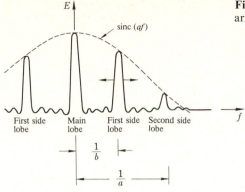

**Fig. 4.17.** Method of designing the antenna array

problem. With this in mind, Fig. 4.17 was drawn as an expanded view of the central portion of Fig. 4.15. The tip of the side lobe follows the envelope of $E_0 \operatorname{sinc}(af)$ when $1/b$ is varied. This is indicated by the dashed line in Fig. 4.17. Therefore the answer is obtained by finding $f$ such that $\operatorname{sinc}(af)$ is $2/\pi$, i.e.

$$\frac{\sin a\pi f}{a\pi f} = \frac{2}{\pi}. \tag{4.26}$$

The solution of (4.26) is $f = 0.5/a$, and the value of $1/b$ is set equal to this value of $f$, namely

$$b = 2a. \tag{4.27}$$

The answer is therefore that the period should be set to twice the width of the aperture of the element antenna.

## 4.4 Video Disk System

An example of the reflection grating is given, followed by an explanation of a video disk system [4.1, 4].

### 4.4.1 Reflection Grating

**Exercise 4.7**    A reflection grating is fabricated by periodically depressing a reflective metal surface. Obtain the far-field pattern of a reflection grating with dimensions as shown in Fig. 4.18. The width of the upper reflecting surface is $a$ and that of the depressed surface is $a'$. The distance from the lower to the higher surface is $h$ and the light is incident normal to the surfaces. Assume the number of periods is infinite.

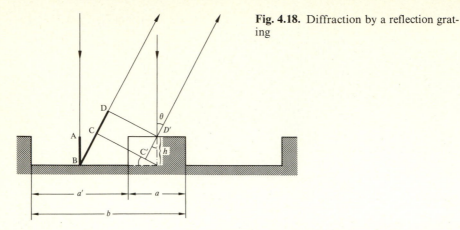

**Fig. 4.18.** Diffraction by a reflection grating

*Solution*  First the radiation pattern of one period is considered. Referring to Fig. 4.18, it is seen that the ray path reflected from the bottom surface is longer by AB + BD than that reflected from the top surface. The path difference $l$ is

$$l = AB + BC + CD = h + \frac{b}{2} \sin \theta + h \cos \theta. \tag{4.28}$$

Hence, the radiation pattern $G_e(f)$ of one period can be obtained by summing the diffraction patterns of the top and bottom surfaces while taking their relative phase differences into consideration:

$$G_e(f) = a \, \text{sinc} \, (a f) + a' \, e^{j\phi} \, \text{sinc} \, (a' f) \tag{4.29}$$

where the value of $\phi$ is from (4.28):

$$\phi = \frac{2\pi}{\lambda} \left[ h(1 + \cos \theta) + \frac{b}{2} \sin \theta \right]. \tag{4.30}$$

The grating array factor $G_a(f)$ with period $b$ is III $(b f)$; hence the radiation pattern $G(f)$ of the entire reflection grating is

$$G(f) = \text{III} \, (b f) \, [a \, \text{sinc} \, (a f) + a' \, e^{j\phi} \, \text{sinc} \, (a' f)] . \tag{4.31}$$

For the special case of $a = a'$ and $\theta \ll 1$, the intensity distribution of the grating is

$$I(f) = 4 a^2 \left[ \text{III} \, (2 a f) \, \text{sinc} \, (a f) \, \cos \frac{\phi}{2} \right]^2 \tag{4.32}$$

where

$$\phi = 2\pi a f + \frac{4\pi}{\lambda} h.$$

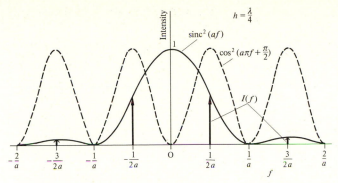

**Fig. 4.19.** Diffraction pattern of a reflection grating with $b = 2a$ for $\theta \ll 1$

The intensity distribution with $h = \lambda/4$ is shown in Fig. 4.19. Notice that the spectral intensity is significantly different from the case of a flat plate. The fact that the diffraction pattern can be controlled by varying geometry is used to store information in the video disk.

*Alternate derivation of* $G_e(f)$. Imagine a fictitious aperture $A$ parallel and close to the reflecting surface in Fig. 4.18. The field distribution of the reflected wave across this aperture is obtained with the help of Fig. 4.18 as

$$g(x_0) = \Pi\left(\frac{x_0}{a}\right) + \Pi\left(\frac{x_0 + b/2}{a'}\right) e^{2jkh},$$

where the point D′ is taken as the origin of the coordinates, and $2kh$ accounts for the phase delay of the bottom surface. The element pattern is given by its Fourier transform $G_e(f)$, with $f = \sin\theta/\lambda$,

$$G_e(f) = a \operatorname{sinc}(af) + a' e^{j\phi_F} \operatorname{sinc}(a'f)$$

with     $$\phi_F = \frac{2\pi}{\lambda}\left(2h + \frac{b}{2}\sin\theta\right).$$

Thus for small $\theta$, the results of this fictitious aperture approach are equivalent to (4.29, 30), which had been derived by tracing the phase delay along the ray path.

### 4.4.2 Principle of the Video Disk System

The video disk is a memory device designed to store pre-recorded television programs using a spiral string of pits on the surface of a spinning disk.

Figure 4.20 shows the play back mechanism of the video disk. The light is focused onto the reflecting surface and the intensity of the reflected beam is modulated according to the pitch length of the pits. The reflected light is then detected by the photodiode which sends the signal to a television set.

Figure 4.21 a shows a photograph of the video disk and Fig. 4.21 b shows an electron microscope photograph of a string of pits on the video disk. Information is encoded in the length and the spacing of the pits along the track [4.5].

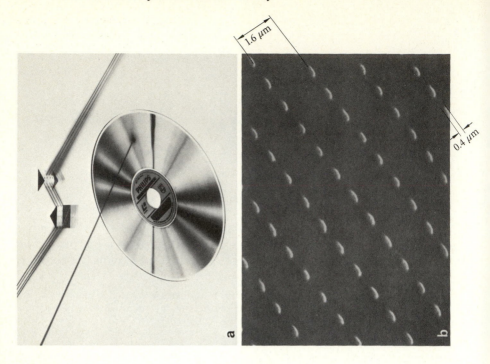

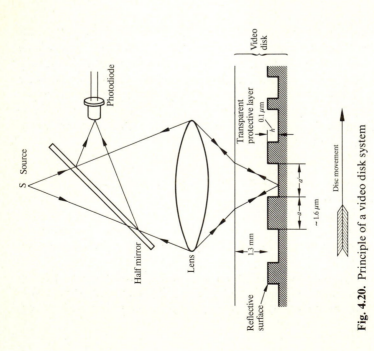

**Fig. 4.20.** Principle of a video disk system

**Fig. 4.21 a, b.** Video disk. (a) Photograph of the video disk. (b) Electron microscope photograph of the disk surface (By courtesy of Philips)

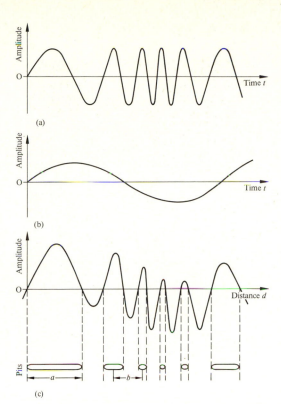

Figure 4.22 illustrates the mode of encoding the pits. The video frequency modulated signal, such as that shown in (a), and the audio frequency, such as that shown in (b), are simultaneously recorded. The $dc$ value of the video signal is changed according to the amplitude of the audio signal. The zero crossing points of the subsequent signal are used to determine the length of the pits. Therefore, the frequency of the video signal is represented by the spacing between the pits, and the audio signal is represented by the length of the pits.

The main features of the video disk system are:

1) There is no contact between the disk and read out system, thus, there is no wearing of the disk.
2) The disks are suitable for mass production.
3) When the protective plastic layer (Fig. 4.20) is made thick, the quality of the signal is hardly affected by dust, finger prints or scratches on the disk surface.
4) The density of the recording is high. For example, the typical spacing between adjacent tracks is 1.6 µm. On a 30 cm diameter disk this results in a linear track length of 34 km or a storage capacity of 54,000 still pictures.

## 4.5 Diffraction Pattern of a Circular Aperture

**Exercise 4.8**   Find the Fraunhofer diffraction pattern of a circular aperture of unit radius, as shown in Fig. 4.23.

*Solution*   First (3.31) is converted into cylindrical coordinates and then the Fourier Bessel transform is performed.

$$u(l) = \frac{1}{j \lambda z_i} \exp\left[j k \left(z_i + \frac{l^2}{2 z_i}\right)\right] \mathcal{B}\{g(r)\}|_{\varrho = l/\lambda z_i \ \ \text{or} \ \ \sin\theta/\lambda}. \tag{4.33}$$

In (4.33), $r$ refers to the radius coordinate in the plane of the aperture, and $l$ refers to the radius coordinate in the plane of the diffraction pattern. Since the input function is

$$g(r) = \text{circ}(r), \tag{4.34}$$

Eq. (2.76) may be used to obtain

$$u(l) = \frac{1}{j \lambda z_i} \exp\left[j k \left(z_i + \frac{l^2}{2 z_i}\right)\right]\left[\frac{J_1(2 \pi \varrho)}{\varrho}\right]\Bigg|_{\varrho = l/\lambda z_i}. \tag{4.35}$$

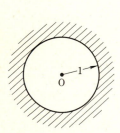

**Fig. 4.23.** Circular aperture

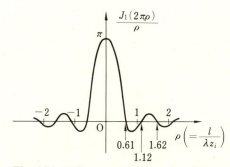

**Fig. 4.24.** Diffraction pattern of a circular aperture

The diffraction pattern is circularly symmetric and the amplitude distribution in the radial direction is shown in Fig. 4.24. The values of $v$ which make the value of $J_1(\pi v)$ zero are $v = 1.22, 2.23, 3.24, \ldots$, and the corresponding values of the $l$'s are $l = 0.61 \lambda z_i, 1.12 \lambda z_i, 1.62 \lambda z_i, \ldots$. The diameter to the first zero of the diffraction pattern is $1.22 \lambda z_i$. This value is compared with that of a square aperture with dimensions $2 \times 2$. The distance between the first zeros of the diffraction pattern of the square is $\lambda z_i$ and is slightly narrower than that of the circular aperture ($1.22 \lambda z_i$).

**Exercise 4.9**   In many optical systems, the cross-sectional amplitude distribution of a light beam is a Gaussian function of the form $\exp(-a^2 r^2)$.

Prove that if the distribution in one plane is Gaussian, then the distribution in a plane at an arbitrary distance away from the first plane is also Gaussian.

*Solution*   Let the first plane be $z = 0$ and the second plane be $z = z_i$. First, the Fourier Bessel transform $G(\varrho)$ of the Gaussian distribution is calculated so that (4.33) may be used. From (2.72), the Fourier Bessel transform is

$$G(\varrho) = 2\pi \int_0^\infty r\, e^{-a^2 r^2}\, J_0(2\pi r \varrho)\, dr. \tag{4.36}$$

Using Bessel's integral formula

$$\int_0^\infty e^{-a^2 x^2}\, x\, J_0(b\,x)\, dx = \frac{1}{2a^2}\, e^{-b^2/4a^2}, \tag{4.37}$$

Eq. (4.36) becomes

$$G(\varrho) = \frac{\pi}{a^2}\, e^{-\pi^2 \varrho^2/a^2} \tag{4.38}$$

and substituting $\varrho = l/(\lambda z_i)$, the distribution $u(l)$ of the light at $z = z_i$ is

$$u(l) = \frac{\pi}{j\, a^2\, \lambda z_i}\, \exp\left[j\, k\left(z_i + \frac{l^2}{2\, z_i}\right)\right] \cdot \exp\left[-\left(\pi/(a\, \lambda\, z_i)\right)^2 l^2\right]. \tag{4.39}$$

Thus it has been proven that the amplitude distribution at $z = z_i$ is again Gaussian.

As a matter of fact, there are only a very limited number of functions whose Fourier Bessel transform can be found in a closed form. The Gaussian distribution function is one of these special functions. Another function of this kind is the delta function.

## 4.6 One-Dimensional Fresnel Zone Plate

When a plane wave is incident upon a mask whose transmission is distributed according to Fig. 4.25a, the transmitted light is focussed to a point. Such a mask behaves like a lens and is called a *Fresnel zone plate* [4.3]. The Fresnel zone plate is one of the most powerful means of forming images in the spectral regions of infra-red light, x-rays, and gamma rays where glass lenses are of no use.

In Exercise 4.5, it was shown that the equal spacing of the periodic slits produced peaks of the diffraction pattern along the $x_i$ axis, but not along the $z$ axis. It will be demonstrated, however, that a quadratic spacing produces peaks along the $z$ axis, but not along the $x_i$ axis.

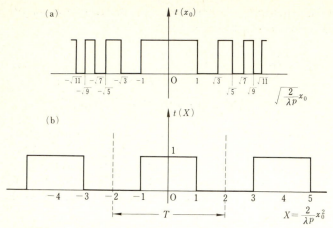

**Fig. 4.25 a, b.** Transmission distribution of a one dimensional zone plate expressed **(a)** as a function of $\sqrt{2/\lambda p}\ x_0$ and **(b)** as a function of $X = (2/\lambda p)\,x_0^2$

Since the transmission of the zone plate shown in Fig. 4.25a is periodic with $x_0^2$, a conversion of variables

$$X = \frac{2}{\lambda p}\,x_0^2 \tag{4.40}$$

is made. Equation (4.40) makes it possible to expand the transmission function into a Fourier series. The expansion coefficients $a_n$ are

$$a_n = \frac{1}{4}\int_{-2}^{2} t(X)\,e^{-j2\pi n(X/4)}\,dX$$

$$= \frac{1}{4}\int_{-1}^{1}\cos\left(n\,\frac{\pi}{2}\,X\right)dX + j\,\frac{1}{4}\int_{-1}^{1}\sin\left(n\,\frac{\pi}{2}\,X\right)dX\ ,$$

where the limits of the integral were changed because $t(x) = 0$ for $1 < |x| < 2$. The integral in the second term is zero because its integrand is an odd function, so that

$$a_n = \frac{\sin n\,\frac{\pi}{2}}{n\,\pi}\ . \tag{4.41}$$

The transmittance function $t(X)$ then can be expressed by

$$t(X) = \sum_{n=-\infty}^{\infty} \frac{\sin n\,\frac{\pi}{2}}{n\,\pi}\,e^{jn(\pi/2)X}\ . \tag{4.42}$$

Changing $n$ to $-n$, and reversing the order of the summation, and using (4.40) to transform back to the variable $x_0$, gives

$$t(x_0) = \sum_{n=-\infty}^{\infty} \frac{\sin n \frac{\pi}{2}}{n \pi} e^{-jn(\pi/\lambda p)x_0^2}. \tag{4.43}$$

This is a series expression for the transmittance distribution of a one-dimensional Fresnel zone plate.

Next, the diffraction pattern of the Fresnel zone plate when illuminated by a parallel beam is examined. The parallel beam is launched along the $z$-axis and the Fresnel zone plate is placed in the plane of $z = 0$, as shown in Fig. 4.26. The distribution function $u(x_i y_i)$ of the diffracted light is obtained by inserting (4.43) into (3.60) and manipulating the equation in a manner similar to that used to arrive at (3.60). The result is

$$u(x_i, z_i) = \frac{1}{\sqrt{z_i \lambda}} \exp\left[j\left(k z_i + k \frac{x_i^2}{2 z_i} - \frac{\pi}{4}\right)\right] \sum_{n=-\infty}^{\infty} \frac{\sin n \frac{\pi}{2}}{n \pi}$$
$$\times \int_{-\infty}^{\infty} \exp\left[j \frac{\pi}{\lambda}\left(\frac{1}{z_i} - \frac{n}{p}\right) x_0^2 - j 2\pi f x_0\right] dx_0 \Bigg|_{f = x_i/\lambda z_i}, \tag{4.44}$$

where the Fresnel zone plate was assumed to extend to infinity. Equation (4.44) is a summation of a series with respect to $n$. The term with $n$ satisfying

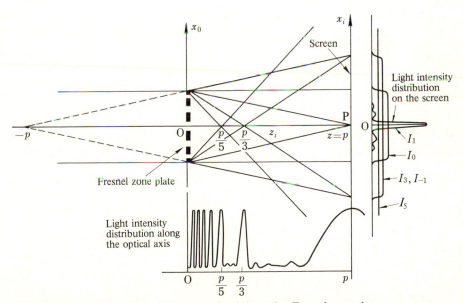

**Fig. 4.26.** Distribution of the light transmitted through a Fresnel zone plate

$$\frac{1}{z_i} - \frac{n}{p} = 0 \tag{4.45}$$

becomes predominantly large, in which case the integral takes on the value

$$\delta\left(\frac{x_i}{\lambda z_i}\right) \tag{4.46}$$

and the intensity peaks up. For a Fresnel zone plate of finite dimension $a$, $t(x_0)$ is replaced by $t(x_0) \, \text{II}(x_0/a)$ and the value corresponding to (4.46) becomes

$$a \, \text{sinc}\,(a \, x_i/\lambda \, z_i).$$

The term with $n = 1$ corresponds to a converged beam focused at $z_i = p$ and $p$ is called the principal focal length. The focused point is indicated by $P$ in Fig. 4.26. Besides the point $P$, the light is focused at the points $z = p_n$ where $p_n = -p, \, -p/3, \, -p/5 \ldots p/n, \, 0, \, p/n \ldots p/5, \, p/3, \, p$. The reason that the peaks with even $n$ are absent is that $\sin(n \, \pi/2)$ in (4.44) becomes zero with even values of $n$. The lower curve in Fig. 4.26 shows the distribution of these amplitude peaks. The curves on the right-hand side of Fig. 4.26 show the relative contribution of each value of $n$ when the screen is placed at the principal focal plane. The curves represent the relative intensities and it should be noted that the resultant intensity distribution is not the mere sum of these intensities, but rather each amplitude should be added taking its phase into consideration, and the intensity should be calculated from this sum. As the position of the screen is moved to another focal point along the $z$ axis, the intensity associated with that particular focal length becomes predominantly high, compared to the intensities associated with other focal lengths. The distribution in Fig. 4.26 is for the case when $n = 1$. If the screen is moved to $z_i = p/3$, then the intensity associated with $n = 3$ becomes high compared to all others. Now that some understanding of the action of the Fresnel zone plate has been gained, it would be instructive to perform the integration in (4.44). The integral is obtained with the aid of (3.38)

$$u\,(x_i; p) = \frac{1}{\sqrt{p\,\lambda}} \exp\left[j\left(k\,p + k\,\frac{x_i^2}{2\,p} - \pi/4\right)\right] \sum_{n=-\infty}^{\infty} \frac{\sin n\frac{\pi}{2}}{n\,\pi}$$

$$\times \sqrt{\frac{p\,\lambda}{1-n}} \exp\left\{j\,\pi\left[1/4 - \frac{x_i^2}{(1-n)\,\lambda\,p}\right]\right\} \tag{4.47}$$

$$= \frac{1}{\sqrt{p\,\lambda}}\, e^{jkp} \sum_{n=-\infty}^{\infty} \frac{\sin n\frac{\pi}{2}}{n\,\pi} \sqrt{\frac{p\,\lambda}{1-n}} \exp\,j\,k\,\frac{x_i^2}{2\,\dfrac{(n-1)}{n}\,p} \,. \tag{4.48}$$

Recall that a cylindrical wavefront measured along a line located a distance $d$ away is

$$A \, e^{jk(x_1^2/2d)} \, .$$    (4.49)

Notice that the $n$th terms of (4.48) have the same form as (4.49), except that the light source is located at a distance $(n-1) \, p/n$ from the screen rather than $d$.

When the Fresnel zone plate is used to converge a parallel beam, not only the contribution from the term with $n = 1$ in (4.48) but also the contributions of all other terms like $n = 0, \pm 3, \pm 5, \pm (2n-1)$ are present. As a result, the converged beam has background illumination and is not as sharp as that made by a convex glass lens. Notice that both positive and negative $n$ terms are included. While positive $n$ corresponds to convex lenses, negative $n$ corresponds to concave lenses, and thus the Fresnel zone plate possesses the properties of both kinds of lenses. Another important point to remember is that the focal length of the Fresnel zone plate varies with the wavelength of light.

## 4.7 Two-Dimensional Fresnel Zone Plate

A Fresnel zone plate can be made two-dimensional, as shown in Fig. 4.27, by changing the array of strips into an array of concentric rings. The radii $R_n$ of the concentric rings are the same as the one-dimensional strip spacing of the zone plate in the previous section, i.e.,

$$R_n = \sqrt{\frac{\lambda p}{2}(2n-1)} \, .$$    (4.50)

The intensity distribution in the principal focal plane can be obtained in a manner similar to that of the one-dimensional Fresnel zone plate. The transmittance distribution $t(r)$ of a two-dimensional Fresnel zone plate is

$$t(r) = \sum_{n=-\infty}^{\infty} \frac{\sin n \frac{\pi}{2}}{n \pi} \, e^{-jn(\pi/\lambda p)r^2} \, .$$    (4.51)

**Fig. 4.27.** Two dimensional Fresnel zone plate

When the Fresnel zone plate is placed in the plane $z = 0$ and illuminated by a beam propagating parallel to the $z$ axis, the diffraction pattern on the screen at $z = z_i$ in cylindrical coordinates is

$$u(l, z_i) = \frac{1}{j \lambda z_i} \exp\left[j k \left(z_i + \frac{l^2}{2 z_i}\right)\right]$$

$$\times 2\pi \int_0^\infty r\, P(r)\, t(r) \exp\left(j k \frac{r^2}{2 z_i}\right) J_0(2\pi \varrho r)\, dr \Bigg|_{\varrho = l/\lambda z_i} \qquad (4.52)$$

where $P(r)$ is the pupil function of the aperture. $P(r)$ is unity inside the aperture and zero outside the aperture. Inserting (4.51) into (4.52) gives

$$u(l, z_i) = \frac{1}{j \lambda z_i} \exp\left[j k \left(z_i + \frac{l^2}{2 z_i}\right)\right] 2 \sum_{n=-\infty}^\infty \frac{\sin n \frac{\pi}{2}}{n}$$

$$\times \int_0^\infty r\, P(r) \exp\left[j\left(\frac{\pi}{\lambda}\right)\left(\frac{1}{z_i} - \frac{n}{p}\right) r^2\right] J_0(2\pi \varrho r)\, dr \Bigg|_{\varrho = l/\lambda z_i}. \qquad (4.53)$$

The term that satisfies the condition

$$\frac{1}{z_i} - \frac{n}{p} = 0 \qquad (4.54)$$

dominates, as in the case of one-dimensional zone plate, and its distribution can be approximated by

$$u(l, p_n) = \frac{1}{j \lambda p_n} \exp\left[j k \left(p_n + \frac{l^2}{2 p_n}\right)\right] \frac{\sin n \frac{\pi}{2}}{n \pi} \mathcal{B}\{P(r)\} \Bigg|_{\varrho = l/\lambda p_n}. \qquad (4.55)$$

From (4.55) it is seen that the distribution of light in the focal plane is a Fourier Bessel transform of the pupil function associated with the two-dimensional Fresnel zone plate.

Up to this point, the distribution of the transmittance across the two-dimensional zone plate was made up of zones of either full transmission or no transmission in a uniform pattern. The transmittance distribution of the zone plate can be further modulated by an arbitrary function $\phi(r)$. The Fresnel zone plate whose transmittance is further modulated in this way is called a *Modulated Zone Plate*, abbreviated MZP. The light distribution in the back focal plane of the MZP is the Hankel transform of $\phi(r)$. By choosing an appropriate function, the distribution of the light at the back focal plane can be formed to a desired shape. By combining the MZP with a high-intensity laser, a cutting tool can be made, which cuts sheet metal into a shape determined by the shape of the focused beam. The MZP is usually made out of thin chromium film. Fig. 4.28 shows two arrays of holes made by MZP [4.6].

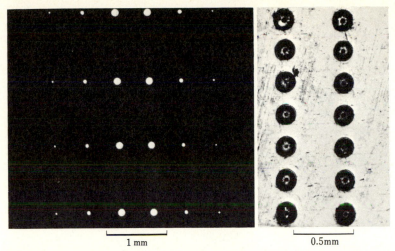

**Fig. 4.28.** The photographs of the machined holes obtained with a laser beam focused by means of MZP [4.6]. (**a**) An array of holes made on a chrome film deposited on a sheet glass. (**b**) Two arrays of blind holes made on a polished surface of steel

As mentioned in Chap. 2, the Hankel transform in cylindrical coordinates is identical to that of the Fourier transform in rectangular coordinates. Therefore, the transform of $\phi(r)$ may be evaluated using whichever transform is the easier. When using rectangular coordinates for the Fourier transform, (4.55) becomes

$$u(x_i, y_i, p_n) = \frac{1}{j\lambda p_n} \exp\left\{ jk\left[ p_n + \frac{(x_i^2 + y_i^2)}{2p_n} \right]\right\} \frac{\sin n\frac{\pi}{2}}{n\pi} \mathscr{F}\{\phi(x,y)\} \qquad (4.56)$$

with $f_x = x_i/\lambda p_n$ and $f_y = y_i/\lambda p_n$. Even though the Fresnel zone plate has such drawbacks as a focal length variation with wavelength and the presence of background light along with the main focused beam, it has the *invaluable advantage that it can be used with types of electromagnetic radiation other than visible light.*

As an example of the application of the Fresnel zone plate to radiation outside the visible light spectrum, the gamma-ray camera is considered. The *gamma-ray camera* uses the Fresnel zone plate to form the image of an isotope administered to a patient for diagnostic purposes. As shown in Fig. 4.29a, the gamma rays from an isotope in a patient strike a Fresnel zone plate made out of concentric lead rings. The pattern of the shadow cast by the zone plate is recorded either by a photographic plate or by a sodium iodine scintillator connected to a processing computer. The shadow recorded by the photographic plate is reduced in size and illuminated by coherent light. As shown Fig. 4.29b, the pattern of the superposed zone plate on the photographic plate converges the light to their respective

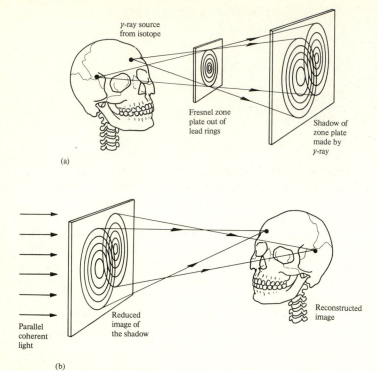

(a)

(b)

**Fig. 4.29 a, b.** Gamma ray camera. (**a**) Shadow of zone plate cast by isotope tracer. (**b**) Reconstruction of the image by coherent light

principal focal points and thus reconstructs the entire shape of the patient's organ. It should be noted that it is *not the diffraction pattern of the gamma rays that forms the image, but rather the diffraction pattern of the light which illuminates the reduced pattern made by the gamma rays.*

The difference between an x-ray and a gamma-ray camera is that the purpose of the former camera is to take a picture of the structure of the organ whereas that of the latter is to examine the function of a particular organ. A comparison of two gamma-ray images taken at a certain time interval shows a change in the distribution of the labelled compound in the patient with respect to time. This change can be used to assess the functioning of that particular organ.

# Problems

**4.1** A grating whose surface is shaped like a saw tooth as shown in Fig. 4.30, is called an echelette grating. Find the optimum height $h$ that

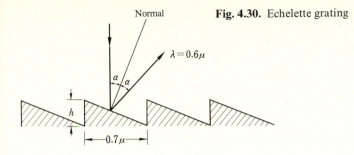

Normal

**Fig. 4.30.** Echelette grating

$\lambda = 0.6\mu$

$\alpha$ $\alpha$

$h$

$\mapsto$—$0.7\mu$—$\dashv$

maximizes the peak intensity of the diffracted light. The wavelength is 0.6 μm (1 μm is $10^{-3}$ mm) and the pitch of the tooth is 0.7 μm.

**4.2** Figure 4.31 shows a light deflector based on the principle of the photo-elastic effect. Liquid in the deflector is periodically compressed by an acoustic wave generated by the transducer at the bottom, thus creating a periodic variation in the index of refraction. Such a structure can be considered as a kind of diffraction grating. The direction of the deflection of the beam is controlled by the frequency.

It is assumed that the phase modulation of the transmitted wave is expressed by

$$\phi(x) = \phi_0 + \phi_m \sin \frac{2\pi}{\lambda_a} x \,, \tag{4.57}$$

when the direction of the incident beam is perpendicular to the wall of the

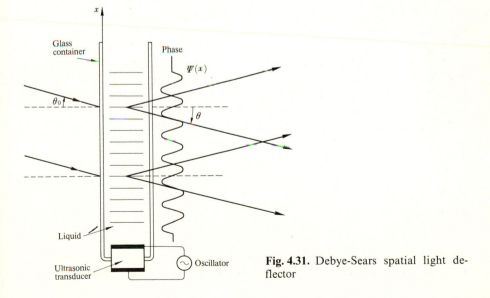

$x$

Glass container

Phase

$\Psi(x)$

$\theta_0$

$\theta$

Liquid

Ultrasonic transducer

Oscillator

**Fig. 4.31.** Debye-Sears spatial light deflector

deflector. Prove that when the angle of incidence of the beam to the wall is $\theta_0$, the angle $\theta$ of the peak of the deflected beam satisfies

$$\sin \theta = \sin \theta_0 - n \frac{\lambda}{\lambda_a}, \tag{4.58}$$

where $\lambda_a$ is the wavelength of the acoustic wave, $\lambda$ is the wavelength of light, and $n$ is an integer. In Fig. 4.31 all the angles are taken as positive when they are measured clockwise from the outward normal of either wall.

**4.3** Referring to Problem 4.2, when the angles are adjusted to satisfy $\theta_0 = - \theta$, (4.58) becomes $2 \sin \theta_B = n(\lambda/\lambda_a)$, which is the formula for Bragg's condition. Explain why the intensity of the peak deflected in this particular direction is substantially larger than all others.

**4.4** A parallel beam is incident upon the modulated zone plate (MZP) shown in Fig. 4.32. Find an expression for the pattern at the back focal plane of this MZP. (A combination of a high intensity laser and such an MZP is used as a cutting tool for thin steel plates.)

**Fig. 4.32.** Modulated zone plate (MZP) for machining tool

**4.5** In the text, the transmittance $t(x_0)$ of the one-dimensional zone plate was made up of strips that were either completely opaque or completely transparent. Show that the semi-transparent mask with the transmittance

$$t(x_0) = 1 + \cos \left( k \frac{x_0^2}{2f} \right) \tag{4.59}$$

also possesses focusing action, and determine its focal length.

# 5. Geometrical Optics

The rigorous way of treating light is to accept its electromagnetic wave nature and solve Maxwell's equations. However the number of configurations for which exact solutions can be found is very limited and most practical cases require approximations.

Based on the specific method of approximation, optics has been broadly divided into two categories, these being *geometrical optics (ray optics)* treated in this chapter and *wave optics (physical optics)* already treated in Chapter 3. The approximation used in geometrical optics puts the emphasis on finding the light path; it is especially useful for tracing the path of propagation in *inhomogeneous* media or in designing optical instruments. The approximation used in physical optics, on the other hand, puts the emphasis on analyzing interference and diffraction and gives a more accurate determination of light distributions.

## 5.1 Expressions Frequently Used for Describing the Path of Light

In this section, mathematical expressions often used for describing the path of light are summarized [5.1, 2]. The mathematical foundation necessary to clearly understand geometrical optics will also be established.

### 5.1.1 Tangent Lines

The tangent line is often used to describe the direction of curvilinear propagation. With the curve *l* shown in Fig. 5.1, an expression for the tangent drawn at P is obtained. An arbitrarily fixed point 0 is taken as the reference point. The vector made up by connecting 0 and P is designated by $R$, and is called the *position vector*. Another point P' on the curve close to P is selected, and a straight line (chord) from P to P' is drawn. The position vector of P' is $R + \Delta R$, $\Delta R$ being the vectorial expression of the chord. As P' approaches P, the chord tends to reach a definite limiting orientation. This limiting orientation is the direction of the tangent to the curve at P. In this definition, the direction of the vector $\Delta R$ indeed

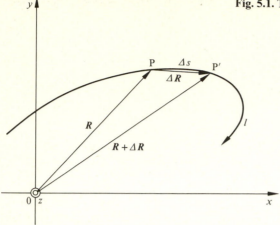

**Fig. 5.1.** Tangent to a curve

coincides with the tangent but its magnitude vanishes, thus making it an inconvenient expression. In order to get around this, the ratio $\Delta R/\Delta s$ is used, $\Delta s$ being the length measured along the curve from P' to P. Now, as P' approaches P both the magnitude $|\Delta R|$ and $\Delta s$ decrease at the same time. The magnitude of $|\Delta R|/\Delta s$ reaches the limiting value of unity. Thus $\Delta R/\Delta s$ represents the *unit vector ŝ of the tangent*. In the following we will interchangeably call ŝ the direction of the light ray, or the unit tangent vector,

$$\hat{s} = \frac{dR}{ds}. \tag{5.1}$$

Using rectangular coordinates, the position vector $R$ is written in terms of its components as

$$R = i\,x + j\,y + k\,z.$$

Substituting this expression for $R$ into (5.1), the unit tangent vector ŝ for rectangular coordinates is

$$\hat{s} = i\,\frac{dx}{ds} + j\,\frac{dy}{ds} + k\,\frac{dz}{ds}. \tag{5.2}$$

Depending on the geometry, some problems are more easily solved using cylindrical rather than rectangular coordinates. Hence, an expression for the unit tangent vector ŝ in cylindrical coordinates is derived. Figure 5.2 shows the projection of a curve $l$ onto the $x$-$y$ plane. The position vector of a point P on the curve $l$ is the sum of the position vector of the point P', which is the projection of P onto the $x$-$y$ plane, and the $z$ component vector of the curve $l$, i.e.,

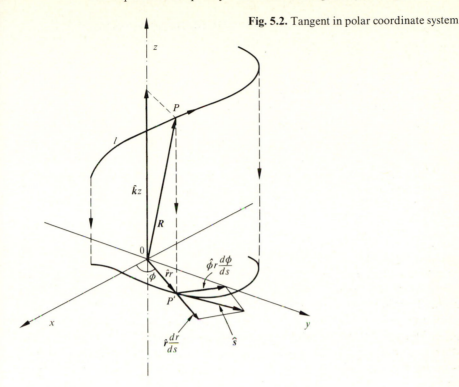

**Fig. 5.2.** Tangent in polar coordinate system

$$R = \hat{r}\,r + \hat{k}\,z . \tag{5.3}$$

The position vector for the projection is expressed as $\hat{r}\,r$, where $r$ is the distance from 0 to P′, and $\hat{r}$ is a unit vector pointing in the direction from 0 to P′. The unit vector $\hat{r}$ is observed to point in the direction of constant $\phi$, where $\phi$ is the angle between the line $\overline{OP'}$ and the $x$ axis. Before inserting (5.3) into (5.1) the unit vector $\hat{\phi}$ is introduced as being a unit vector in the direction of constant radius. These unit vectors can be decomposed into components parallel to the rectangular coordinates as

$$\hat{r} = \quad i\cos\phi + j\sin\phi, \quad \hat{\phi} = -i\sin\phi + j\cos\phi \tag{5.4}$$

Inserting (5.3) into (5.1) gives

$$\hat{s} = \hat{r}\,\frac{dr}{ds} + r\,\frac{d\hat{r}}{ds} + \hat{k}\,\frac{dz}{ds} . \tag{5.5}$$

It is important to realize that *both $\hat{r}$ as well as $r$ change as point P moves.* From (5.4), the derivative of $\hat{r}$ in (5.5) is

$$\frac{d\hat{r}}{ds} = (-i\sin\phi + j\cos\phi)\,\frac{d\phi}{ds} = \hat{\phi}\,\frac{d\phi}{ds} . \tag{5.6}$$

Thus, the *tangent expressed in cylindrical coordinates* becomes

$$\hat{s} = \hat{r}\frac{dr}{ds} + \hat{\phi}\,r\,\frac{d\phi}{ds} + \hat{k}\frac{dz}{ds}\,. \tag{5.7}$$

### 5.1.2 Curvature of a Curve

Light changes its direction of propagation when it encounters an inhomogeneity in the medium. The curvature of the path is used to quantify this change of direction. This curvature is defined as the ratio of the change in the direction of propagation to the length measured along the curved path, i.e., the curvature is $d\hat{s}/ds$. The *radius of curvature $\varrho$ of a curve* is

$$\frac{1}{\varrho} = \left| \frac{d\hat{s}}{ds} \right|\,, \tag{5.8}$$

or from (5.1),

$$\frac{1}{\varrho} = \left| \frac{d^2 R}{ds^2} \right|\,. \tag{5.9}$$

Thus, the *inverse of the radius of curvature is the magnitude of the first derivative of the tangent vector, or the second derivative of the position vector.*
Next, the relationship,

$$\frac{d\hat{s}}{ds} = \frac{\hat{N}}{\varrho} \tag{5.10}$$

known as the *Frenet-Serret formula*, will be proven with the aid of Fig. 5.3. The vector $\hat{N}$ appearing in (5.10) is the unit normal vector to the curve. A circle which shares a common tangent with the curve at a point P is drawn. The radius of such a circle is designated by $\varrho$. The path length $ds$ common with the circle can be represented by

$$ds = \varrho\,d\phi\,. \tag{5.11}$$

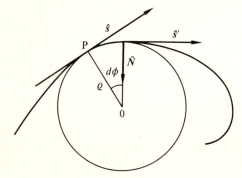
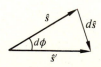

**Fig. 5.3.** Geometry used to prove Frenet-Serret formula

The unit tangent vector $\hat{s}$ does not change its magnitude but changes its direction over the path. The direction changes from $\hat{s}$ to $\hat{s}'$, as illustrated in Fig. 5.3. Since $|\hat{s}| = 1$, it follows from the figure that

$$|d\hat{s}| = d\phi. \tag{5.12}$$

The scalar product of $\hat{s}' \cdot \hat{s}'$ is

$$\hat{s}' \cdot \hat{s}' = 1 = (\hat{s} + d\hat{s}) \cdot (\hat{s} + d\hat{s}) = |\hat{s}|^2 + 2\hat{s} \cdot d\hat{s} + |d\hat{s}|^2 = 1 + 2\hat{s} \cdot d\hat{s} + |d\hat{s}|^2.$$

When $|d\hat{s}|^2$ is much smaller than the other terms,

$$\hat{s} \cdot d\hat{s} = 0.$$

This means that $d\hat{s}$ is perpendicular to $\hat{s}$ and is normal to the curve. In Fig. 5.3, the change from $\hat{s}$ to $\hat{s}'$ is in a downward direction corresponding to a section of the curve which is concave down. The direction of the normal $N$ is also downward in this case. The direction of $N$ is the same as that of the change in $\hat{s}$.

Thus $d\hat{s}$ is a vector whose magnitude is $d\phi$, from (5.12), and whose direction is $N$,

$$d\hat{s} = \hat{N} d\phi. \tag{5.13}$$

Finally, combining (5.11, 13), the Frenet-Serret formula is derived, namely,

$$\frac{d\hat{s}}{ds} = \hat{N} \frac{1}{\varrho}.$$

This formula will be used in describing the optical path in an inhomogeneous medium.

### 5.1.3 Derivative in an Arbitrary Direction and Derivative Normal to a Surface

While most of the groundwork for a discussion of geometrical optics has now been laid, there remains yet one more important concept to be introduced before proceeding. This concept involves the rate of change of a multivariable function along a specified direction. In dealing with multivariable functions, the partial derivative is commonly used to describe the rate of change of the function along a particular coordinate axis direction. However, there is no reason for being restricted to the coordinate axis directions, and a more general expression for the derivative along an arbitrary direction is sought. Only three variable functions are considered here, although generalizations to a greater number of variables can easily be made.

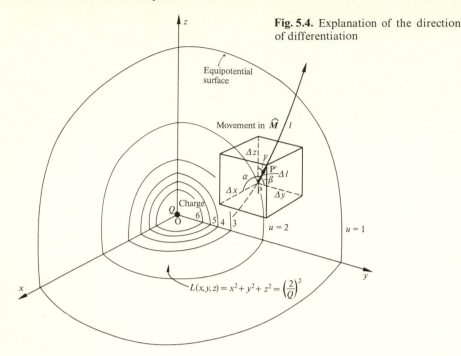

**Fig. 5.4.** Explanation of the direction of differentiation

Let us begin with a specific example. The electrostatic potential $u$ observed at a distance $r$ from a point charge $Q$ is $u = Q/r$. The expression for an equipotential surface of a specific value, let us say $u = 2$, is

$$x^2 + y^2 + z^2 = (Q/2)^2. \tag{5.14}$$

This equipotential surface is a sphere, as shown in Fig. 5.4. All other equipotential surfaces can be drawn in a similar manner. In general,

$$L(x, y, z) = C \tag{5.15}$$

represents a surface which is called the *level surface* associated with the value $C$.

Referring to Fig. 5.4, the change in $L$ with a movement of the point P of observation from one equipotential surface to the other will be investigated. Let the length of the movement be $\Delta l$. The direction of its movement is arbitrary. The movement $\Delta l$ consists of a movement of $\Delta x$ in the $x$ direction, and $\Delta y$ in the $y$ direction, and $\Delta z$ in the $z$ direction, such that $\Delta l = \sqrt{\Delta x^2 + \Delta y^2 + \Delta z^2}$. The increment $\Delta L$ associated with the movement from P $(x, y, z)$ to P′ $(x + \Delta x, y + \Delta y, z + \Delta z)$ is obtained using the multivariable Taylor's formula,

$$\Delta L = \frac{\partial L}{\partial x} \Delta x + \frac{\partial L}{\partial y} \Delta y + \frac{\partial L}{\partial z} \Delta z + \text{infinitesimals of higher order.} \tag{5.16}$$

The directional derivative $\nabla_s L$ is

$$\nabla_s L = \lim_{\Delta l \to 0} \frac{\Delta L}{\Delta l} = \frac{\partial L}{\partial x} \frac{\Delta x}{\Delta l} + \frac{\partial L}{\partial y} \frac{\Delta y}{\Delta l} + \frac{\partial L}{\partial z} \frac{\Delta z}{\Delta l} \tag{5.17}$$

+ infinitesimals of higher order.

As $\Delta l$ decreases, the infinitesimals of higher order in (5.17) tend to zero. Note that $\Delta x/\Delta l$, $\Delta y/\Delta l$, $\Delta z/\Delta l$ are the direction cosines $\cos \alpha$, $\cos \beta$, $\cos \gamma$ of the movement $\Delta l$, as shown in Fig. 5.4. Using the direction cosines, $\nabla_s L$ is written as

$$\nabla_s L = \frac{dL}{dl} = \frac{\partial L}{\partial x} \cos \alpha + \frac{\partial L}{\partial y} \cos \beta + \frac{\partial L}{\partial z} \cos \gamma . \tag{5.18}$$

The directional derivative $\nabla_s L$ is of the form of a scalar product of $\nabla L$ and $\hat{M}$,

$$\nabla L = i \frac{\partial L}{\partial x} + j \frac{\partial L}{\partial y} + k \frac{\partial L}{\partial z} \tag{5.19}$$

$$\hat{M} = i \cos \alpha + j \cos \beta + k \cos \gamma \tag{5.20}$$

where $\hat{M}$ is the unit vector in the direction of the movement of the point P. Thus,

$$\nabla_s L = (\nabla L) \cdot \hat{M} \qquad \text{or} \tag{5.21}$$

$$\Delta L = (\nabla L) \cdot \hat{M} \, dl = (\nabla L) \cdot dl \tag{5.22}$$

where $dl$ is the vector form of $dl$. For a given $L$, the direction of $\nabla L$ is determined by (5.19) but the value of $\nabla_s L$ varies with the choice of the direction of the movement. According to (5.22), the change in $L$ becomes a maximum when the movement is selected in the same direction as that of $\nabla L$. Conversely, it can be said that the direction of $\nabla L$ is the direction that gives the maximum change in $L$ for a given length of movement. The direction which gives the maximum change in $L$ with the minimum length of movement is the direction of the normal to the equi-level surface because it connects the two adjacent equi-level surfaces with the shortest distance.

The unit vector representing the *normal N to the equi-level surface* is therefore

$$N = \frac{\nabla L}{|\nabla L|} . \tag{5.23}$$

In optics, (5.23) is a particularly useful formula because it determines the optical path from the equi-phase surface.

With these preparations completed, geometrical optics will be introduced in the next sections.

## 5.2  Solution of the Wave Equation in Inhomogeneous Media by the Geometrical-Optics Approximation

A vector $E(x, y, z)$ representing a light wave needs to satisfy the wave equation

$$(\nabla^2 + \omega^2 \mu \, \varepsilon) \, E(x, y, z) = 0 \tag{5.24}$$

where $\omega$ is the radian frequency of the light, $\mu$ the permeability of the medium, and $\varepsilon$ the dielectric constant of the medium. In a rectangular coordinate system, the equations for $E_x$, $E_y$, and $E_z$ are identical if the medium is isotropic[1], so that solving for one of them is sufficient to obtain a general solution. The light wave vector may then be replaced by a scalar function $u(x, y, z)$ and the wave equation becomes

$$(\nabla^2 + [k \, n \, (x, y, z)]^2) \, u(x, y, z) = 0 , \tag{5.25}$$

where $\omega^2 \mu \, \varepsilon = k^2 \, n^2 \, (x, y, z)$, $k$ is the free space propagation constant, and $n$ is refractive index of the medium. When the medium is homogeneous and $n(x, y, z)$ is constant, the solutions are well known trigonometric functions. However, for an inhomogeneous medium in which $n(x, y, z)$ is a function of location, the rigorous solution becomes complicated and some kind of approximation usually has to be employed. The method of approximation treated here is the geometrical-optics approximation [5.3 – 5] which is valid for large $k$.

The solution of (5.25) is assumed to be of the form

$$u(x, y, z) = A(x, y, z) \, e^{j[k \, L(x, y, z) - \omega t]} . \tag{5.26}$$

Whether or not (5.26) is correct can be verified by substituting it back into the differential equation. The functions $A(x, y, z)$ and $L(x, y, z)$ are unknown and have to be determined in such a way that (5.26) satisfies (5.25). Computing the derivative of (5.26) with respect to $x$ gives

$$\frac{\partial u}{\partial x} = \left( \frac{\partial A}{\partial x} + j \, k \, A \, \frac{\partial L}{\partial x} \right) e^{j(k L - \omega t)} .$$

Equation (5.25) is obtained by differentiating once more and adding a term $n^2 \, k^2 \, u$. Similar operations are performed with respect to $y$ and $z$, giving

$$n^2 k^2 u + \frac{\partial^2 u}{\partial x^2} = e^{j(k L - \omega t)} \left\{ k^2 \left[ n^2 - \left( \frac{\partial L}{\partial x} \right)^2 \right] A + \frac{\partial^2 A}{\partial x^2} + j \, k \, A \, \frac{\partial^2 L}{\partial x^2} + j 2 k \, \frac{\partial A}{\partial x} \cdot \frac{\partial L}{\partial x} \right\} = 0$$

$$\downarrow \qquad\qquad\qquad \downarrow \qquad \downarrow \qquad \downarrow \qquad \downarrow \quad \dots \tag{5.27}$$

$$n^2 k^2 u + \nabla^2 u = e^{j(k L - \omega t)} \left\{ k^2 \, [n^2 - |\nabla L|^2] A + \nabla^2 A + j \, k \, A \, \nabla^2 L + j 2 k \, (\nabla A) \cdot (\nabla L) \right\} = 0 ,$$

where $|\nabla L|^2$ means the sum of the squares of the $i, j, k$ components of $\nabla L$.

---

[1] A medium is said to be isotropic if the physical properties at each point in the medium are independent of the direction of measurement.

Multiplying (5.27) by $1/k^2$ gives

$$(n^2 - |\nabla L|^2) A + \frac{1}{k^2} \nabla^2 A + \frac{j}{k} [A \nabla^2 L + 2 (\nabla A) \cdot (\nabla L)] = 0. \qquad (5.28)$$

If the wavelength of light is much shorter than the dimensions of the associated structure, then the second and third terms are very small compared to the first term and can be ignored. Equation (5.28) then becomes

$$|\nabla L|^2 = n^2 . \qquad (5.29)$$

Expressed in rectangular coordinates, (5.29) is written as

$$\left(\frac{\partial L}{\partial x}\right)^2 + \left(\frac{\partial L}{\partial y}\right)^2 + \left(\frac{\partial L}{\partial z}\right)^2 = n^2 . \qquad (5.30)$$

The function $u(x, y, z)$ can be obtained by inserting the solution of (5.30) into (5.26). In this way, (5.25) does not have to be solved directly. It should be noted, however, that the approximations applied in obtaining (5.29) implied not only that $k$ was very large, but also that the values of $\nabla^2 L$, $\nabla^2 A$, and $\nabla A \cdot \nabla L$ were very small. In other words, the variation of both $L$ and $A$ has to be very small. Referring to (5.22), the variation $\Delta L$ due to a movement in an arbitrary direction $dl$ is $(\nabla L) \cdot dl$ and therefore

$$L = \int_{\substack{\text{arbitrary} \\ \text{direction}}} (\nabla L) \cdot dl . \qquad (5.31)$$

If the movement is restricted to the $\nabla L$ direction which is the direction of the normal to the equi-level surface of $L$, then (5.29) gives

$$L = \int_{\substack{\text{along} \\ \text{normal to} \\ \text{equi-level} \\ \text{surface}}} |\nabla L| \, ds = \int_{\substack{\text{along} \\ \text{normal to} \\ \text{equi-level} \\ \text{surface}}} n \, ds . \qquad (5.32)$$

At first glance, it appears as if the differential equation (5.29) has been solved directly by the integration of (5.32). However, it is not quite so simple, because, in order to perform the integration, the direction of the normal has to be known, and in order to know the direction of the normal, $L$ has to be known, and $L$ is the quantity one is trying to determine by the integration of (5.32) in the first place. There is no way to avoid solving the differential equation (5.29).

Equation (5.29) is an important equation and is called the *eikonal equation*. The wave front $L$ itself is called the *eikonal or optical path*. The origin of the word "eikonal" is Greek, meaning "image". As seen from (5.26), the eikonal is a quantity which specifies the phase front.

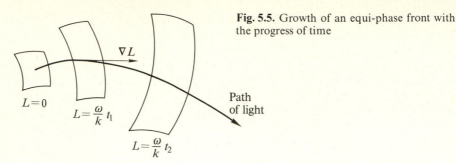

**Fig. 5.5.** Growth of an equi-phase front with the progress of time

The movement of a particular phase surface, say the zero phase surface, is considered. From (5.26), such a phase front is on a surface represented by

$$L(x, y, z) = \frac{\omega}{k} t \, .$$

As time elapses, the zero radian phase front advances, as shown in Fig. 5.5. Recall from (5.23) that the *direction of the normal to this equi-phase surface is* $\nabla L / |\nabla L|$. The direction of the normal is called the *"wave normal"* and coincides with the direction of propagation if the refractive index is isotropic, and is the same in all directions.

Since the velocity of light in a medium with refractive index $n$ is $v = c/n$, the time $T$ that the light takes to travel between the two points $P_1$ and $P_2$ in this medium is

$$T = \int_{P_1}^{P_2} \frac{ds}{v} = \int_{P_1}^{P_2} \frac{n}{c} \, ds = \frac{L}{c} \qquad \text{where} \tag{5.33}$$

$$L = \int_{P_1}^{P_2} n \, ds \, . \tag{5.34}$$

The path of integration in (5.34) coincides with the optical path which is the direction of the normal to the equi-phase surface, so that (5.32) can be used. One interpretation of $L = cT$ is that the *eikonal is the distance that the light would travel in vacuum during the time that the light travels between the two specified points in the medium.* Another interpretation is that *the ratio of the eikonal to the speed of light $c$ represents the time it takes for the light ray to travel from* $P_1$ *to* $P_2$. 

The value of $A(x, y, z)$ has still to be determined. Since the first term of (5.28) is zero from (5.29), and since the term with $1/k^2$ can be assumed negligibly small, (5.28) becomes

$$A \nabla^2 L + 2 (\nabla A) \cdot (\nabla L) = 0 \, . \tag{5.35}$$

In general, it is difficult to solve (5.35) for $A$ except for special cases, one of which is considered in Problem 5.1. The importance of geometrical optics lies in the solution for $L$ rather than for $A$.

**Exercise 5.1**   A glass slab is the fundamental building block of many kinds of integrated micro-optic circuits. Consider a glass slab whose index of refraction is variable in the $x$ direction but is constant in both the $y$ and $z$ directions. The index of refraction is denoted by $n(x)$. Find a solution $L(x, y, z)$ of the eikonal equation, and find the geometrical optics solution $u(x, y, z)$ of the light propagating in this medium.

*Solution*   $L(x, y, z)$ is assumed to be separable so that it can be expressed by the sum

$$L(x, y, z) = f(x) + g(y) + h(z) \tag{5.36}$$

where $f(x)$ is a function solely of $x$, $g(y)$ is a function solely of $y$ and $h(z)$ is a function solely of $z$. Whether or not a solution of this form is correct can be verified by inserting it back into (5.29). The inserted result is

$$\{[f'(x)]^2 - [n(x)]^2\} + [g'(y)]^2 + [h'(z)]^2 = 0. \tag{5.37}$$

The first term is a function of $x$ only, and the second and the third terms are functions only of $y$ and $z$, respectively. In order that (5.37) be satisfied at any point in space, in other words, for any possible combination of the values of $x, y$ and $z$, each term has to be constant,

$$\left.\begin{aligned} &f'(x)^2 - [n(x)]^2 = a^2 \\ &[g'(y)]^2 = b^2 \\ &[h'(z)]^2 = c^2 \\ &a^2 + b^2 + c^2 = 0. \end{aligned}\right\} \tag{5.38}$$

Integrating (5.38) gives

$$f(x) = \pm \int_0^x \sqrt{[n(x)]^2 - (b^2 + c^2)}\ dx \tag{5.39}$$

$$g(y) = \pm b\, y + m_1 \tag{5.40}$$

$$h(z) = \pm c\, z + m_2. \tag{5.41}$$

Inserting all of these terms into (5.36), $L(x, y, z)$ becomes

$$L(x, y, z) = \pm \int_m^x \sqrt{[n(x)]^2 - (b^2 + c^2)}\ dx \pm b\, y \pm c\, z. \tag{5.42}$$

The constants $m_1$ and $m_2$ in (5.40, 41) are absorbed in the lower limit of the integral of the first term of (5.42). The values of $b$ and $c$ are determined by the boundary conditions, such as launching position and angle. The plus or minus signs are determined by the direction of propagation. Taking the direction of propagation to be positive for all three $x$, $y$ and $z$ directions, and inserting (5.42) into (5.26) finally gives

$$u(x, y, z) = A \exp\left\{ j k \left[ \int_m^x \sqrt{n^2(x) - (b^2 + c^2)}\, dx + b\, y + c\, z \right] - j\, \omega\, t \right\}.$$

(5.43)

The solution for $A$ is left as a problem.

## 5.3 Path of Light in an Inhomogeneous Medium

The direction of light propagation in an isotropic medium is the direction of the wave normal [5.3, 4]. The unit vector $N$ of the wave normal is from (5.23, 29)

$$N = \frac{\nabla L}{n}.$$

(5.44)

The wave normal is in the same direction as that of the unit tangent vector $\hat{s}$ to the light path. Combining (5.1 and 44) gives

$$\frac{dR}{ds} = \frac{\nabla L}{n}.$$

(5.45)

Equation (5.45) is a differential equation that is used for determining the light path. In a rectangular coordinate system, (5.45) is

$$n\frac{dx}{ds} = \frac{\partial L}{\partial x}, \qquad n\frac{dy}{ds} = \frac{\partial L}{\partial y}, \qquad n\frac{dz}{ds} = \frac{\partial L}{\partial z}.$$

(5.46)

Using (5.46), a study will be made of the optical path of light launched into a medium whose refractive index varies only in the $x$ direction. The results of Exercise 5.1 are applicable to this situation. Inserting (5.42) into (5.46) gives

$$n(x)\frac{dx}{ds} = \sqrt{[n(x)]^2 - (b^2 + c^2)}$$

(5.47)

$$n(x)\frac{dy}{ds} = b$$

(5.48)

$$n(x)\frac{dz}{ds} = c.$$

(5.49)

From (5.48, 49), one obtains

$$\frac{dy}{dz} = \frac{b}{c}$$

and therefore

$$y = \frac{b}{c} z + d. \tag{5.50}$$

There exists a unique plane perpendicular to the $y$-$z$ plane defined by (5.50). A ray entering a medium characterized by an $x$ dependent refractive index will always remain in this plane regardless of launching conditions. The projection of the launched ray to the $y$-$z$ plane is, of course, a straight line. The constants are determined by the launching point and angle. For example, let us assume that the launching point is $(0, 0, C_0)$, and that the ray is launched at an angle $\phi_0$ with respect to the $y$ axis and $\theta_0$ with respect to the $x$ axis. The expression for the projected line becomes

$$z = y \tan \phi_0 + C_0. \tag{5.51}$$

The expressions for $y$ and $z$ are obtained from (5.47 – 49),

$$y = \int \frac{b \, dx}{\sqrt{[n(x)]^2 - (b^2 + c^2)}} \tag{5.52}$$

$$z = \int \frac{c \, dx}{\sqrt{[n(x)]^2 - (b^2 + c^2)}} . \tag{5.53}$$

Referring to Fig. 5.6, $dy$ and $dz$ can be written as

$$ds \sin \theta \cos \phi_0 = dy, \qquad ds \sin \theta \sin \phi_0 = dz \tag{5.54}$$

where $\theta$ is the angle between $ds$ and the $x$ axis at any point on the path. Combining (5.54) with (5.48, 49) gives

$$n(x) \sin \theta \cos \phi_0 = b, \quad n(x) \sin \theta \sin \phi_0 = c. \tag{5.55}$$

Eliminating $\phi_0$ from the expressions in (5.55) gives

$$n(x) \sin \theta = \sqrt{b^2 + c^2} . \tag{5.56}$$

This leads to the important conclusion, that

$$n(x) \sin \theta = \text{constant}, \tag{5.57}$$

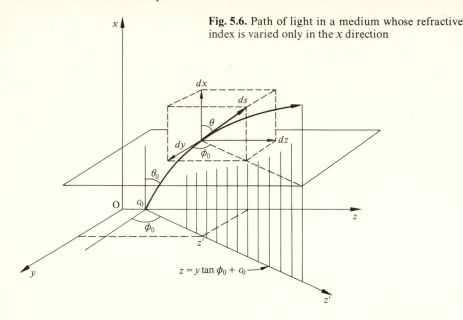

**Fig. 5.6.** Path of light in a medium whose refractive index is varied only in the $x$ direction

which holds true throughout the trajectory. Equation (5.57) is *Snell's law* for a one-dimensional stratified medium. At the launching point $x = 0$, the index of refraction is $n_0$ and the launching angle $\theta_0$ satisfies

$$n_0 \sin \theta_0 = \sqrt{b^2 + c^2} \ . \tag{5.58}$$

Inserting (5.55, 58) into (5.52, 53) yields

$$y = \int_0^x \frac{n_0 \sin \theta_0 \cos \phi_0}{\sqrt{[n(x)]^2 - n_0^2 \sin^2 \theta_0}} \, dx \quad \text{and} \tag{5.59}$$

$$z - C_0 = \int_0^x \frac{n_0 \sin \theta_0 \sin \phi_0}{\sqrt{[n(x)]^2 - n_0^2 \sin^2 \theta_0}} \, dx . \tag{5.60}$$

The values of the lower limits of (5.59, 60) are selected such that $y = 0$ and $z = C_0$ at $x = 0$. If the $z'$ axis which starts off at $(0, 0, C_0)$ at angle $\phi_0$, as shown in Fig. 5.6, is taken as a new axis, then (5.60) can be written as

$$z' = \int_0^x \frac{n_0 \sin \theta_0}{\sqrt{[n(x)]^2 - n_0^2 \sin^2 \theta_0}} \, dx . \tag{5.61}$$

It is worthwhile to examine (5.61) closely. When the quantity inside of the square root becomes negative, the value of $z'$ becomes imaginary and the light does not propagate. The light will reflect at the point where $[n(x)]^2 - n_0^2 \sin^2 \theta_0$ becomes negative, i.e.,

$$n(x) = n_0 \sin \theta_0. \qquad (5.62)$$

For the case of a refractive index in which the value of $n(x)$ is decreasing monotonically with $x$, the light will not propagate beyond $x = x_0$, where $n(x_0) = n_0 \sin \theta_0$, and total reflection takes place at this point. The location of total reflection is a function of the launching angle $\theta_0$, so that the smaller $\theta_0$ is, the further away the point of reflection is. If the launching angle is decreased to such an extent that $n_0 \sin \theta_0$ is smaller than any value of $n(x)$ within the medium, then the light penetrates through the medium. This behavior is illustrated in Fig. 5.7. In a medium with refractive index always larger than that at the launching point, namely $n(x) > n_0$, no total reflection takes place regardless of the distribution of refractive index.

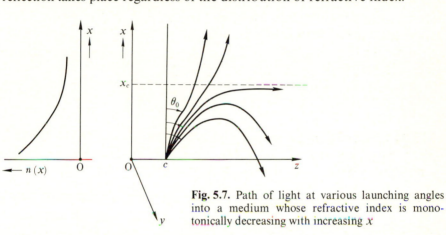

**Fig. 5.7.** Path of light at various launching angles into a medium whose refractive index is monotonically decreasing with increasing $x$

For the special case of a refractive index which is a maximum on the $x$-axis and then decreases both in the positive and negative $x$ directions, total reflection takes place at both upper and lower limits, so that, under the proper launching conditions, the light is confined to propagate within these limits. This kind of medium has practical significance for guiding light and is used in fiber optical communication. Propagation in this particular type of medium will be discussed in more detail next. A typical distribution which is frequently employed in practical devices is the parabolic distribution

$$n^2 = n_c^2 (1 - \alpha^2 x^2). \qquad (5.63)$$

A glass slab with such a variation of refractive index is frequently referred to by the tradename of Selfoc.

The light path is derived by inserting (5.63) directly into (5.61)

$$z' = \int \frac{a\,dx}{\sqrt{n_c^2 - a^2 - \alpha^2 n_c^2 x^2}} \qquad (5.64)$$

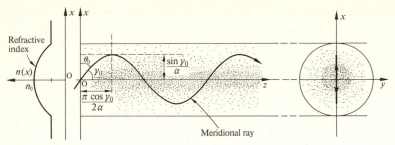

**Fig. 5.8.** Propagation path of light inside a one dimensional Selfoc slab

where $a = n_0 \sin \theta_0$. When the launching point is at the origin and the launching direction is in the $y = 0$ plane, as indicated in Fig. 5.8, then $a = n_c \sin \theta_0$ and (5.64) simplifies to

$$z = \frac{\sin \theta_0}{\alpha} \int_0^x \frac{dx}{\sqrt{\left(\dfrac{\cos \theta_0}{\alpha}\right)^2 - x^2}}$$

which upon integrating becomes

$$z = \frac{\sin \theta_0}{\alpha} \sin^{-1}\left(\frac{\alpha}{\cos \theta_0}\right) x .$$

In optical communications, the angle $\gamma_0 = 90° - \theta_0$ is conventionally used instead of $\theta_0$ (Fig. 5.8). Expressed in terms of $\gamma_0$, and solving for $x$, the result becomes

$$x = \frac{\sin \gamma_0}{\alpha} \sin\left(\frac{\alpha}{\cos \gamma_0}\right) z . \tag{5.65}$$

The optical path in such a medium is sinusoidal with an oscillation amplitude equal to $(\sin \gamma_0)/\alpha$ and a one-quarter period equal to $\pi (\cos \gamma_0)/2 \alpha$, as shown in Fig. 5.8. As the incident angle $\gamma_0$ increases, the amplitude grows and the period shortens.

Now, compare the ray having a larger incident angle with that having a smaller incident angle. Even though the ray with a larger incident angle travels along a longer path, it travels mostly in an outer region where the refractive index is lower and the velocity is higher. On the other hand, the ray with a smaller incident angle travels along a shorter path but in an inner region where the refractive index is higher and the velocity is lower. As a result, the dispersion or the difference in the travel time of the two rays is small. Hence, the dispersion with the Selfoc slab is smaller than that of a step index slab (a uniform core glass sandwiched by a uniform cladding glass of lower refractive index). Similarly, the dispersion of a Selfoc fiber is less than that of a step index fiber. Therefore the Selfoc fiber has the merit of having a larger information carrying capability than does a step index fiber. Dispersion will be discussed further in Sect. 5.9.

## 5.4 Relationship Between Inhomogeneity and Radius of Curvature of the Optical Path

Whenever light encounters an inhomogeneity of the refractive index, the direction of propagation changes. In this section, the relationship between the inhomogeneity of the refractive index and the radius of curvature associated with the change in direction of the optical path is studied [5.6].

Equation (5.32) can be written in terms of the derivative *in the direction of the light path* which is along the normal to the equi-level surface as

$$\frac{dL}{ds} = n. \tag{5.66}$$

Taking the gradient of both sides of (5.66) gives

$$\frac{d}{ds} \nabla L = \nabla n \tag{5.67}$$

where the order of applying the gradient and the derivative on the left-hand side was exchanged. Combining (5.45) with (5.67) yields

$$\frac{d}{ds} \left( n \frac{d\mathbf{R}}{ds} \right) = \nabla n, \tag{5.68}$$

called the *Euler-Lagrange formula*. This formula specifies the relationship between the inhomogeneity of the medium and the change in the optical path. It gives the change in the optical path directly without going through $L$. The more quantitative evaluation of the change in the light path can be found by using the Frenet-Serret formula. Performing the differentiation of (5.68)

$$\frac{dn}{ds} \frac{d\mathbf{R}}{ds} + n \frac{d\hat{s}}{ds} = \nabla n$$

and using (5.10), $\nabla n$ is expressed as

$$\frac{dn}{ds} \hat{s} + n \hat{N} \frac{1}{\varrho} = \nabla n.$$

Taking the scalar product of both sides of the equation with $\hat{N}$, and making use of the fact that $\hat{N}$ is perpendicular to $\hat{s}$, yields

$$n\frac{1}{\varrho} = \hat{N} \cdot \nabla n. \tag{5.69}$$

Now, the meaning of (5.69) is examined. Since $\varrho$ was defined as a positive quantity in Sect. 5.1 and since $n$ is always positive, the left-hand side of (5.69) is always positive, so that $\hat{N} \cdot \nabla n$ must also be a positive quantity. This means that the light ray always bends toward the direction of the maximum incremental increase in $n$. This relationship is often helpful for finding the direction of the changes of the optical path in a complicated optical system. For example, in Fig. 5.8, $\hat{N}$ points down in the region of $x > 0$ and points up in the region of $x < 0$. The vector $\hat{N}$ always points toward the direction of $\nabla n$.

Next, it will be proven that when light is incident upon a medium whose refractive index distribution is spherically symmetric, the light path is always confined to a plane containing the launching point and the center of the symmetry. Take the point of symmetry as being the origin of the coordinate system. Consider a plane containing both the position vector $\boldsymbol{R}$ originating from the point of symmetry and the direction $\hat{s}$ of the incident ray at the point of entry. This plane naturally contains the origin. The normal vector $\boldsymbol{K}$ to this plane is expressed by

$$\boldsymbol{K} = \boldsymbol{R} \times (n\,\hat{s}).$$

As soon as the ray $\hat{s}$ starts to deviate from this plane, the normal vector defined above starts to change its direction. Conversely, as long as $d\boldsymbol{K}/ds = 0$, the ray stays in this initial plane. Now, using (5.1), gives

$$\frac{d\boldsymbol{K}}{ds} = \frac{d\boldsymbol{R}}{ds} \times \left( n\frac{d\boldsymbol{R}}{ds} \right) + \boldsymbol{R} \times \frac{d}{ds}\left( n\frac{d\boldsymbol{R}}{ds} \right).$$

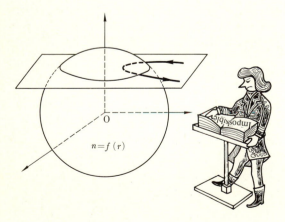

**Fig. 5.9.** Such an optical path is impossible if the medium is spherically symmetric

The first term is obviously zero and the second term can be rewritten using (5.68) as

$$\frac{d\boldsymbol{K}}{ds} = \boldsymbol{R} \times (\nabla n).$$

If the refractive index distribution is spherically symmetric, $\nabla n$ points in the radial direction, hence $d\boldsymbol{K}/ds = 0$. Thus, the ray is always confined in the initial plane, and the situation depicted in Fig. 5.9 is impossible.

## 5.5 Path of Light in a Spherically Symmetric Medium

The light path in a medium whose refractive index distribution is spherically symmetric is examined further [5.3, 4]. The expression for $\nabla L$ in spherical coordinates is

$$\nabla L = \hat{r}\frac{\partial L}{\partial r} + \hat{\phi}\frac{1}{r\sin\phi}\frac{\partial L}{\partial\phi} + \hat{\theta}\frac{1}{r}\frac{\partial L}{\partial\theta}. \tag{5.70}$$

As proven at the end of Sect. 5.4, the path of light is always in a plane containing the origin, regardless of launching angle, if the refractive index is spherically symmetric. If the $z$ axis is taken in this plane, then the path direction (or $\nabla L$) does not have a $\hat{\phi}$ component. Thus, if the medium is spherically symmetric, the eikonal equation becomes

$$|\nabla L|^2 = \left(\frac{\partial L}{\partial r}\right)^2 + \left(\frac{1}{r}\frac{\partial L}{\partial\theta}\right)^2 = [n(r)]^2. \tag{5.71}$$

Applying the method of separation of variables, a solution of the eikonal equation of the form

$$L(r, \theta) = R(r) + \Theta(\theta) \tag{5.72}$$

is assumed. Inserting (5.72) into (5.71) gives

$$r^2\{[R'(r)]^2 - [n(r)]^2\} + [\Theta'(\theta)]^2 = 0. \tag{5.73}$$

The first term of (5.73) is a function of $r$ only and the second is a function of $\theta$ only. In order that (5.73) be satisfied for any combination of $r$ and $\theta$, the first and second terms each have to be constant, i.e.,

$$[\Theta'(\theta)]^2 = a^2 \tag{5.74}$$

$$r^2\{[R'(r)]^2 - [n(r)]^2\} = -a^2. \tag{5.75}$$

The solution of (5.74) is

$$\Theta(\theta) = \pm a\theta + m_1$$

and that of (5.75) is

$$R(r) = \pm \int_{m_1}^{r} \sqrt{[n(r)]^2 - \left(\frac{a}{r}\right)^2}\, dr + m_2. \tag{5.76}$$

Inserting both results back into (5.72) yields

$$L(r, \theta) = a\theta + \int_{m}^{r} \sqrt{[n(r)]^2 - \left(\frac{a}{r}\right)^2}\, dr, \tag{5.77}$$

where $m_1$ and $m_2$ are included in the lower limit of the integral, and where the propagation is assumed to be in the direction of positive $r$ and $\theta$.

Since the eikonal $L(r, \theta)$ has been found, the light path can be calculated by inserting $L(r, \theta)$ into (5.45). However (5.45) must first be expressed in polar coordinates. Using relationships similar to those that led to (5.7), $d\mathbf{R}/ds$ is written as

$$\mathbf{R} = r\,\hat{r} \tag{5.78}$$

$$\frac{d\mathbf{R}}{ds} = \hat{r}\frac{dr}{ds} + \hat{\theta}r\frac{d\theta}{ds}. \tag{5.79}$$

Thus, the components of (5.45) become

$$n\frac{dr}{ds} = \frac{\partial L}{\partial r} \tag{5.80}$$

$$n\,r\frac{d\theta}{ds} = \frac{1}{r}\frac{\partial L}{\partial \theta}. \tag{5.81}$$

Inserting (5.77) into (5.80, 81) gives

$$n\frac{dr}{ds} = \sqrt{[n(r)]^2 - \left(\frac{a}{r}\right)^2} \tag{5.82}$$

$$n\,r\frac{d\theta}{ds} = \frac{1}{r}a. \tag{5.83}$$

Combining (5.82) and (5.83), one obtains

$$\theta = \int_{m}^{r} \frac{a\,dr}{r\,\sqrt{[r\,n(r)]^2 - a^2}} \tag{5.84}$$

where integration constants are included in the lower limit of the integral.

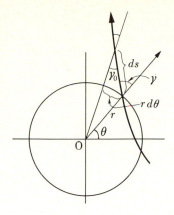

**Fig. 5.10.** Initial launching condition into a spherically symmetric medium

Examination of Fig. 5.10 shows that, at an arbitrary point $(r, \theta)$, we have

$$r \frac{d\theta}{ds} = \sin \gamma \tag{5.85}$$

where $\gamma$ is the angle which the ray makes with the radial vector. Inserting (5.85) into (5.83) gives

$$r\, n(r) \sin \gamma = a. \tag{5.86}$$

It is worth mentioning that, based on (5.86), points with the same $r$ have the same angle $\gamma$ or $(180° - \gamma)$, as shown in Fig. 5.11 a.
[Note that $\sin \gamma = \sin(180° - \gamma)$.]
    The constant $a$ is now determined. At the launching point $(r_0, \theta_0)$ with $\gamma = \gamma_0$, (5.86) becomes

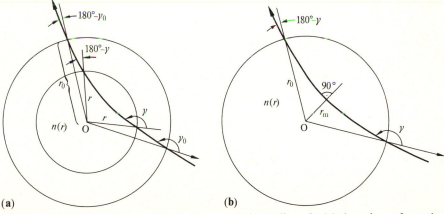

**(a)**                                    **(b)**

**Fig. 5.11 a, b.** Light path in a spherically symmetric medium. In (**a**) the values of $\gamma$ at the same $r$ are identical. (**b**) shows the value of $r_m\, n(r_m)$ for a given launching condition

$$r_0 \, n(r_0) \sin \gamma_0 = a. \tag{5.87}$$

It is also interesting to examine what happens at $\gamma = 90°$ and $r = r_m$, as indicated in Fig. 5.11 b. Equations (5.86, 87) become

$$r_m \, n(r_m) = r_0 \, n(r_0) \sin \gamma_0. \tag{5.88}$$

This equation means, for a given launching condition, the value of $r_m \, n(r_m)$ is immediately known.

**Exercise 5.2**  Find the light path in a medium with a spherically symmetric refractive index given by

$$n(r) = \frac{1}{\sqrt{r}}. \tag{5.89}$$

The launching position and the angle are $(r_0, \theta_0)$ and $\gamma_0$ respectively.

*Solution*  Inserting (5.89) into (5.84), $\theta$ becomes

$$\theta = \int \frac{a \, dr}{r \sqrt{r - a^2}}. \tag{5.90}$$

The square root term is real only for $r \geq a^2$, so that $r = a^2$ is a point of total reflection, and there is no propagation in the region of $r < a^2$. From a table of integrals [Ref. 5.7, p. 40],

$$\theta = \cos^{-1}\left(\frac{2a^2 - r}{r}\right) + \alpha,$$

which can be rewritten as

$$r = 2a^2 - r \cos(\theta - \alpha). \tag{5.91}$$

From the initial condition that the ray starts at $r = r_0$, $\theta = \theta_0$, the value of $\alpha$ can be determined,

$$\alpha = \theta_0 - \cos^{-1}\left(\frac{2a^2}{r_0} - 1\right).$$

The value of $a$, from (5.87), is

$$a = \sqrt{r_0} \, \sin \gamma_0.$$

Equation (5.91) is the equation of a parabola. The point of total reflection is $(a^2, \alpha)$. The focus is at the origin. If new axes $X$ and $Z$ are chosen which are rotated by $\alpha$ from the original axes, then the directrix, which is a line used as a reference to generate the locus of the parabola, is at $Z = 2a^2$, as shown

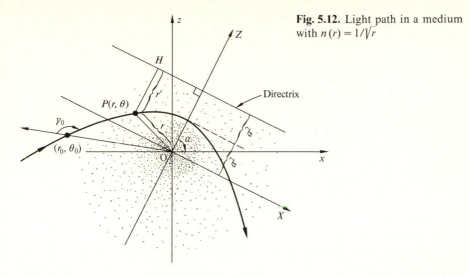

**Fig. 5.12.** Light path in a medium with $n(r) = 1/\sqrt{r}$

in Fig. 5.12. The distance from $P(r, \theta)$ to $H$ in Fig. 5.12 is $r = r'$. Another way of identifying (5.91) as a parabola is to convert it into rectangular co-ordiantes using $Z = r \cos(\theta - \alpha)$, and $X = -r \sin(\theta - \alpha)$. In these coordinates (5.91) becomes $X^2 = 4a^2(a^2 - Z)$.

## 5.6  Path of Light in a Cylindrically Symmetric Medium

Many optical components have cylindrical symmetry with respect to the optical axis [5.8−12]. It is, therefore, important to be able to treat the formulae in cylindrical coordinates.

First, the eikonal equation is solved in cylindrical coordinates using the method of separation of variables. The expression for $\nabla L$ in cylindrical coordinates is

$$\nabla L = \hat{r} \frac{\partial L}{\partial r} + \hat{\phi} \frac{1}{r} \frac{\partial L}{\partial \phi} + \hat{k} \frac{\partial L}{\partial z} . \tag{5.92}$$

The eikonal equation in cylindrical coordinates is

$$\left( \frac{\partial L}{\partial r} \right)^2 + \left( \frac{1}{r} \frac{\partial L}{\partial \phi} \right)^2 + \left( \frac{\partial L}{\partial z} \right)^2 = n^2 . \tag{5.93}$$

Let the solution of (5.93) be

$$L(r, \phi, z) = R(r) + \Phi(\phi) + Z(z) \tag{5.94}$$

where $R(r)$, $\Phi(\phi)$ and $Z(z)$ are functions solely of $r$, $\phi$, and $z$, respectively. Inserting (5.94) into (5.93) gives

$$(R')^2 + \left(\frac{1}{r}\,\Phi'\right)^2 + (Z')^2 = n^2. \tag{5.95}$$

If $n$ is assumed to be cylindrically symmetric, the solution is simplified considerably,

$$\underbrace{\left\{R'^2 - [n(r)]^2 + \left(\frac{\Phi'}{r}\right)^2\right\}}_{-a^2} + \underbrace{(Z')^2}_{a^2} = 0. \tag{5.96}$$

The first term is a function of the independent variables $r$ and $\phi$ only, and the second term is a function of the independent variable $z$ only. Since for all possible combinations of $r$, $\phi$, and $z$, the sum of the two terms has to be constant, each term itself has to be constant. Thus, (5.96) becomes

$$(Z')^2 = a^2 \tag{5.97}$$

$$R'^2 - [n(r)]^2 + \left(\frac{\Phi'}{r}\right)^2 = -a^2. \tag{5.98}$$

Equation (5.98) is further separated in terms of the functions $R$ and $\Phi$,

$$\underbrace{r^2\left\{R'^2 - [n(r)]^2 + a^2\right\}}_{-c^2} + \underbrace{\Phi'^2}_{c^2} = 0. \tag{5.99}$$

Similarly, one obtains

$$\Phi'^2 = c^2 \tag{5.100}$$

$$R'^2 = [n(r)]^2 - \frac{c^2}{r^2} - a^2. \tag{5.101}$$

Finally, (5.94, 97, 100, 101) are combined to form the eikonal in a cylindrically symmetric medium

$$L(r, \phi, z) = \int_m^r \sqrt{[n(r)]^2 - \frac{c^2}{r^2} - a^2}\ dr + c\,\phi + a\,z \tag{5.102}$$

where all integration constants are absorbed in the lower limit of the integral.

Next, the expression for the path of light is derived. Using (5.7, 92) in (5.45) and breaking the equation into its components gives

$$n \frac{dr}{ds} = \frac{\partial L}{\partial r} \qquad \hat{r} \text{ component} \tag{5.103}$$

$$n r \frac{d\phi}{ds} = \frac{1}{r} \frac{\partial L}{\partial \phi} \qquad \hat{\phi} \text{ component} \tag{5.104}$$

$$n \frac{dz}{ds} = \frac{\partial L}{\partial z} \qquad \hat{k} \text{ component.} \tag{5.105}$$

These differential equations are used to derive the path of light from the eikonal in cylindrical coordinates. In particular, when the medium is cylindrically symmetric, the value of $L$ has already been obtained by (5.102) and the above formulae become

$$n(r) \frac{dr}{ds} = \sqrt{[n(r)]^2 - \frac{c^2}{r^2} - a^2} \tag{5.106}$$

$$n(r) r \frac{d\phi}{ds} = \frac{1}{r} c \tag{5.107}$$

$$n(r) \frac{dz}{ds} = a. \tag{5.108}$$

Division of (5.107) and (5.108) by (5.106), followed by integration, gives

$$\phi = \int \frac{c \, dr}{r^2 \sqrt{[n(r)]^2 - \left(\frac{c^2}{r^2} + a^2\right)}} \tag{5.109}$$

$$z = \int \frac{a \, dr}{\sqrt{[n(r)]^2 - \left(\frac{c^2}{r^2} + a^2\right)}}. \tag{5.110}$$

Equation (5.109, 110) are general solutions for the optical path in a medium whose refractive index distribution is cylindrically symmetric. The constants $a$ and $c$ are to be determined from the initial conditions. Referring to Fig. 5.13, $dz/ds$ in (5.108) is the direction cosine $\cos \gamma$ of the path with respect to the $\hat{k}$ direction and $r \, d\phi/ds$ in (5.107) is the direction cosine $\cos \delta$ of the path with respect to the $\hat{\phi}$ direction. Let the launching condition be

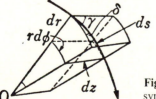

Fig. 5.13. Geometry of the optical path in a cylindrically symmetric medium (direction cosines)

$$r = r_0$$

$$\phi = \phi_0$$

$$z = 0 \tag{5.111}$$

$$n(r_0) = n_0.$$

Let the launching direction cosines be $\cos \gamma_0$ and $\cos \delta_0$, then

$$a = n_0 \cos \gamma_0 \tag{5.112}$$

$$c = n_0 r_0 \cos \delta_0. \tag{5.113}$$

The integral in (5.109, 110) simplifies considerably when $c = 0$, i.e. when $\delta_0 = 90°$, or in other words, when the launching path is in a plane containing the $z$ axis. The integrand of (5.109) is then zero and the path stays in a plane containing the $z$ axis. Such a ray is called a *meridional ray*. The integral (5.110) for this case becomes identical to (5.53) of the one-dimensional case. The ray that does not stay in this plane, due to non-zero $c$, rotates around the $z$ axis and is called a *skew ray*. As long as the medium is perfectly cylindrically symmetric, *only the initial launching condition determines whether the path will be meridional or skew.* However, if the medium contains some inhomogeneities, the meridional ray will become a skew ray as soon as the ray moves out of the meridional plane. It is interesting to recognize that any ray launched at the center of the medium is always meridional because $c$ in (5.107) is zero.

The general behavior of the light path is discussed by considering the sign of the quantity inside of the square root in (5.109, 110). Let $\eta_1$ and $\eta_2$ be functions of $r$ defined as follows,

$$\eta_1 = n^2(r) \tag{5.114}$$

$$\eta_2 = \frac{c^2}{r^2} + a^2. \tag{5.115}$$

Note that $\eta_1$ is dependent only on the refractive index distribution, and that the value of $\eta_2$ is dependent only on the launching conditions. Consider the special case for which the curve of $\eta_1$ is bell shaped (Fig. 5.14). For a skew ray, the function $\eta_2$ is monotonically decreasing with respect to $r$, and light can propagate only in the region $r_{\min} < r < r_{\max}$, where $r_{\max}$ and $r_{\min}$ are the intersections of $\eta_1$ and $\eta_2$. Outside this region, e.g. along the $z$ axis, the quantity inside the square root in (5.109, 110) becomes negative, and light cannot propagate. As the curve $\eta_2$ is raised, the region of propagation is narrowed. In the limit that $\eta_1$ and $\eta_2$ just touch each other, propagation is possible only on this particular radius, and the projection of the path becomes a circle. For a meridional ray, $c$ is zero and $\eta_2 = a^2$, so that the region of propagation is $r < r_c$, which includes propagation along the $z$ axis.

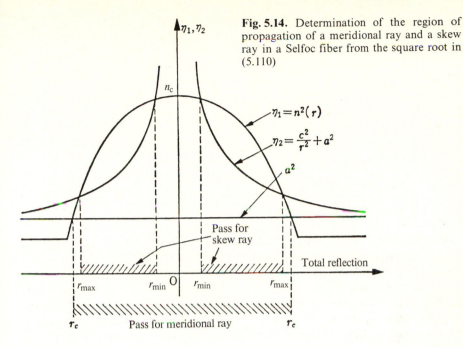

**Fig. 5.14.** Determination of the region of propagation of a meridional ray and a skew ray in a Selfoc fiber from the square root in (5.110)

## 5.7 Selfoc Fiber

In Sect. 5.3, the one dimensional Selfoc in slab shape was discussed. In this section, the Selfoc in fiber form will be investigated [5.8, 12]. The distribution of the *refractive index of the Selfoc fiber* is

$$n^2 = n_c^2 (1 - \alpha^2 r^2).$$

(5.116)

With this distribution, light is confined inside the fiber and propagates for a long distance with very little loss, so that the Selfoc fiber plays an important role in fiber-optical communication. In the following, the optical path in the Selfoc fiber will be analyzed separately for meridional ray propagation and skew ray propagation.

### 5.7.1 Meridional Ray in Selfoc Fiber

The expression for the meridional ray in the Selfoc fiber is derived. Referring to the equations in Sect. 5.6, it follows that for the case of the meridional ray, $c = 0$ and the integrand of (5.109) is zero. Equation (5.110) with (5.116) becomes

$$z = \int \frac{a \, dr}{\sqrt{n_c^2 - a^2 - \alpha^2 n_c^2 r^2}} \, .$$

(5.117)

Performing the integration in (5.117) gives

$$r = \frac{\sqrt{n_c^2 - a^2}}{\alpha n_c} \sin \frac{\alpha n_c}{a} (z - z_0).$$ (5.118)

This ray path is sinusoidal with an oscillation amplitude equal to $\sqrt{n_c^2 - a^2} / \alpha n_c$ and a period equal to $2\pi a / \alpha n_c$.

When the launching point is at the center of the fiber and $r_0 = z_0 = 0$, $n_0$ becomes $n_c$ and the expression with the help of (5.112) simplifies to

$$r = \frac{\sin \gamma_0}{\alpha} \sin \left( \frac{\alpha}{\cos \gamma_0} \right) z.$$ (5.119)

Figure 5.15 shows the path of light. Comparing (5.119) with (5.65) or even comparing (5.117) with (5.64), one finds that as far as the meridional ray is concerned, the path in the fiber is identical to that in the slab. The result of the one-dimensional treatment is often applicable to the meridional ray in the fiber. The physical reason for this is that, as far as the small section of circumference that reflects the meridional ray is concerned, the plane of incidence is always perpendicular to this circumference (Fig. 5.15 b), just as the plane of incidence in the slab is perpendicular to the boundaries.

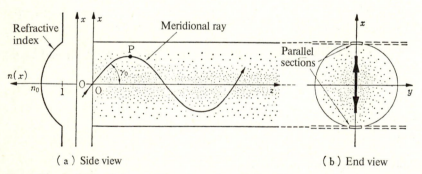

(a) Side view          (b) End view

**Fig. 5.15 a, b.** Path of the meridional ray in a Selfoc fiber; (**a**) side view, (**b**) end view. As seen in the end view, the plane of incidence is perpendicular to the circumference, just as the ray in the slab is perpendicular to the boundary

### 5.7.2 Skew Ray in Selfoc Fiber

When $c \neq 0$, the integrations of (5.109, 110) are more complex. A short manipulation after inserting (5.116) into (5.110) gives

$$z = \frac{a}{2 n_c \alpha} \int \frac{dt}{\sqrt{q^2 - (t - p)^2}} \quad \text{where}$$ (5.120)

$$t = r^2 \tag{5.121}$$

$$p = \frac{n_c^2 - a^2}{2 n_c^2 \, \alpha^2}, \quad q^2 = p^2 - \left(\frac{c}{n_c \, \alpha}\right)^2. \tag{5.122}$$

The result of the integration in (5.120) is

$$t = p + q \sin\left[\frac{2 n_c \, \alpha}{a} (z - m)\right]. \tag{5.123}$$

Having evaluated (5.110), the next step is to evaluate (5.109) and obtain an expression for $\phi$. By comparing (5.109) with (5.110), one finds that the numerator differs in the constants $c$ and $a$, and the denominator of (5.109) contains an $r^2$ factor which the denominator of (5.110) does not. Making use of the similarities in (5.109, 110), the former can be written as

$$\phi = \frac{c}{2 n_c \, \alpha} \int \frac{dt}{t \sqrt{q^2 - (t - p)^2}}. \tag{5.124}$$

The equation below, taken from a table of indefinite integrals [Ref. 5.7, p. 47], is used to evaluate (5.124),

$$\int \frac{dx}{(c^2 x + a b) \sqrt{b^2 - c^2 x^2}} = \frac{1}{b c \sqrt{a^2 - c^2}} \sin^{-1} \frac{a c x + b c}{c^2 x + a b}, \tag{5.125}$$

where, $a$, $b$, and $c$ are arbitrary constants not to be confused with $a$ and $c$ in previous equations.

Making the following substitutions for $x$, $a$, $b$, and $c$ in (5.125),

$$\dot{x} = t - p \qquad b = q$$
$$a = p/q \qquad c = 1,$$

the result of the integration in (5.124) becomes

$$\phi = \frac{c}{2 n_c \, \alpha} \frac{1}{\sqrt{p^2 - q^2}} \sin^{-1}\left(\frac{p x + q^2}{q (x + p)}\right) + \phi_0. \tag{5.126}$$

Writing $t - p$ in place of $x$, and making use of (5.121, 122), (5.126) becomes

$$\frac{1}{r^2}\left(\frac{c}{n_c \, \alpha}\right)^2 = p - q \sin 2(\phi - \phi_0). \tag{5.127}$$

If $p$ is replaced by $p[\cos^2(\phi - \phi_0 - \pi/4) + \sin^2(\phi - \phi_0 - \pi/4)]$ and the relationship $\sin(\phi - \phi_0) = \cos 2(\phi - \phi_0 - \pi/4)$ is used, (5.127) can be further rewritten as

$$1 = \frac{r^2 \cos^2 (\phi - \phi_0 - \pi/4)}{\left(\dfrac{c}{n_c \, \alpha \, \sqrt{p-q}}\right)^2} + \frac{r^2 \sin^2 (\phi - \phi_0 - \pi/4)}{\left(\dfrac{c}{n_c \, \alpha \, \sqrt{p+q}}\right)^2} \quad \text{or} \tag{5.128}$$

$$1 = \frac{X^2}{A^2} + \frac{Y^2}{B^2} . \tag{5.129}$$

Equation (5.128) is the equation of an ellipse with its major axis $A = c(n_c \, \alpha \, \sqrt{p-q})^{-1}$ and minor axis $B = c(n_c \, \alpha \, \sqrt{p+q})^{-1}$ tilted by $\phi_0 + \pi/4$ radians. Inserting all the constants of (5.122) into (5.128), the lengths of the major and minor axes become

$$A = \frac{1}{\sqrt{2} \, n_c \, \alpha} \sqrt{(n_c^2 - a^2) + \sqrt{(n_c^2 - a^2)^2 - 4 \, n_c^2 \, \alpha^2 \, c^2}} \tag{5.130}$$

$$B = \frac{1}{\sqrt{2} \, n_c \, \alpha} \sqrt{(n_c^2 - a^2) - \sqrt{(n_c^2 - a^2)^2 - 4 \, n_c^2 \, \alpha^2 \, c^2}} . \tag{5.131}$$

Thus, the projection of the optical path is found to be an ellipse (Fig. 5.16). It can easily be proven that the solutions $r_{min}$ and $r_{max}$ obtained from $\eta_1 = \eta_2$ in (5.114, 115) are identical to the lengths of the major and minor axes expressed by (5.130, 131).

In short, the *skew ray takes the path of an elliptic helix*, as shown in Fig. 5.16. Since the cross section is an ellipse, the radial distance from the $z$ axis in the cross sectional plane modulates as the ray advances. One complete turn around the circumference passes two maxima; hence, the pitch of the helix is found from (5.123) to be

$$z_p = \frac{2\pi a}{n_c \, \alpha} \tag{5.132}$$

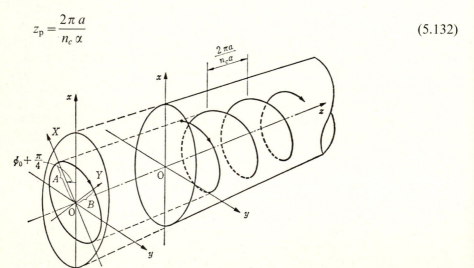

Projection                                    **Fig. 5.16.** Skew ray in a Selfoc fiber

which is identical to the pitch of the meridional ray. This is because the skew ray travels longer but stays in the region of lower refractive index.

A Selfoc micro lens is a cylindrical glass rod with a parabolic refractive index distribution. The diameter of the Selfoc micro lens, however, is several millimeters and is much larger than that of the Selfoc fiber. The Selfoc micro lens is used to transfer the image on the front surface to the back surface. The reason this device can transfer the image is due to the fact that the pitches of the skew and meridional rays are identical so that the image is not blurred.

When the innermost square root of (5.130, 131) is zero, the cross section is a circle. It is also seen from these equations that $A/B$ increases with a decrease in $c$, and the skew ray starts to convert into a meridional ray.

## 5.8 Quantized Propagation Constant

### 5.8.1 Quantized Propagation Constant in a Slab Guide

It will be demonstrated that only those rays that are incident at certain discrete angles can propagate into the fiber [5.9]. In order to simplify matters, a rectangular slab guide such as the one shown in Fig. 5.17 is taken as a first example. This slab guide (step-index guide) is made of three layers. The inner layer (core glass) has a refractive index $n_1$, which is larger than the refractive index $n_2$ of the two outer layers (cladding glass). If the angle of incidence $\theta$ is small enough, total reflection takes place at $R_2$ and $R_3$ on the boundaries and the incident ray $R_1 S$ will take the zigzag path shown in Fig. 5.17, eventually reaching the point T. The equi-phase fronts of the zigzag ray are indicated by the dotted lines drawn perpendicular to the optical path, assuming a plane wave. In addition to this zigzag ray, there is another wave front indicated by the dashed line that propagates directly from S to T. The field at T is the sum of these two wave fronts. The two wave fronts add constructively at T and are accentuated when the incident angle $\theta$ is at discrete angles $\theta_N$ such that the two wave fronts are in phase at T. The field at T is attenuated when the incident angle $\theta$ is not at such angles. Here, just one section $\overline{ST}$ of the zigzag path was considered but the wave fronts originating in the sections preceding this section also contribute to the field at T. The phases of the wave fronts from all of these other sections varies over the entire range of zero to $2\pi$ radians, and the resultant field becomes zero, unless the incident angle $\theta$ is at such $\theta_N$ specified above [5.13].

These discrete angles $\theta_N$ will be calculated by using the geometry in Fig. 5.17. The total length of the zigzag path $\overline{ST}$ is equal to the length of the straight line $\overline{SP}$, and the phase of the zigzag wave is $n_1 k \overline{SP}$. The phase of the directly propagated wave at T is the same as that at Q, so that the phase

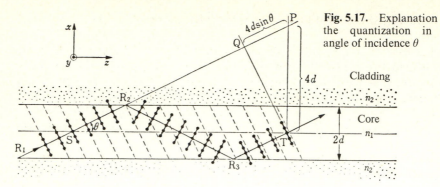

Fig. 5.17. Explanation of the quantization in the angle of incidence $\theta$

of this wave at T is $n_1 \, k \, \overline{SQ}$. The difference between these phases is

$$n_1 \, k \, (\overline{SP} - \overline{SQ}) = n_1 \, k \, \overline{QP} \, .$$

The length $\overline{QP}$ from the geometry of Fig. 5.17 is $4 \, d \sin \theta$. Thus $\theta$ can take on only such discrete values as specified by

$$4 n_1 \, k \, d \sin \theta_N = 2N \, \pi \tag{5.133}$$

where $N$ is an integer.

Next, (5.133) is modified slightly. Since $n_1 \, k \sin \theta_N$ is the propagation constant $\beta_x$ in the vertical $x$ direction, (5.133) can be rewritten as

$$\psi_x = 2 d \beta_x = N \, \pi \tag{5.134}$$

which means that the *phase difference $\psi_x$ between the upper and lower boundaries of the core glass can take only discrete values of an integral multiple of $\pi$ radians.*

The $z$ component of the propagation constant is $n_1 \, k \cos \theta_N$ and is normally designated by $\beta_N$. From (5.133), the propagation constant is expressed as

$$\beta_N = n_1 \, k \, \sqrt{1 - \left( \frac{N \, \pi}{2 n_1 \, k \, d} \right)^2} \, . \tag{5.135}$$

Thus, it is seen that the values of $\beta_N$ are discrete. The so called *Goos-Hänchen phase shift* [5.14], which appears at each reflection on the boundary due to a small spread in incident angle, has been ignored here. The reason that the quantization phenomenon is not readily observed with the thick slab geometry is that the difference $\theta_N - \theta_{N-1}$ in (5.133) is too minute for quantization to be observed.

Although, the above treatment applies to a slab waveguide, the same results hold true for a meridional ray in an optical fiber because, in both cases, the planes containing the rays are perpendicular to the boundary surfaces (Fig. 5.15b).

### 5.8.2 Quantized Propagation Constant in Optical Fiber

For the case of a skew ray, the explanation of the quantization is more complicated than that of the meridional ray. Quantizations in both the $\phi$ and $r$ directions have to be considered. The phase factor $\psi$ of a skew ray is obtained from (5.26, 102) as

$$\psi = k \int \sqrt{n^2 - (c/r)^2 - a^2} \, dr + k \, c \, \phi + k \, a \, z. \tag{5.136}$$

The propagation constants $v$, $\omega$ and $\beta$ in the $r$, $\phi$, and $z$ directions, respectively, are obtained from $\psi$ as follows:

$$v = \partial \psi / \partial r, \qquad \omega = r^{-1} \, \partial \psi / \partial \phi, \qquad \beta = \partial \psi / \partial z.$$

Applying these differentiations to $\psi$ in (5.136) gives

$$v = k \sqrt{n^2 - (c/r)^2 - a^2} \tag{5.137}$$

$$\omega = k \frac{c}{r} \tag{5.138}$$

$$\beta = k \, a. \tag{5.139}$$

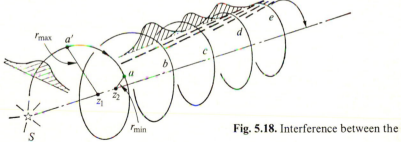

**Fig. 5.18.** Interference between the light rays

The quantization in the $\phi$ direction will be considered first. Figure 5.18 is a sketch of the path and cross-sectional field distributions of the wave front of a skew ray. The cross-sectional distribution of the ray has tails. The tail of the cross-sectional distribution of the first turn of the spiral path is overlapping with that of the second turn. If the phases of the overlapping wave front are random, then the larger the number of contributing wave fronts is, the smaller the field intensity becomes. However, when their phase differences become an integral multiple of $2\pi$ radians, the field grows with the number of participating rays. This condition is satisfied when

$$\int_0^{2\pi} \omega \, r \, d\phi = 2v \, \pi \tag{5.140}$$

where $v$ is an integer. Inserting (5.138) into (5.140) gives

$$v = k \, c. \tag{5.141}$$

Thus, the value of $k \, c$ is quantized and the condition of quantization is (5.141).

Next, quantization in the $r$ direction is considered. The drawing in Fig. 5.19 is a comprehension aid. A piece of soft noodle is wrapped around a piece of pencil. Before wrapping the noodle, a wavy line is drawn on the side of the noodle. If the shape and the period of the wavy line is properly selected, the line becomes an elliptic helix of a skew ray when it is wrapped around the pencil.

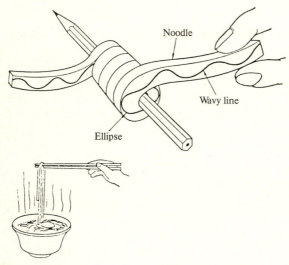

Noodle

Wavy line

Ellipse

**Fig. 5.19.** A wavy line is drawn on a side of a piece of noodle, and then wrapped around a piece of pencil. The line can represent the path of a skew ray

Noodle optics is
very easy to swallow

Before wrapping, the pattern of the wavy lines resembles the one-dimensional zigzag light path shown in Fig. 5.17. As mentioned earlier, the condition of quantization in the one-dimensional zigzag path was that the phase difference between the upper and lower boundaries has to be a multiple of $\pi$ radians. This condition *that the phase difference between the inner and outer boundaries be a multiple of $\pi$ radians* approximately applies to the *quantization in the $r$ dimension of the cylindrical guide*. Thus, this condition gives

$$\mu \, \pi = \int_{r_{max}}^{r_{min}} v \, dr \tag{5.142}$$

where $\mu$ is an integer. In order to make it easier to perform the integral, (5.142) is rewritten. The ratio of (5.106) to (5.108), and (5.137) gives

$$\frac{dr}{dz} = \frac{v}{k\,a},$$

and with (5.139),

$$\mu\,\pi = \frac{1}{\beta} \int_{z_1}^{z_2} v^2\,dz. \tag{5.143}$$

From (5.137, 139, 141), the condition for quantization in the $r$ direction becomes

$$\mu\,\pi = \frac{1}{\beta} \int_{z_1}^{z_2} \left[ k^2\,n^2\,(r) - \left(\frac{v}{r}\right)^2 - \beta^2 \right] dz. \tag{5.144}$$

Thus, the path of a skew ray is discrete and is controlled by the two integers $(\mu, v)$.

The integration of (5.144) will be performed for the case of a Selfoc fiber, and the quantized propagation constant $\beta$ will be derived. First, the integration constant $m$ in (5.123) is determined such that at $z = z_1$, the maximum value of $r$ appears. The maximum value of $r$ appears when $2 n_c \alpha (z_1 - m)/a = \pi/2$. Putting $m$ back into (5.123) gives

$$r^2 = p + q \cos\left[2 n_c \alpha (z - z_1)/a\right] \tag{5.145}$$

so that, the position $z_2$ of the adjacent minimum is

$$z_2 = z_1 + \frac{\pi\,a}{2 n_c\,\alpha}. \tag{5.146}$$

Inserting (5.116, 122, 145, 146) into (5.144) and using the integration formula

$$\int \frac{dx}{p + q \cos b\,x} = \frac{1}{b\sqrt{p^2 - q^2}}\cos^{-1}\frac{q + p \cos b\,x}{p + q \cos b\,x} \tag{5.147}$$

gives the final result

$$\beta_{\mu,v} = n_c\,k\,\sqrt{1 - 2\alpha\left(\frac{2\mu + v}{n_c\,k}\right)}. \tag{5.148}$$

When the end of an optical fiber is excited by a light source, only those rays propagate that are incident at discrete angles for which the propagation constant in the $z$ direction satisfies (5.148). Each ray that can propagate is designated by the integers $\mu$ and $v$, and is called the $(\mu, v)$ mode (of propagation).

Next, the average of the radius of the skew path is considered. From (5.145), the average radius $r_{\text{ave}}^2$ is

$$r_{\text{ave}}^2 = p.$$ (5.149)

Combining (5.122, 139, 148) gives

$$r_{\text{ave}}^2 = \frac{2\mu + v}{\alpha\, n_c\, k}.$$ (5.150)

The conclusion is that the higher the mode order is, i.e. the larger $2\mu + v$ is, the further the ray will travel into the outer regions of the optical fiber.

In all the above discussions, the plus sign has been employed in (5.102), but the minus sign is equally legitimate. The minus sign indicates a spiraling of the skew ray in the opposite direction. Actually, two skew rays spiraling in opposite directions are present in the fiber. *The two waves propagating in opposite directions create a standing-wave pattern. These standing-wave patterns are nothing but the so-called mode patterns.* Some of these mode patterns are shown in Fig. 5.20.

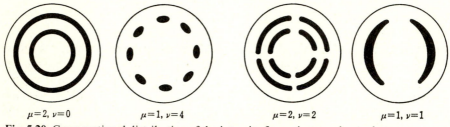

$\mu=2,\ v=0$          $\mu=1,\ v=4$          $\mu=2,\ v=2$          $\mu=1,\ v=1$

**Fig. 5.20.** Cross-sectional distribution of the intensity for various mode numbers

Recall that in the case of a standing-wave pattern made by the interference of two plane waves travelling in opposite directions, the intensity goes through two extrema for a distance that corresponds to the advance of the phase by $2\pi$ radians. For one complete turn in circumference, the phase goes through $2\pi v$ radians as indicated by (5.140) and hence the intensity passes through $2v$ extrema. According to (5.142) the phase advances by $\mu\pi$ radians in the radial direction, so that the intensity goes through $\mu$ extrema across the radius.

Next, the behavior of $\beta$, which is the propagation constant in the $z$ direction, is examined more closely. Using (5.148), $\beta$ is plotted with respect to $k\,n_c$ with $\mu$ and $v$ as parameters in Fig. 5.21. If $\beta_{\mu,v}$ is imaginary instead of real (propagating), the wave decays rapidly in the $z$ direction. Whether or not the ray propagates is determined by whether the quantity inside the square root in (5.148) is positive or negative. The condition

$$1 - 2\alpha\,\frac{2\mu + v}{n_c\, k} = 0$$ (5.151)

is called the *cut-off condition.*

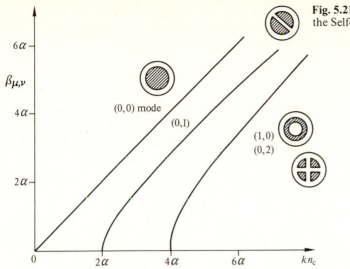

Fig. 5.21. $k - \beta$ diagram of the Selfoc fiber

Among the modes, the $(0, 0)$ mode is very special. Note that, from either (5.151) or Fig. 5.21, the $(0, 0)$ mode has its cut-off at $k\, n_c = 0$ and can propagate throughout the frequency band. The cutoff for the next higher mode $(0, 1)$ is $k\, n_c = 2\alpha$, so that the condition that only the $(0, 0)$ mode is excited is that $k\, n_c < 2\alpha$. The optical fiber which is designed to propagate only one mode is called a *single-mode fiber*. Some important characteristics of single-mode fibers are mentioned in the next section.

## 5.9 Group Velocity

The speed of propagation of the light energy is called the *group velocity* $v_g$ of the light [5.9]. Its inverse $\tau = v_g^{-1}$ is the time it takes for the light energy to travel a unit distance and is called the group delay. The group velocity may be obtained from the propagation constant $\beta$ using the relationship

$$v_g = \frac{d\omega}{d\beta} = c\, \frac{dk}{d\beta} \,. \tag{5.152}$$

Alternatively, the time needed for the light to travel along the helical path in Fig. 5.16 may be calculated using (5.33) (Problem 5.8). Here, the former method is used. Differentiating (5.148) with respect to $k$ gives

$$\frac{d\beta_{\mu,\nu}}{dk} = \frac{n_c^2\, k}{\beta_{\mu,\nu}} \left( 1 - \alpha\, \frac{2\mu + \nu}{n_c\, k} \right). \tag{5.153}$$

Equation (5.148) is again used to find an expression for $(2\mu + v)$ in terms of $\beta_{\mu,v}$. This expression for $(2\mu + v)$ is substituted into (5.153) and the result is used in (5.152) to obtain

$$v_g = \frac{c}{n_c} \frac{2\,(\beta_{\mu,v}/n_c\,k)}{1+(\beta_{\mu,v}/n_c\,k)^2} \,. \tag{5.154}$$

The lower-order modes have a higher group velocity than the higher-order modes. When light enters into a multimode fiber, several fiber modes may be excited, depending on the launch conditions, and each of these excited modes will support the propagation of light. For an input signal which is composed of a series of light pulses, the difference in the time of arrival at the receiver for the various fiber modes causes a spreading of the individual light pulses. The difference in arrival time between the rays of the various modes limits the time interval between the pulses, and hence, the rate of pulses that can be sent. Following similar reasoning, if the input light signal is amplitude modulated, the difference in group velocity causes a distortion of the received waveform. The spreading of a pulse or distortion of an amplitude modulated light signal due to the different velocities between the modes is called *mode dispersion*. Obviously an isotropic[2] single mode fiber has the merit of being free from mode dispersion. An additional advantage of a single-mode fiber is that $\beta$ is a linear function of frequency $(k\,n_c)$, as seen from Fig. 5.21, so that the distortion due to the difference in the velocity of transmission among the different frequency components of the signal is eliminated. Because of these special characteristics, a single-mode fiber plays an important role in high density information transmission in the field of fiberoptical communication, as will be discussed in Chapter 12.

---

2 A single-mode fiber actually allows the propagation of two orthogonal polarization modes. If the fiber is isotropic, these two polarization modes propagate with the same velocity. However, in real fibers, there is usually some anisotropy, and the two polarization modes propagate at slightly different velocities. This effect should be taken into account when using polarization sensitive optical devices, or when designing long-haul, high-bandwidth fiber-optical communications links. There are two methods of controlling polarization mode dispersion. In one approach, a large controlled anisotropy is purposely introduced, by, for example, stressing the fiber as it is drawn, so that this controlled anisotropy is much larger in magnitude than any random anisotropy. If polarization of the incident light is aligned with the direction of the stress, the polarization is stabilized. In the other approach, the geometry of the fiber is modified in such a way that one of the two polarization modes is cutoff. An example of such a geometry is a fiber in which the refractive index of two diametrically opposite sections of the cladding is increased.

It is important to realize that, in the first approach, both polarization modes will propagate if they are excited. This means that for polarization stabilized operation, the incident light must be launched correctly so that only one polarization mode is excited. In the second approach, an improper launching polarization increases the losses in the cutoff polarization mode, but the output state of polarization does not change.

This section concludes by giving an alternate expression to (5.154) for the group velocity. As discussed previously, $\beta$ is the propagation constant in the $z$ direction and $n_c\,k$ is the propagation constant along the helical path, so that $(\beta/(n_c\,k))$ is $\cos\gamma$ in Fig. 5.13. Equation (5.154) can be written as

$$v_g = \tau^{-1} = \frac{c}{n_c}\,\frac{2\cos\gamma}{1+\cos^2\gamma}\,. \tag{5.155}$$

## Problems

**5.1**  By solving the differential equation (5.35), find the value of $A$ for the one-dimensional problem; $n = n(x)$, $\partial A/\partial y = \partial A/\partial z = 0$.

**5.2**  Prove that the optical path is circular if light is launched in a direction parallel to the $z$ axis in a medium such as that shown in Fig. 5.22. The refractive index varies only in the $x$-direction and is expressed by

$$n_\pm = \frac{b}{a \pm x} \tag{5.156}$$

where $a > 0$, $b > 0$ and $x < a$. The launching point is $(x_0, 0, 0)$.

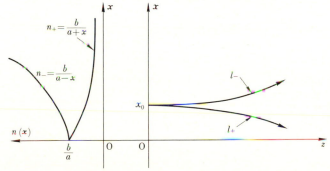

**Fig. 5.22.** The light path in a medium with $n = b/(a \pm x)$ is an arc of a circle

**5.3**  In Exercise 5.2, an expression was derived for the optical path in a spherically symmetric medium having the refractive index $n = 1/\sqrt{r}$. Calculate the optical path for the refractive index $n = 1/r$.

**5.4**  Using the relationship between the direction cosines shown in Fig. 5.23,

$$\cos^2\alpha + \cos^2\delta + \cos^2\gamma = 1$$

derive (5.106) directly from (5.107, 108).

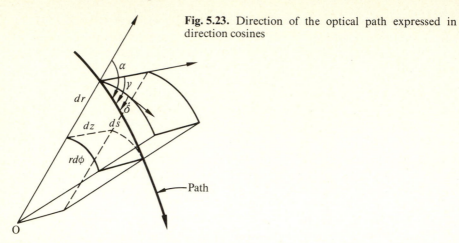

**Fig. 5.23.** Direction of the optical path expressed in direction cosines

**5.5** Find the length of the major axis $A$ and that of the minor axis $B$ of a skew ray in the Selfoc fiber by a method other than mentioned in the text. (*Hint:* For $r = r_{max}$ and $r = r_{min}$, $dr/ds = 0$).

**5.6** What is the relationship between $\phi$ and $z$ of a skew ray?

**5.7** Prove that (5.118) for a meridional ray is a special case of (5.145) for a skew ray.

**5.8** Derive the group velocity of a skew ray inside a Selfoc fiber by calculating the time it takes light to travel along the helical path using (5.33).

# 6. Lenses

Lenses are among the most used components in optical systems. The functions that lenses perform include the convergence or divergence of light beams, and the formation of virtual or real images. An additional interesting property of lenses is that they can be used for optically performing the two-dimensional Fourier transform.

In this chapter, a special emphasis is placed on the detailed explanation of the Fourier transformable properties of a lens. A special feature of the optical Fourier transform is that the location of the Fourier transformed image is not influenced by that of the input image. This feature is widely used in optical signal processing. The effect of the finite size of a lens on the quality of the output image is also investigated.

## 6.1 Design of Plano-Convex Lens

The contour of a plano-convex lens is made in such a way that light from a point source becomes a parallel beam after passing through the lens. In other words, it is designed so that the optical paths $F-P-Q$ and $F-L-R$ in Fig. 6.1 are identical. Let H be the point of projection from P to the optical axis. Since $\overline{PQ} = \overline{HR}$, the optical paths will be identical if $\overline{FP}$ equals $\overline{FH}$. The optical paths are expressed as follows

Optical path of $\overline{FP} = \varrho$

Optical path of $\overline{FH} = f + n\,(\varrho \cos \theta - f)$

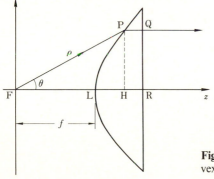

Fig. 6.1. Design of the curvature of a plano convex lens

where $n$ is the index of refraction of the glass, and $f$ is the focal length $\overline{\text{FL}}$. When the optical path of $\overline{\text{FP}}$ is set equal to that of $\overline{\text{FH}}$, the formula for the contour of the plano-convex lens is found to be

$$\varrho = \frac{(n-1)\,f}{n\cos\theta - 1}. \tag{6.1}$$

Next, (6.1) is converted into rectangular coordinates. For now, consider only the $y = 0$ plane. Taking the origin as the focus of the lens, the coordinates of the point P are

$$\left.\begin{array}{l} \varrho = \sqrt{x^2 + z^2} \\[2mm] \cos\theta = \dfrac{z}{\sqrt{x^2 + z^2}}. \end{array}\right\} \tag{6.2}$$

Equation (6.2) is inserted into (6.1), and a little algebraic manipulation gives the result

$$\frac{(z-c)^2}{a^2} - \frac{x^2}{b^2} = 1 \qquad \text{where}$$

$$a = \frac{f}{n+1}, \quad b = \sqrt{\frac{n-1}{n+1}}\,f, \quad c = \frac{n}{n+1}\,f. \tag{6.3}$$

It can be seen from (6.3) that the contour of the lens is a hyperbola.

Even though light from a point source located at the focus becomes a parallel beam after passing through the lens, the *intensity distribution across this beam is not uniform*. In Fig. 6.2 the solid angle $\angle\,\text{AFB}$ is made equal to solid angle $\angle\,\text{A}'\text{FB}'$. The light beam inside the latter angle is spread more and is therefore weaker than that of the former. This inhomogeneity becomes a problem for large lens diameters.

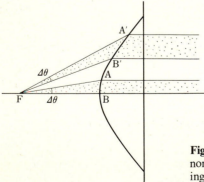

**Fig. 6.2.** Illustration of the cause of the intensity non-uniformity existing in a parallel beam emerging from a plano convex lens

## 6.2 Consideration of a Lens from the Viewpoint of Wave Optics

In the previous section, the lens was considered from the viewpoint of geometrical optics while in this section the viewpoint of wave optics is adopted. The phase distribution of the transmitted light beam through a lens is examined. As shown in Fig. 6.3, the incident beam is incident parallel to the lens axis $z$. When the lens is thin, (6.3) can be simplified. The larger $b$ in (6.3) is, the smaller the variation of $z$ with respect to $x$ becomes, which leads to the thin lens condition

$$\left(\frac{x}{b}\right)^2 \ll 1 . \tag{6.4}$$

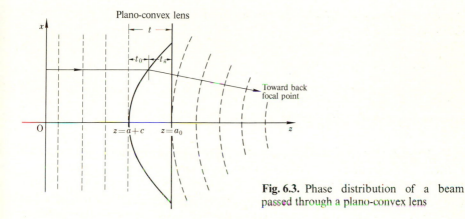

**Fig. 6.3.** Phase distribution of a beam passed through a plano-convex lens

When (6.4) is satisfied, an approximate formula for (6.3) is obtained by using the binomial expansion

$$z \cong a + c + \frac{a}{2b^2}\, x^2 . \tag{6.5}$$

If the plane surface of the lens is at $z = a_0$, the thickness $t_n$ of the glass at $x$ is

$$t_n = a_0 - \left(a + c + \frac{a}{2b^2}\, x^2\right), \tag{6.6}$$

and the phase of the plane wave passed through the lens at $x$ is

$$\phi(x) = k\left[(a_0 - a - c) - t_n\right] + n\, k\, t_n . \tag{6.7}$$

Inserting (6.6) into (6.7) gives

$$\phi(x) = k\left[n(a_0 - a - c) - \frac{a(n-1)}{2b^2}x^2\right].$$

Again, inserting the values for $a, b, c$ from (6.3), the phase becomes

$$\phi(x) = \phi_0 - k\frac{x^2}{2f} \qquad \text{where} \tag{6.8}$$

$$\phi_0 = k\,n\,(a_0 - f).$$

Since an analogous relationship holds in the $y$ direction, the phase can be written as

$$\phi(x, y) = \phi_0 - k\left(\frac{x^2 + y^2}{2f}\right) = \phi_0 - k\frac{r^2}{2f}, \tag{6.9}$$

where $r$ is the distance from the lens axis to the point $(x, y)$.

In conclusion, the lens creates a phase distribution whereby the phase advances with the square of the radius from the optical axis, and the rate of phase advance is larger for smaller $f$.

## 6.3 Fourier Transform by a Lens

Lenses play an important role in optical signal processing, and in particular, *Fourier transformability* is one of the most frequently used properties of lenses other than image formation. The results of the Fourier transform depend critically on the relative positions of the input object, the lens and the output screen. In the following subsections, every combination will be described [6.1−3].

### 6.3.1 Input on the Lens Surface

As shown in Fig. 6.4, the input transparency is pressed against the lens. Let the input image be represented by the function $g(x_0, y_0)$. When a parallel beam of unit amplitude is incident along the optical axis of the lens, the distribution of light just behind the lens is

$$g(x_0, y_0)\,\exp\left(-j\,k\,\frac{x_0^2 + y_0^2}{2f} + j\,\phi_0\right)$$

where the diameter of the lens is assumed infinite. The $\phi_0$ is insignificant for most of the analysis below; it will be assumed as zero.

After propagation to the screen at $z = z_i$ located in the Fresnel region, the pattern according to (3.36) is

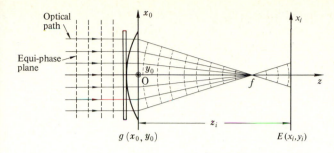

**Fig. 6.4.** A pattern projected onto a screen through a convex lens when the input transparency $g(x_0, y_0)$ is pressed against the lens and the screen is placed at $z = z_i$

$$E(x_i, y_i, z_i) = \frac{1}{j \lambda z_i} \exp\left[j k \left(z_i + \frac{x_i^2 + y_i^2}{2 z_i}\right)\right]$$

$$\times \mathscr{F}\left\{g(x_0, y_0) \exp\left(-j k \frac{x_0^2 + y_0^2}{2 f} + j k \frac{x_0^2 + y_0^2}{2 z_i}\right)\right\} \qquad (6.10)$$

with $f_x = x_i/\lambda z_i$ and $f_y = y_i/\lambda z_i$.

It is unfortunate that the same symbol $f$ is used for both the lens focal length and for the Fourier transform variable. As a guideline for distinguishing between the two, note that the Fourier transform variable always appears with a lettered subscript, whereas the focal length is written subscriptless or with a numbered subscript.

For the special case in which the screen is placed at the focal plane $z_i = f$, (6.10) becomes

$$E(x_i, y_i, f) = \frac{1}{j \lambda f} \exp\left[j k \left(f + \frac{x_i^2 + y_i^2}{2 f}\right)\right] G\left(\frac{x_i}{\lambda f}, \frac{y_i}{\lambda f}\right) \qquad (6.11)$$

where

$$G(f_x, f_y) = \mathscr{F}\{g(x_0, y_0)\} .$$

This is a very important result. Except for the factor

$$\exp\left[j k \left(f + \frac{x_i^2 + y_i^2}{2 f}\right)\right],$$

the pattern on the screen placed at the focal plane is the Fourier transform $G(x_i/\lambda z_i, y_i/\lambda z_i)$ of the input function $g(x_0, y_0)$. This *Fourier transform property of a lens is used often in optical signal processing* [6.3, 4]. The phase factor

$$\exp\left(j k \frac{x_i^2 + y_i^2}{2 f}\right)$$

can be eliminated by moving the input function to the front focal plane. This will be dealt in the next subsection.

### 6.3.2 Input at the Front Focal Plane

As shown in Fig. 6.5, the input function is now located a distance $d_1$ in front of the lens, and the screen is at the back focal plane. It will be easier to separate the propagation distance into two: propagation from the object to the lens and propagation from the lens to the screen.

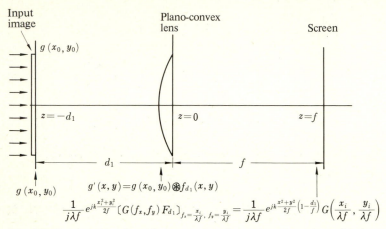

$$g'(x,y) = g(x_0,y_0) \circledast f_{d_1}(x,y)$$

$$\frac{1}{j\lambda f} e^{jk\frac{x_i^2+y_i^2}{2f}} \left[ G(f_x,f_y) F_{d_1} \right]_{f_x=\frac{x_i}{\lambda f}, f_y=\frac{y_i}{\lambda f}} = \frac{1}{j\lambda f} e^{jk\frac{x^2+y^2}{2f}\left(1-\frac{d_1}{f}\right)} G\left(\frac{x_i}{\lambda f}, \frac{y_i}{\lambda f}\right)$$

**Fig. 6.5.** Fourier transform by a convex lens; the case when the input transparency $g(x_0,y_0)$ is located a distance $d_1$ in front of the lens and the screen is at the back focal plane

The field distribution $g'(x,y)$ of the input $g(x_0,y_0)$ which has propagated to the front surface of the lens is, from (3.34),

$$g'(x,y) = g(x,y) * f_{d_1} \tag{6.12}$$

where

$$f_{d_1} = \frac{1}{j\lambda d_1} \exp\left[jk\left(d_1 + \frac{x^2+y^2}{2d_1}\right)\right].$$

Again, it is unfortunate that the letter $f$ appears in the symbol $f_{d_1}$ which represents the point-source transfer function. The symbol $f$ for both the point-source transfer function and the Fourier transform variable appears with a lettered subscript, however, the former can be distinguished by the fact that its subscript refers to distance along the $z$ direction. Note also that

$$\mathscr{F}\{f_{d_1}\} = F_{d_1}(f_x, f_y) = \exp\left[jk\,d_1 - j\pi\lambda\,d_1\,(f_x^2 + f_y^2)\right]. \tag{6.13}$$

The pattern at the back focal plane can be obtained by replacing $g(x_0,y_0)$ in (6.11) with $g'(x,y)$.

$$E(x_i, y_i, f) = \frac{1}{j\lambda f} \exp\left[j k\left(f + \frac{x_i^2 + y_i^2}{2f}\right)\right] G'\left(\frac{x_i}{\lambda f}, \frac{y_i}{\lambda f}\right) \tag{6.14}$$

where

$$G'(f_x, f_y) = \mathscr{F}\{g'(x, y)\} = G(f_x, f_y) \cdot F_{d_1}(f_x, f_y).$$

Inserting (6.13) into (6.14) gives

$$\begin{aligned}
E(x_i, y_i, f) &= \frac{1}{j\lambda f} \exp\left[j k\left(d_1 + f + \frac{x_i^2 + y_i^2}{2f}\right)\right] \\
&\quad \times G\left(\frac{x_i}{\lambda f}, \frac{y_i}{\lambda f}\right) \exp\left\{-j\pi\lambda d_1\left[\left(\frac{x_i}{\lambda f}\right)^2 + \left(\frac{y_i}{\lambda f}\right)^2\right]\right\} \tag{6.15} \\
&= \frac{1}{j\lambda f} e^{jk(f+d_1)} \exp\left[j k\left(\frac{x_i^2 + y_i^2}{2f}\right)\left(1 - \frac{d_1}{f}\right)\right] G\left(\frac{x_i}{\lambda f}, \frac{y_i}{\lambda f}\right).
\end{aligned}$$

When the input is placed at the front focal plane $d_1 = f$, the second exponential factor becomes unity and

$$E(x_i, y_i, f) = \frac{e^{j2kf}}{j\lambda f} G\left(\frac{x_i}{\lambda f}, \frac{y_i}{\lambda f}\right). \tag{6.16}$$

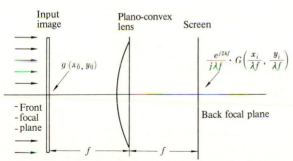

**Fig. 6.6.** Fourier transform by a convex lens; the case when the input transparency $g(x_0, y_0)$ is placed in the front focal plane. When the input transparency is placed at the front focal plane the exact Fourier transform $G(x_i/\lambda f, y_i/\lambda f)$ is observed at the back focal plane

Thus, the field distribution $G(x_i/\lambda f, y_i/\lambda f)$ on the screen becomes *exactly the Fourier transform* of the input function $g(x_0, y_0)$ if the input function is moved to the front focal plane and the screen is located at the back focal plane. The difference between this case and the previous case is that (6.16) does not contain the factor $\exp[j k(x_i^2 + y_i^2)/2f]$. Figure 6.6 summarizes the results of this subsection.

### 6.3.3 Input Behind the Lens

As shown in Fig. 6.7, the input is now located behind the lens. The positions of the lens and the screen are fixed, and only the input image is moved. Let the input image be placed at $d_1$ behind the lens. Again, the region of the propagation is separated into two: the region between the lens and the input image and that between the input image and the screen.

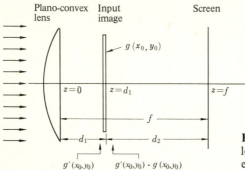

Fig. 6.7. Fourier transform by a convex lens; in the case when the input transparency $g(x_0, y_0)$ is placed behind the lens

By using (3.34), the light distribution $g'(x_0, y_0)$ illuminating the input image is given by

$$g'(x_0, y_0) = \exp\left(-jk\frac{x_0^2 + y_0^2}{2f}\right) * f_{d_1}(x_0, y_0) \tag{6.17}$$

where

$$f_{d_1}(x, y) = \frac{1}{j\lambda d_1}\exp\left[jk\left(d_1 + \frac{x^2 + y^2}{2d_1}\right)\right]$$

$$\mathscr{F}\{f_{d_1}(x, y)\} = F_{d_1}(f_x, f_y) = \exp\left[jkd_1 - j\pi\lambda d_1(f_x^2 + f_y^2)\right]$$

$$\mathscr{F}\left\{\exp\left(-jk\frac{x^2 + y^2}{2f}\right)\right\} = -j\lambda f\exp\left[j\pi\lambda f(f_x^2 + f_y^2)\right].$$

The expression for $g'(x_0, y_0)$ can be simplified by the successive operations of Fourier transform and inverse Fourier transform,

$$g'(x_0, y_0) = \mathscr{F}^{-1}\mathscr{F}\{g'(x_0, y_0)\}$$
$$= -j\lambda f\, e^{jkd_1}\mathscr{F}^{-1}\{\exp[-j\pi\lambda(d_1 - f)(f_x^2 + f_y^2)]\}$$

therefore

$$g'(x_0, y_0) = \frac{f}{f - d_1}\exp\left[jk\left(d_1 + \frac{x_0^2 + y_0^2}{2(d_1 - f)}\right)\right].$$

The field distribution immediately behind the lens is $g'(x_0, y_0)\, g(x_0, y_0)$. The image $E(x_i, y_i, f)$ that will be produced after this image propagates the distance $d_2$ can be obtained from (3.36)

$$E(x_i, y_i, f) = \frac{1}{j \lambda d_2} \exp\left[j k \left(f + \frac{x_i^2 + y_i^2}{2 d_2}\right)\right] \left(\frac{f}{f - d_1}\right) \qquad (6.18)$$

$$\times \mathscr{F}\left\{g(x_0, y_0) \exp\left[j k \left(\frac{1}{2(d_1 - f)}\right.\right.\right.$$

$$\left.\left.\left. + \frac{1}{2 d_2}\right)(x_0^2 + y_0^2)\right]\right\}\Big|_{\substack{f_x = x_i/\lambda d_2, \\ f_y = y_i/\lambda d_2}}$$

From Fig. 6.7

$$d_2 = f - d_1,$$

and (6.18) becomes

$$E(x_i, y_i, f) = \frac{1}{j \lambda d_2} \exp\left[j k \left(f + \frac{x_i^2 + y_i^2}{2 d_2}\right)\right] \left(\frac{f}{d_2}\right) G\left(\frac{x_i}{\lambda d_2}, \frac{y_i}{\lambda d_2}\right). \qquad (6.19)$$

As seen from (6.19), the size of the output image increases with an increase in $\lambda d_2$ and becomes the largest when $d_2 = f$. In short, the *size of the image is very easily controlled by moving the input image*. With the arrangements discussed in previous subsections, unless one uses lenses of different focal lengths, the size of the output image could not be changed. The ease of adjustment obtained with the present arrangement is a handy advantage. It should be noted that this case contains a phase factor similar to those found in (6.11, 15) and that the intensity of the image varies as $(f/d_2^2)^2$.

### 6.3.4  Fourier Transform by a Group of Lenses

So far, the Fourier transform was performed using a single lens. In this subsection it will be demonstrated that *the Fourier transform of the image can be obtained as long as the input image is illuminated by a converging beam, and provided that the screen is placed at the converging point of the beam*.

Assume that the light beam converges to the point P after passing through a series of lenses, as shown in Fig. 6.8. Since the wave front of a converging beam is spherical, the wave incident on the input image plane can be expressed by

$$E(x_0, y_0) = A \exp\left(-j k \frac{x_0^2 + y_0^2}{2 d_2}\right).$$

The negative sign in the exponent denotes a converging beam. This spherical wave is transmitted through the input image and further propa-

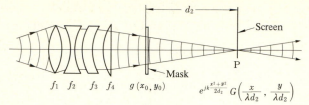

**Fig. 6.8.** Fourier transform performed by a converging beam of light. The Fourier transform is observed in a plane containing the point $P$ of convergence. It does not matter how the convergent beam is generated

gates over the distance $d_2$ to reach the screen. The light distribution on the screen is

$$E(x_i, y_i, d_2) = A \exp\left[j\,k\left(d_2 + \frac{x_i^2 + y_i^2}{2\,d_2}\right)\right] \mathscr{F}\left\{g(x_0, y_0)\right.$$

$$\left. \times \exp\left(-j\,k\,\frac{x_0^2 + y_0^2}{2\,d_2}\right) \exp\left(j\,k\frac{x_0^2 + y_0^2}{2\,d_2}\right)\right\}_{\substack{f_x = x_i/\lambda d_2,\\ f_y = y_i/\lambda d_2}}$$

$$= A \exp\left[j\,k\left(d_2 + \frac{x_i^2 + y_i^2}{2\,d_2}\right)\right] G\left(\frac{x_i}{\lambda\,d_2}, \frac{y_i}{\lambda\,d_2}\right), \tag{6.20}$$

which is proportional to the Fourier transform of the input image.

This is a rather roundabout explanation, but if one considers the group of lenses as one compound convex lens, the analysis would be identical to that of the previous subsection.

### 6.3.5 Effect of Lateral Translation of the Input Image on the Fourier-Transform Image

Figure 6.9a shows the light path when the input object is placed in the center. The parallel incident rays converge to the center of the back focal plane. The beam scattered by the input object has a similar tendency to converge to the center, but it has some spread in the field distribution. *This field spread is nothing but the Fourier transform of the input object.*

Figure 6.9b shows the case when the input object is lowered from the center. The scattered beam starts from a lower location in the input plane, but by the converging power of the lens, the scattered field converges again to the center of the back focal plane.

The property that the *Fourier transform always appears near the optical axis regardless of the location of the input object*, is, as mentioned earlier, one of the most attractive features of Fourier transforming by a lens. This is especially true when the location of the input object is unpredictable. It

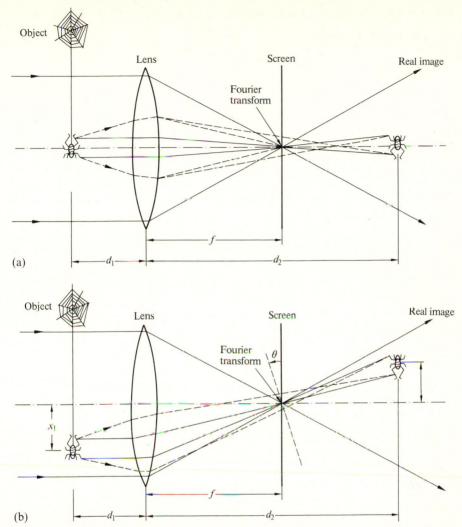

**Fig. 6.9 a, b.** Effect of the lateral translation on the Fourier transform. **(a)** The object is in the center; **(b)** the object is laterally translated. Note that Fourier transform always stays on the optical axis

should, however, be realized that there is a tilt in the Fourier transform image by $\theta = \sin^{-1}(x_1/f)$ when the input image is shifted by $x_1$ (Problem 6.2).

## 6.4 Image Forming Capability of a Lens from the Viewpoint of Wave Optics

It is well known that the condition for forming a real image is

$$\frac{1}{d_1} + \frac{1}{d_2} = \frac{1}{f} , \tag{6.21}$$

where $d_1$ is the distance from the input to the lens, and $d_2$ is the distance from the lens to the image (Fig. 6.10). Using a slightly different approach, this formula will be derived from the viewpoint of wave optics. The propagation distance is separated into two regions as before: the region from the input to the back focal plane of the lens and that from the back focal plane to the screen.

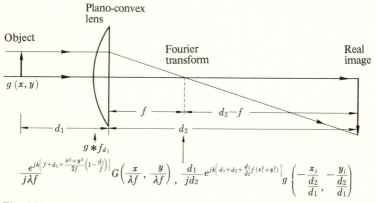

Fig. 6.10. The image forming capability of a lens is examined from the viewpoint of wave optics

As for the propagation from the input to the back focal plane, it was treated already in Sect. 6.3.2 and the result was (6.15). It is rewritten here for convenience, i.e.,

$$E(x, y, f) = \frac{1}{j \lambda f} e^{jk(f+d_1)} \exp\left[j k \left(\frac{x^2 + y^2}{2f}\right)\left(1 - \frac{d_1}{f}\right)\right] G\left(\frac{x}{\lambda f}, \frac{y}{\lambda f}\right)$$

where the plane at $z = f$ is taken as the $x\,y$ plane.

The above distribution further propagates over the distance $d_2 - f$. From (3.36), the field distribution is

$$E(x_i, y_i \cdot d_2) = \frac{1}{j \lambda (d_2 - f)} \exp\left[j k \left(d_1 + d_2 + \frac{x_i^2 + y_i^2}{2(d_2 - f)}\right)\right]$$

$$\times \frac{1}{j\lambda f} \mathscr{F} \left\{ \exp\left[ j k \frac{x^2+y^2}{2f} \left(1 - \frac{d_1}{f}\right)\right]\right.$$

$$\left. \times G\left(\frac{x}{\lambda f}, \frac{y}{\lambda f}\right) \exp\left[j k \frac{x^2+y^2}{2(d_2-f)}\right]\right\} {\scriptstyle f_x = x_i/\lambda(d_2-f), \atop f_y = y_i/\lambda(d_2-f)} \cdot$$

(6.22)

The exponential term inside the curly brackets of the Fourier transform is treated separately by denoting it by $\phi$

$$\phi = j k \frac{x^2+y^2}{2}\left[\frac{1}{f}\left(1-\frac{d_1}{f}\right) + \frac{1}{d_2-f}\right]$$

$$= j k \frac{x^2+y^2}{2}\frac{d_1 d_2}{f(d_2-f)}\left(\frac{1}{d_1} + \frac{1}{d_2} - \frac{1}{f}\right).$$

(6.23)

The value of (6.23) becomes zero when the imaging condition (6.21) is satisfied, and (6.22) becomes

$$E\,(x_i, y_i \cdot d_2) = \frac{-f}{d_2-f}\exp\left[j k \left(d_1 + d_2 + \frac{x_i^2+y_i^2}{2(d_2-f)}\right)\right]$$

$$\times g\left(-\frac{f x_i}{d_2-f}, -\frac{f y_i}{d_2-f}\right).$$

(6.24)

The relationship

$$\frac{f}{d_2-f} = \frac{d_1}{d_2}$$

obtained from (6.21) is inserted in (6.24) giving

$$E\,(x_i, y_i, d_2) = -\frac{d_1}{d_2}\exp\left[j k \left(d_1 + d_2 + \frac{d_1}{d_2}\frac{x_i^2+y_i^2}{2f}\right)\right]$$

$$\times g\left(-\frac{x_i}{d_2/d_1}, -\frac{y_i}{d_2/d_1}\right).$$

(6.25)

Equation (6.25) means that an image which is $d_2/d_1$ times the original image is formed at the location set by (6.21). The negative signs in $x_i, y_i$ indicate that the image is inverted.

It is noteworthy that *both the Fourier transform and the image are simultaneously obtained with only one lens*. It would appear that only one lens is needed to perform the same function as the two lens arrangement shown in Fig. 6.11. However, the difference is that, in the case of the single lens, the phase factor is present in both the Fourier transform and the image as seen in (6.11, 15, 19, 25). For a single lens, the field distribution gives a correct representation of the intensity, but not of the phase. However, if two

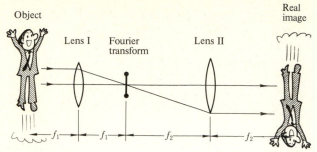

**Fig. 6.11.** Formation of the image using two successive Fourier transforms. The output image obtained in this manner is identical with the input image not only in magnitude but also in phase

lenses are arranged in the manner shown in Fig. 6.11, both the correct amplitude and the correct phase are retained.

## 6.5 Effects of the Finite Size of the Lens

So far, the finite physical extent of the lens has been ignored and all the light scattered from the object was assumed to be intercepted by the lens. In this section, we discuss how the finiteness of the lens affects the quality of the image. Figure 6.12 illustrates the spatial distribution of the signal scattered from the object relative to the lens. The lens is placed in the far-field region of the scattered field and the one-dimensional case is considered for simplicity. The input is a transparency $g(x)$ illuminated by a parallel beam from behind. The image of the input is to be formed by a lens with diameter $D$.

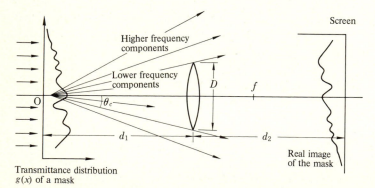

**Fig. 6.12.** The output image through a lens of a finite size. Some of the higher spatial frequency components do not intercept the lens resulting in a loss of the higher frequency component in the output image

The diffracted wave from the input transparency radiates in all directions, but the portion which contributes to the formation of the image is limited by the diameter of the lens. The lens intercepts only the light which falls inside the angle subtended by the lens at the input plane. The input $g(x)$ can be expanded into spatial frequency components by the Fourier transform. At first, only the component with the spatial frequency $f_0$ is considered. By using (3.31), the diffraction pattern of this component is given by

$$U(\theta) = K_0 \mathscr{F} \{e^{j2\pi f_0 x}\}_{f=\sin\theta/\lambda} = K_0 \delta \left(\frac{\sin\theta}{\lambda} - f_0\right), \qquad (6.26)$$

where $K_0$ is the phase factor in (3.31). The direction of the diffracted wave is $\theta = \sin^{-1}(f_0 \lambda)$. The higher the spatial frequency $f_0$ is, the larger the diffraction angle $\theta$ is.

In the case shown in Fig. 6.12, the highest spatial frequency $f_c$ which can be intercepted by the lens is

$$f_c \simeq \frac{D}{2\lambda d_1}, \qquad (6.27)$$

where $d_1$ is the distance between the input image and the lens. Therefore, any spatial frequency component higher than $f_c$ does not participate in forming the output image. The higher frequency components which are needed to fill in the fine details of the output image are missing, and hence the output image is degraded. The degree of degradation with respect to the diameter of the lens is considered in the next section.

### 6.5.1 Influence of the Finite Size of the Lens on the Quality of the Fourier Transform

The relationship between the size of the lens aperture and the quality of the Fourier transformed image is studied [6.5]. As shown in Fig. 6.13, the input

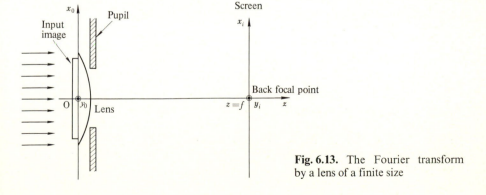

**Fig. 6.13.** The Fourier transform by a lens of a finite size

transparency is pressed against the lens and the Fourier transformed image is obtained at the back focal plane. The aperture of the lens can be expressed by the pupil function $P(x, y)$ defined in Sect. 4.6. The focal length of the lens is $f$. The result of Sect. 6.3.1 may be used, the only difference being that $g(x_0, y_0) P(x_0, y_0)$ is used instead of $g(x_0, y_0)$. From (6.11), the field distribution on the screen is

$$E(x_i, y_i, f) = \frac{1}{j\lambda f} \exp\left[jk\left(f + \frac{x_i^2 + y_i^2}{2f}\right)\right] G\left(\frac{x_i}{\lambda f}, \frac{y_i}{\lambda f}\right) * \bar{P}\left(\frac{x_i}{\lambda f}, \frac{y_i}{\lambda f}\right)$$

$$(6.28)$$

where the bar denotes Fourier transform, and

$$\mathscr{F}\{P(x, y)\} = \bar{P}(f_x, f_y) .$$

By comparing (6.28) with (6.11), the difference between the finite and infinite lens sizes is the effect of the convolution operation of the Fourier transform with $\bar{P}(x_i/\lambda f, y_i/\lambda f)$. If the image is convolved with $\bar{P}(x_i/\lambda f, y_i/\lambda f)$, the image is blurred.

A quantitative "degree of blurring" will be calculated for the case of a one-dimensional aperture. The pupil function for a lens of lateral dimension $D$ and its Fourier transform are

$$P(x) = \Pi\left(\frac{x}{D}\right),$$

$$(6.29)$$

$$\bar{P}\left(\frac{x_i}{\lambda f}\right) = D \operatorname{sinc}\left(D \frac{x_i}{\lambda f}\right).$$

$$(6.30)$$

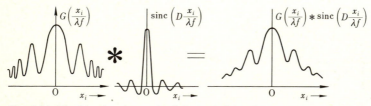

**Fig. 6.14.** The blurring due to the finiteness of the diameter of the lens

Figure 6.14 illustrates the convolution of $G(x_i/\lambda f)$ with $D \operatorname{sinc}(D x_i/\lambda f)$. The "degree of blurring" is determined by the width $f\lambda/D$ of the main lobe of $\operatorname{sinc}(D x_i/\lambda f)$. The smallest size $\Delta x_i$ that is meaningful in the Fourier transform image formed by this lens is approximately the width of the main lobe, i.e.,

$$\Delta x_i = 2 \frac{f}{D} \lambda .$$

$$(6.31)$$

For instance, for $D = 10$ mm, $\lambda = 0.6 \times 10^{-3}$ mm and $f = 50$ mm, we get $\Delta x_i = 6 \times 10^{-3}$ mm.

When the aperture is circular in shape with diameter $D$, the smallest resolvable dimension $\Delta l$ is, from the results in Sect. 4.5,

$$\Delta l = 2.44 \frac{f}{D} \lambda . \tag{6.32}$$

### 6.5.2 Influence of the Finite Size of the Lens on the Image Quality

When a lens is used for the purposes of image formation, it is useful to know the relationship between the lens size and the resolution of the output image. As shown in Fig. 6.15, a real image is formed on a screen placed at a distance $d_2$ behind a lens, whose focal length is $f$, and pupil function is $P(x, y)$. The input image $g(x_0, y_0)$ is placed at a distance $d_1$ in front of the lens. The propagation path is divided into two: from input to the aperture of the lens, and from the aperture of the lens to the screen. From (3.36), the distribution $g_l(x, y)$ at the aperture is

$$g_l(x, y) = \tag{6.33}$$

$$\underbrace{\underbrace{\frac{1}{j\lambda d_1} \exp\left[jk\left(d_1 + \frac{x^2 + y^2}{2d_1}\right)\right] \mathcal{F}\left\{g(x_0, y_0)\exp\left(jk\frac{x_0^2 + y_0^2}{2d_1}\right)\right\}}_{\text{just in front of the lens}} \exp\left(-jk\frac{x^2 + y^2}{2f}\right) P(x, y).}_{\text{just behind the lens}}$$

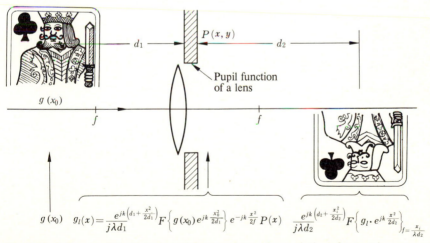

$$g(x_0) \quad g_l(x) = \frac{e^{jk\left(d_1 + \frac{x^2}{2d_1}\right)}}{j\lambda d_1} F\left\{g(x_0)e^{jk\frac{x_0^2}{2d_1}}\right\} e^{-jk\frac{x^2}{2f}} P(x) \quad \frac{e^{jk\left(d_2 + \frac{x_i^2}{2d_2}\right)}}{j\lambda d_2} F\left\{g_l \cdot e^{jk\frac{x^2}{2d_2}}\right\}_{f = \frac{x_i}{\lambda d_2}}$$

**Fig. 6.15.** Influence of the finiteness of the lens size on the resolution of output image

Note that the equation given in Fig. 6.15 is one-dimensional because of limitations on printing space available. Rewriting (6.33) yields

$$g_l(x, y) = \frac{1}{j \lambda d_1} \exp\left\{j k \left[d_1 + \frac{1}{2}\left(\frac{1}{d_1} - \frac{1}{f}\right)(x^2 + y^2)\right]\right\}$$

$$\times G_{d_1}\left(\frac{x}{\lambda d_1}, \frac{y}{\lambda d_1}\right) P(x, y) \qquad \text{where} \tag{6.34}$$

$$g_{d_1}(x, y) = g(x, y) \exp\left(j k \frac{x^2 + y^2}{2 d_1}\right), \tag{6.35}$$

$$\mathscr{F}\{g_{d_1}(x, y)\} = G_{d_1}(f_x, f_y).$$

By using (3.36) the field distribution resulting from the propagation from the aperture to the screen is given by

$$E(x_i, y_i) = \frac{1}{j \lambda d_2} \exp\left[j k \left(d_2 + \frac{x_i^2 + y_i^2}{2 d_2}\right)\right]$$

$$\times \mathscr{F}\left\{g_l(x, y) \exp\left(j k \frac{x^2 + y^2}{2 d_2}\right)\right\}_{\substack{f_x = x_i/\lambda d_2, \\ f_y = y_i/\lambda d_2}} \tag{6.36}$$

Equations (6.34, 35) are inserted into (6.36) to yield

$$E(x_i, y_i) = -\frac{1}{\lambda^2 d_1 d_2} \exp\left[j k \left(d_1 + d_2 + \frac{x_i^2 + y_i^2}{2 d_2}\right)\right]$$

$$\times \mathscr{F}\left\{\exp\left[j \frac{k}{2}\left(\frac{1}{d_1} - \frac{1}{f} + \frac{1}{d_2}\right)(x^2 + y^2)\right]\right.$$

$$\times \left. G_{d_1}\left(\frac{x}{\lambda d_1}, \frac{y}{\lambda d_1}\right) P(x, y)\right\}_{\substack{f_x = x_i/\lambda d_2, \\ f_y = y_i/\lambda d_2}} \tag{6.37}$$

If the imaging condition is satisfied, (6.37) becomes

$$E(x_i, y_i) = -\frac{1}{M} \exp\left[j k \left(d_1 + d_2 + \frac{x_i^2 + y_i^2}{2 d_2}\right)\right]$$

$$\times \left\{\left[g\left(-\frac{x_i}{M}, -\frac{y_i}{M}\right) \exp\left(j k \frac{(x_i^2 + y_i^2)}{2 M d_2}\right)\right] * \bar{P}\left(\frac{x_i}{\lambda d_2}, \frac{y_i}{\lambda d_2}\right)\right\} \tag{6.38}$$

where $M = d_2/d_1$ is the magnification. The operation of the convolution is performed, which yields

$$E(x_i, y_i) = -\frac{1}{M} \exp\left\{j k \left[d_1 + d_2 + \frac{x_i^2 + y_i^2}{2 d_2}\left(1 + \frac{1}{M}\right)\right]\right\}$$

$$
\times \int\!\!\int_{-\infty}^{\infty} \left\{ g\left(\frac{x_i - \xi}{M}, \frac{y_i - \eta}{M}\right) \exp\left[-j \frac{2\pi}{\lambda d_2 M}(x_i\, \xi + y_i\, \eta) + j \frac{\pi}{\lambda} \frac{(\xi^2 + \eta^2)}{d_2\, M}\right]\right.
$$

$$
\left. \times \bar{P}\left(\frac{\xi}{\lambda d_2}, \frac{\eta}{\lambda d_2}\right)\right\} d\xi\, d\eta .
\tag{6.39}
$$

The computation of (6.39) is messy but can be simplified if the value of $\bar{P}(x_i/\lambda d_2, y_i/\lambda d_2)$ is sharply peaked at certain values of $x_i$ and $y_i$. This method of simplification is demonstrated for a one-dimensional aperture using typical numbers. Let the aperture width be $D$, so that

$$
P(x, y) = \Pi\left(\frac{x}{D}\right) \qquad \text{and}
\tag{6.40}
$$

$$
\bar{P}(x, y) = D \operatorname{sinc}\left(D \frac{x_i}{\lambda d_2}\right).
\tag{6.41}
$$

A typical set of numbers such as

$$
D = 60 \text{ mm},
$$
$$
d_1 = d_2 = 100 \text{ mm}, \qquad \text{and}
\tag{6.42}
$$
$$
\lambda = 0.6 \times 10^{-3} \text{ mm}
$$

is inserted into (6.39)

$$
E(x_i, y_i) = -\frac{1}{M} \exp\left[j\, k\left(d_1 + d_2 + \frac{(x_i^2 + y_i^2)}{2\, M\, f}\right)\right]
$$

$$
\times \int_{-\infty}^{\infty} \left\{ \exp\left[-j\, 100\,(x_i\, \xi + y_i\, \eta) + j\, 50\,(\xi^2 + \eta^2)\right]\right.
\tag{6.43}
$$

$$
\left. \times g\left(-\frac{d_1}{d_2}(x_i - \xi), -\frac{d_1}{d_2}(y_i - \eta)\right) \operatorname{sinc}\left(\frac{\xi}{10^{-3}}\right)\right\} d\xi .
$$

The value of $\operatorname{sinc}(\xi/10^{-3})$ becomes negligible as soon as $|\xi|$ exceeds $10^{-3}$, so that the error involved in changing the limits of the integration to $-10^{-3}$, $10^{-3}$ from $-\infty, \infty$, would be very small. In the region $|\varepsilon| < 10^{-3}$, the value of $\exp[-j\, 100\,(x_i\, \xi + y_i\, \eta) + j\, 50\,(\xi^2 + \eta^2)]$ can be considered to be unity as long as $|x_i| + |y_i| < 10$. Under these conditions, (6.43) can be approximated by

$$
E(x_i, y_i) = -\frac{1}{M} \exp\left[j\, k\left(d_1 + d_2 + \frac{(x_i^2 + y_i^2)}{2\, M\, f}\right)\right]
$$

$$
\times \left[g\left(-\frac{x_i}{M}, -\frac{y_i}{M}\right) * \operatorname{sinc}\left(D \frac{x_i}{\lambda d_2}\right)\right].
\tag{6.44}
$$

Making a general inference from this typical example, (6.38) can be approximated as

$$E(x_i, y_i) = -\frac{1}{M} \exp \left[ j k \left( d_1 + d_2 + \frac{(x_i^2 + y_i^2)}{2 M f} \right) \right]$$

$$\times \left[ g \left( -\frac{x_i}{M}, -\frac{y_i}{M} \right) * \bar{P} \left( \frac{x_i}{\lambda d_2}, \frac{y_i}{\lambda d_2} \right) \right]. \tag{6.45}$$

It can be seen that if the diameter of the lens is infinite, $\bar{P}(x_i/\lambda d_2, y_i/\lambda d_2)$ becomes a $\delta$ function, and a perfect image can be obtained provided the lens has no aberration. However, *if the size of the lens is finite, a perfect image can never be obtained even if the lens is aberration free.* The resolution is limited by the convolution with $\bar{P}(x_i/\lambda d_2, y_i/\lambda d_2)$ which is associated with the diffraction from the aperture. A system is called *diffraction limited* if the system is perfect and the diffraction effect is the sole cause of limiting the resolution. In astronomy, the number of the stars one can observe is often limited by the incident light energy. In this case, the system is power limited. The university laboratory is often money limited.

Next, by again using the one-dimensional aperture as an example, the phrase "degree of blur" will be further qualified. When the system has an aperture width $D$, the output image is given by (6.44). The convolution of the image and $\text{sinc}[D(x_i/\lambda d_2)]$ can be obtained in the manner illustrated in Fig. 6.14. If one assumes that the smallest dimension $\Delta x_i$ in the output image that can be resolved is more or less the width of the main lobe, $\Delta x_i$ is

$$\Delta x_i = 2 \frac{d_2}{D} \lambda. \tag{6.46}$$

The value given by (6.46) is the resolution in the output image. This resolution can be converted into that of the input image by multiplying by $1/M = d_1/d_2$

$$\Delta x_0 = 2 \frac{d_1}{D} \lambda. \tag{6.47}$$

The relationship between the highest spatial frequency $f_c$ and the limit of the resolution is

$$f_c \cong \frac{1}{\Delta x_0}.$$

It is seen that this result agrees with that of (6.27).

# Problems

**6.1**   The front surface $v$ of a meniscus lens is spherical, as shown in Fig. 6.16. The center of the sphere is at the focus of the lens.

Prove that the expression for the back surface $v'$ of the lens is

$$r = \frac{f(n-1)}{n - \cos \theta} \tag{6.48}$$

where $f$ is the distance between the focus and the front vertex, $n$ the index of refraction of the glass, and $\theta$ the angle between the lens axis and a line connecting the focus and a point $S_2$ on the back surface $v'$ of the lens.

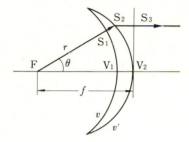

**Fig. 6.16.** The design of the surface of a meniscus lens

**6.2**   Referring to Fig. 6.9 b, find the angle of the tilt of the Fourier transform image due to the translational shift $x_1$ of the input object.

**6.3**   One wants to take a highly reduced picture of a diagram as shown in Fig. 6.17 by using a lens whose aperture stop is $F = 1.2$ and the focal length $f = 50$ mm. ($F = f/D$, $D$ being the diameter and $f$ the focal length of the lens.) The highest spatial frequency of the diagram is 1 line/mm. Derive the maximum obtainable reduction ratio (size of the picture/size of the object). The wavelength of the illuminating light is $\lambda = 0.555\ \mu m$, and the camera is assumed to be diffraction limited.

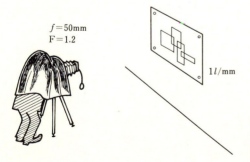

**Fig. 6.17.** Maximum obtainable reduction ratio of a camera with a lens of a finite size

**6.4**  As shown in Fig. 6.18, by using a convex lens with focal length $f$, the Fourier transform of the input $g(x_0, y_0)$ is to be obtained on the screen placed at the back focal plane of the lens. In order to eliminate the phase factor associated with the Fourier transform, as seen in (6.19), convex lens is placed on the surface of the screen. What should the focal length of the convex lens be?

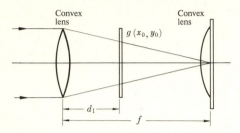

**Fig. 6.18.** Elimination of the phase factor associated with the Fourier transform using a convex lens

**6.5**  As shown in Fig. 6.19, the input images $t_1(x_0, y_0)$ and $t_2(x_1, y_1)$ were placed on the surface of the lens $L_1$ and $L_2$. The focal length of the lens $L_1$ is $f_1 = 2a$ and that of $L_2$ is $f_2 = a$. The spacings between $L_1$ and $L_2$ and the screen are all $2a$. Derive an expression for the light distribution on the screen. What are the applications of such an arrangement?

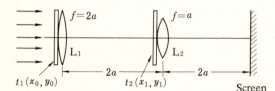

**Fig. 6.19.** Optical computing using two convex lenses

**6.6**  The cylindrical lens has the property of Fourier transforming only in one dimension. Describe the nature of the output image when a convex lens and a cylindrical lens are arranged in the manner shown in Fig. 6.20. What are the possible applications?

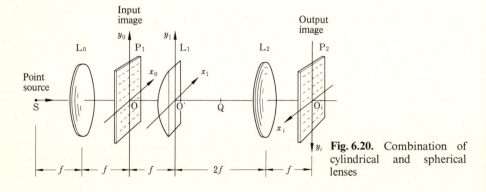

**Fig. 6.20.** Combination of cylindrical and spherical lenses

**6.7**   Optics and communication theory have a lot in common. Figure 6.21 is meant to be an equivalent circuit for representing the image-forming system shown in Fig. 6.15. Insert the relevant mathematical expressions in the blanks in Fig. 6.21 [1].

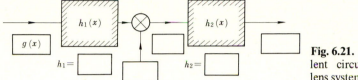

$h_1 =$

$h_2 =$

**Fig. 6.21.** Electrical equivalent circuit of an optical lens system

1  In optics, it is quite easy to reduce the size of the image because a lens is available for this purpose. The pulse width of a radar echo can be reduced to improve the range resolution of the radar in a similar manner by electronically realizing a circuit, such as is shown in Fig. 6.21.

# 7. The Fast Fourier Transform (FFT)

The Fast Fourier Transform (FFT) is one of the most frequently used mathematical tools for digital signal processing. Techniques that use a combination of digital and analogue approaches have been increasing in numbers. This chapter is for establishing the basis of this combined approach in dealing with computer tomography, computer holography and hologram matrix radar.

## 7.1 What is the Fast Fourier Transform?

One can immediately see from a table of Fourier transforms that only a limited number of functions can be transformed into closed analytic forms. When a transform is not available in an analytic form, it must be estimated by numerical computation. The numerical approximation to the Fourier transform integral, which can be obtained by summing over small sections of the integrand, takes time even when a high speed computer is used. In 1965, *Cooley* and *Tukey* published a new algorithm which substantially reduced the computation time [7.1]. For instance, an 8,192 point Discrete Fourier Transform (DFT), which takes about 30 minutes of computer time when conventional integration programming is used, can be computed in less than 5 seconds with the algorithm. Although first known as the Cooley-Tukey algorithm, it has since become practically synonymous with FFT [7.2−4]. Figure 7.1 shows the ratio of the computation time between

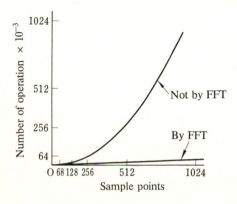

**Fig. 7.1.** Substantial reduction in number of computations is demonstrated; One curve with FTT the other with ordinary computation

normal programming and FFT algorithm programming. Besides the economic advantages of FFT, there are certain applications where high-speed processing is essential. Real-time radar-echo processing is one such example.

The Fourier transform pair is defined as

$$G(f) = \int_{-\infty}^{\infty} g(x) \, e^{-j2\pi fx} \, dx \tag{7.1}$$

$$g(x) = \int_{-\infty}^{\infty} G(f) \, e^{j2\pi fx} \, df. \tag{7.2}$$

Similarly, the Discrete Fourier Transform (DFT) is defined as

$$G_l = \sum_{k=0}^{N-1} g_k \, e^{-j2\pi kl/N} \tag{7.3}$$

$$g_k = \frac{1}{N} \sum_{l=0}^{N-1} G_l \, e^{j2\pi kl/N}. \tag{7.4}$$

Notice that both (7.3) representing the transform and (7.4) representing the inverse transform are quite similar mathematically. The only difference is whether the exponent contains a positive or negative sign, and whether or not the normalizing factor $N$ is present. Because the two expressions are basically equivalent, the discussion will proceed solely with (7.3). This equation is rewritten in a simpler form as

$$G_l = \sum_{k=0}^{N-1} g_k \, W^{kl} \quad \text{where} \quad l = 0, 1, 2, 3, \ldots, N-1 \tag{7.5}$$

$$W = e^{-j2\pi/N}. \tag{7.6}$$

In the analysis to follow, it will be convenient to use matrix notation. The expression (7.5) for the DFT, is written in matrix form below.

$$
\begin{pmatrix} G_0 \\ G_1 \\ G_2 \\ G_3 \\ \cdot \\ \cdot \\ \cdot \end{pmatrix} =
\begin{pmatrix}
W^0 & W^0 & W^0 & W^0 & \cdots & W^0 \\
W^0 & W^1 & W^2 & W^3 & \cdots & W^{N-1} \\
W^0 & W^2 & W^4 & W^6 & \cdots & W^{2(N-1)} \\
W^0 & W^3 & W^6 & W^9 & & \\
\cdot & & & & & \\
\cdot & & & & & \\
W^0 & W^{N-1} & & & & W^{(N-1)(N-1)}
\end{pmatrix}
\begin{pmatrix} g_0 \\ g_1 \\ g_2 \\ g_3 \\ \cdot \\ \cdot \\ g_{N-1} \end{pmatrix}. \tag{7.7}
$$

Thus the DFT can be thought of as a kind of linear transform. A normal Fourier transform can be considered as the special case of $N$ being infinity. The number of $G$'s which can be uniquely determined is identical to the number of $g$'s, namely, if there are only $N$ sampled points then there are only $N$ spectrum points. All the elements of the $N \times N$ matrix in (7.7) are derived from (7.6) and have peculiar characteristics which can be used for simplifying the matrix. The FFT is, in fact, a systematic method of matrix simplification.

The principle of FFT will be explained by taking an example of an 8 point DFT. Inserting $N = 8$ into (7.5, 7) gives

$$G_0 = W^0 g_0 + W^0 g_1 + W^0 g_2 + W^0 g_3 + W^0 g_4 + W^0 g_5 + W^0 g_6 + W^0 g_7$$
$$G_1 = W^0 g_0 + W^1 g_1 + W^2 g_2 + W^3 g_3 + W^4 g_4 + W^5 g_5 + W^6 g_6 + W^7 g_7$$
$$G_2 = W^0 g_0 + W^2 g_1 + W^4 g_2 + W^6 g_3 + W^8 g_4 + W^{10} g_5 + W^{12} g_6 + W^{14} g_7$$
$$G_3 = W^0 g_0 + W^3 g_1 + W^6 g_2 + W^9 g_3 + W^{12} g_4 + W^{15} g_5 + W^{18} g_6 + W^{21} g_7$$
$$G_4 = W^0 g_0 + W^4 g_1 + W^8 g_2 + W^{12} g_3 + W^{16} g_4 + W^{20} g_5 + W^{24} g_6 + W^{28} g_7$$
$$G_5 = W^0 g_0 + W^5 g_1 + W^{10} g_2 + W^{15} g_3 + W^{20} g_4 + W^{25} g_5 + W^{30} g_6 + W^{35} g_7$$
$$G_6 = W^0 g_0 + W^6 g_1 + W^{12} g_2 + W^{18} g_3 + W^{24} g_4 + W^{30} g_5 + W^{36} g_6 + W^{42} g_7$$
$$G_7 = W^0 g_0 + W^7 g_1 + W^{14} g_2 + W^{21} g_3 + W^{28} g_4 + W^{35} g_5 + W^{42} g_6 + W^{49} g_7$$

$$(7.8)$$

In its present form, (7.8) requires 64 multiplications and 56 additions. Since computer multiplication is essentially the repetition of additions, the key factor to reducing the computer time is to reduce the number of multiplications that must be performed.

The value of $W^{kl}$ ranges from $W^0$ to $W^{49}$ but they all can be reduced to one of 8 values between $W^0$ and $W^7$. This is shown in Fig. 7.2 by using a phasor diagram in the complex plane. The magnitude of $W^{kl} = \exp(-j\,2\pi\,l/N)$ is unity, and for $N = 8$, the phase angle associated with $W^{kl}$ is an integral multiple of $\pi/4$ radians. In the phasor diagram, $W^{kl}$ is a vector whose end point lies on a unit circle. Starting at $kl = 0$ which corresponds to $W^0 = 1$, the vector is seen to rotate clockwise by $\pi/4$ radians each time $kl$ is increased by one. In going from $W^7$ to $W^8$ the vector arrives once again at the starting point and the cycle is repeated so that there are only 8 uniquely

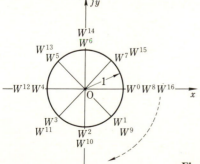

Fig. 7.2. Representation of $W^k$ on the complex plane

determined points on the circle. Notice that the $W^{kl}$'s diametrically opposite each other on the circle differ only in their sign.

Referring to Fig. 7.2, equivalent values of $W^{kl}$ are factored out of the terms in the DFT. This saves on the number of multiplications as can be seen by comparing $G_2$ and $G_3$ below with those from (7.8), i.e.,

$$G_2 = (g_0 + g_4) + W^2(g_1 + g_5) + W^4(g_2 + g_6) + W^6(g_3 + g_7)$$
$$G_3 = W(g_0 - g_4) + W^3(g_1 - g_5) + W^6(g_2 - g_6) + W^9(g_3 - g_7).$$

The results can be summarized as follows

$$\begin{pmatrix} G_0 \\ G_2 \\ G_4 \\ G_6 \\ G_1 \\ G_3 \\ G_5 \\ G_7 \end{pmatrix} = \begin{pmatrix} 1 & 1 & 1 & 1 & 0 & 0 & 0 & 0 \\ 1 & W^2 & W^4 & W^6 & 0 & 0 & 0 & 0 \\ 1 & W^4 & W^8 & W^{12} & 0 & 0 & 0 & 0 \\ 1 & W^6 & W^{12} & W^{18} & 0 & 0 & 0 & 0 \\ 0 & 0 & 0 & 0 & 1 & W^1 & W^2 & W^3 \\ 0 & 0 & 0 & 0 & 1 & W^3 & W^6 & W^9 \\ 0 & 0 & 0 & 0 & 1 & W^5 & W^{10} & W^{15} \\ 0 & 0 & 0 & 0 & 1 & W^7 & W^{14} & W^{21} \end{pmatrix} \begin{pmatrix} g_0 + g_4 \\ g_1 + g_5 \\ g_2 + g_6 \\ g_3 + g_7 \\ g_0 - g_4 \\ g_1 - g_5 \\ g_2 - g_6 \\ g_3 - g_7 \end{pmatrix}. \quad (7.9)$$

In the above matrix, only two of the $4 \times 4$ submatrices are non-zero, and the number of multiplication operations in (7.9) is one half of (7.8). Taking the reduction process one step further, a similar method may be applied to reduce each of these non-zero $4 \times 4$ submatrices into two $2 \times 2$ non-zero matrices. Systematic reduction in the number of multiplication operations is thus made, which is the goal of FFT. The FFT can be divided into two major methods, and each of them will be explained in the following sections.

## 7.2   FFT by the Method of Decimation in Frequency

Suppose that the sampled values are $g_0, g_1, \ldots g_k, \ldots, g_{N-1}$ and that the DFT of these values is sought. First, these sampled values are split into two groups; one group containing values $g_0$ through $g_{N/2-1}$, and the other group containing values $g_{N/2}$ through $g_{N-1}$. These two groups are designated as follows:

$$f_k = g_k$$
$$h_k = g_{k+N/2} \quad \text{where} \quad k = 0, 1, 2, 3, \ldots N/2 - 1. \quad (7.10)$$

Equation (7.10) is inserted into (7.5),

$$G_l = \sum_{k=0}^{N/2-1} g_k W^{kl} + \sum_{k=0}^{N/2-1} g_{k+N/2} W^{(N/2+k)l}$$

$$= \sum_{k=0}^{N/2-1} g_k \, W^{kl} + \sum_{k=0}^{N/2-1} h_k \, W^{kl} \, W^{Nl/2} \qquad \text{and}$$

$$W^{Nl/2} = \exp\left(-j\,\frac{2\pi}{N}\,\frac{N}{2}\,l\right) = e^{-j\pi l} = (-1)^l$$

and therefore

$$G_l = \sum_{k=0}^{N/2-1} [g_k + (-1)^l \, h_k] \, W^{kl}. \tag{7.11}$$

The second term inside the bracket of (7.11) keeps changing its sign depending on whether $l$ is even or odd. Thus, first of all, the $G_l$'s with even number $l$ are pulled out and calculated, and then those with odd number $l$ are calculated. This explains why this method is called *method of decimation in frequency*.

The $G_l$'s with even $l$ are given by

$$G_{2l} = \sum_{k=0}^{N/2-1} (g_k + h_k)(W^2)^{kl} \tag{7.12}$$

where $l$ ranges from 0 to $N/2 - 1$. Carefully comparing (7.12) with (7.5, 6), it can be seen that (7.12) is again in a form of DFT but the number of sampling points has been reduced to $N/2$. Next, the $G_l$'s with odd $l$ are considered. From (7.11) one obtains

$$G_{2l+1} = \sum_{k=0}^{N/2-1} (g_k - h_k) \, W^k (W^2)^{kl} \tag{7.13}$$

where $l$ ranges from 0 to $N/2 - 1$. Equation (6.13) is again in the form of a DFT with $N/2$ sampling points. So far, the $N$-point DFT has been split into two $N/2$-point DFT's. If this process of reducing the sampling points is repeated, each $N/2$-point DFT is further reduced to two $N/4$-point DFT's. If one repeats the procedure $n$ times, the $N$ point DFT is reduced to several $N/2^n$ point DFT's. The number of repetitions of the procedure necessary to reduce to the 1 point DFT is

$$\frac{N}{2^n} = 1. \tag{7.14}$$

Therefore the number of necessary repetitions is

$$n = \log_2 N. \tag{7.15}$$

The signal flow graph, which is a convenient guideline for programming the FFT algorithm, will be explained next. Again, take as an example the case $N = 8$. Figure 7.3 is a signal flow graph representing the calculation of

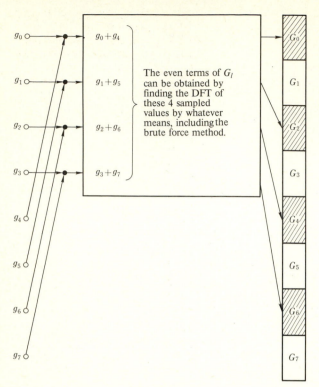

**Fig. 7.3.** Signal flow graph representing the calculation of (7.12). Black dot represents the summing of the two numbers transferred by the arrows. A number written on the side of a line means the written number is multiplied by the number brought by the line.

(7.12). An arrow denotes transfer of the number to a new location. A black dot denotes the summing of the two numbers transferred by the arrow. A number written on the side of a line means that the written number should be multiplied by the number brought by the line. A dotted line denotes transfer of the number after changing its sign. This symbolism will be better understood by looking at a specific example. The black dot to the right of $g_0$ in the upper left part of Fig. 7.3 has two arrows pointing at it. This black dot signifies the operation of $g_0 + g_4$. In Fig. 7.4, the black dot and $W^3$ to the right of $g_7$ signify the operation of the product $(g_3 - g_7) \, W^3$.

Figure 7.3 shows that $G_0$, $G_2$, $G_4$ and $G_6$ can be obtained by performing the 4 point DFT of $(g_0 + g_4)$, $(g_1 + g_5)$, $(g_2 + g_6)$, and $(g_3 + g_7)$. The method of performing this 4-point DFT is not specified, but is instead represented by a plain square box in Fig. 7.3. Figure 7.4 is a similar diagram except for the odd $l$ of $G_l$'s.

Suppose that the square boxes of Figs. 7.3, 4 are calculated without using special methods to reduce the number of multiplications. Each of the square boxes has four inputs. Therefore, the number of multiplications needed for the $G_l$ with even $l$ is $(4)^2 = 16$ and those with odd $l$ is again $(4)^2 = 16$. The total number of multiplications therefore, is 32. This is half the amount of the direct transform using (7.8) which is $8^2 = 64$.

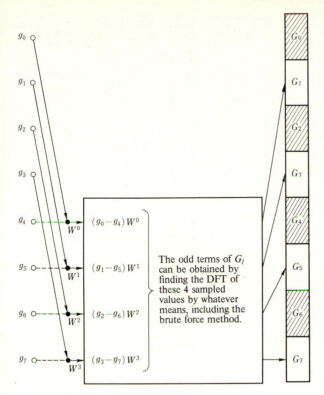

**Fig. 7.4.** Same as Fig. 7.3 but for (7.13).
A dotted line means transfer the number after changing its sign

In order to further reduce the number of operations, each 4 point DFT is split into two 2-point DFT's. Figure 7.5 illustrates this method. In order to simplify the expression, the $g$'s and $G$'s are labeled as follows

$$q_0 = g_0 + g_4 \qquad Q_0 = G_0$$
$$q_1 = g_1 + g_5 \qquad Q_1 = G_2$$
$$q_2 = g_2 + g_6 \qquad Q_2 = G_4$$
$$q_3 = g_3 + g_7 \qquad Q_3 = G_6,$$

namely, the 4 point DFT of the $q_k$'s is expressed by the $Q_l$'s. As before, the $Q_l$'s are separated into even $l$'s and odd $l$'s.

$Q_0$ and $Q_2$ are the 2-point DFT of $q_0 + q_2$ and $q_1 + q_3$, $Q_1$ and $Q_3$ are the 2 point DFT of $(q_0 - q_2) W^0$ and $(q_1 - q_3) W^2$. As seen from (7.5, 6) with $N = 2$, the 2 point DFT is quite simple and is given by the sum and difference of the inputs. The sequence of operations for obtaining the 4-point DFT is pictured in Fig. 7.5. Thus, the operation in the square box in Fig. 7.4 can be completed in a similar manner and results in the diagram shown in Fig. 7.6.

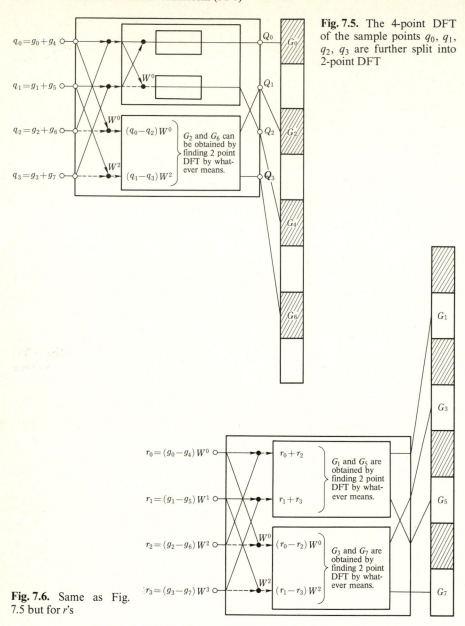

**Fig. 7.5.** The 4-point DFT of the sample points $q_0$, $q_1$, $q_2$, $q_3$ are further split into 2-point DFT

**Fig. 7.6.** Same as Fig. 7.5 but for $r$'s

The combination of Figs. 7.5, 6 results in Fig. 7.7. If the bent lines in Fig. 7.7 are straightened and the square frames removed, the signal flow graph shown in Fig. 7.8 is obtained. After close observation, one would find that not only does the signal flow graph in Fig. 7.8 reduce the amount of computation, but it also has a special property that plays a key role in saving

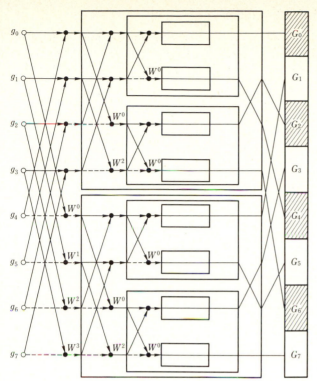

**Fig. 7.7.** Signal flow graph of 8-point DFT by a method of decimation in frequency

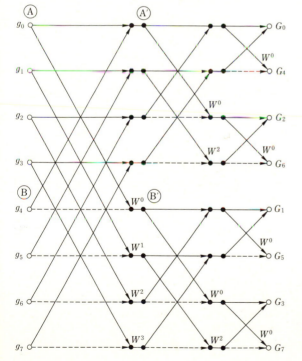

**Fig. 7.8.** Signal flow graph obtained by removing the frames and straightening the lines of Fig. 7.7 (method of decimation in frequency)

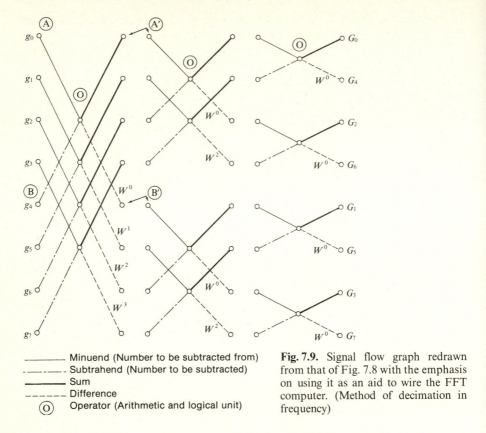

———— Minuend (Number to be subtracted from)
–·——·– Subtrahend (Number to be subtracted)
**——** Sum
– – – – Difference
Ⓞ Operator (Arithmetic and logical unit)

**Fig. 7.9.** Signal flow graph redrawn from that of Fig. 7.8 with the emphasis on using it as an aid to wire the FFT computer. (Method of decimation in frequency)

memory space. For instance, the value of $g_0$ which is stored at the location Ⓐ and that of $g_4$ stored at the location Ⓑ are added, the sum is stored at the location Ⓐ and the difference between $g_0$ and $g_4$ is stored at the location Ⓑ′. The rule is that the computed results are always stored at the memories located on the same horizontal level as the location of the inputs. Every computation follows this rule.

Figure 7.9 is a signal flow graph drawn to emphasize this point. The signal flow graph in Fig. 7.9 can be conveniently used at the time the computer circuits are wired. The points Ⓐ Ⓑ Ⓐ′ Ⓑ′ in Fig. 7.9 correspond to those in Fig. 7.8. A fine line (——) and a broken line (–·——·) both represent the inputs. The heavy line (**——**) represents the sum of the two inputs and a dashed line (- - -) represents the difference of the two inputs. O indicates an arithmetic unit which calculates the sum and the difference. The $W^k$ is always associated with the difference which is represented by the dashed line, and the product of $W^k$ and the difference is made. The reason that the inputs are designated by two different kinds of lines is that it is necessary to indicate which of the two inputs is subtracted from the other. The input represented by the broken line is subtracted from that represented by the fine line. For example, the connection by the broken line in the upper left of Fig. 7.9 means $g_0 - g_4$ and not $g_4 - g_0$.

The input to the computer is first put into the input register and the two numbers, e.g., $g_0$ stored at the memory location Ⓐ and $g_4$ stored at the memory location Ⓑ, are used to make the sum and the difference. The result $g_0 + g_4$ is stored at Ⓐ and the $g_0 - g_4$ is stored at Ⓑ′. After a similar operation is repeated three times, the values of all $G_i$'s are obtained.

At first, it may seem as if four columns of memory storage locations are needed, but in fact *only one column of memory can manage all the computations.* This is a major advantage in building an FFT computer. This advantage will be explained by taking the example of the operation at Ⓐ and Ⓑ. The value $g_0$ stored at Ⓐ is used to calculate $g_0 + g_4$ and $g_0 - g_4$; after that, there is no more need to use $g_0$ by itself. The same holds true for $g_4$ stored at Ⓑ. Hence $g_0$ can be erased from the memory location Ⓐ and be replaced by $g_0 + g_4$. Similarly $g_4$ can be erased and replaced by $g_0 - g_4$. This means that there is no need for the memory locations at Ⓐ′ and Ⓑ′. In fact, all the memory locations expect the first column can be eliminated, and the construction of the FFT computer is simplified.

One may notice that the order of the $G_i$'s is mixed up in Fig. 7.9. The $G_i$'s as they appear in Fig. 7.9 are $G_0$, $G_4$, $G_2$, $G_6$, $G_1$, $G_5$, $G_3$, $G_7$. Expressing the subscripts as binary numbers while maintaining the same order as above gives

000, 100, 010, 110, 001, 101, 011, 111 .

Taking each of these binary numbers individually and writing them backwards, as for example, rewriting 110 as 011, gives

000, 001, 010, 011, 100, 101, 110, 111 .

The decimal numbers corresponding to these binary numbers are 0, 1, 2, 3, 4, 5, 6, 7 and are in proper order. In the computer, by issuing the control signal of "*bit reversal*", the output can be read out in proper order.

Next, the total number of operations is counted. In Fig. 7.9 there are 3 columns of operations. In one column there are 4 additions, 4 subtractions and 4 multiplications. This is a substantial reduction in the total number of operations. Recall the direct computation of (7.8) required 56 additions and 64 multiplications.

In general, the number of repetitions of operations needed to reach the 1 point DFT was calculated to be (7.15), and in each repetition there are $N$ additions and subtractions and $N/2$ multiplications. The total number of operations for the FFT is

$$(3/2) N \log_2 N. \qquad (7.16)$$

Either from (7.16) or its graph in Fig. 7.1, it is seen that a remarkable reduction in computation is achieved for a large number $N$ of sampling points.

## 7.3 FFT by the Method of Decimation in Time

In this section, the other method of FFT is used. Let the $N$ sampled values be $g_0, g_1, g_2, g_3, \ldots, g_{N-1}$ and their DFT spectrum be $G_0, G_1, G_2, G_3, \ldots, G_{N-1}$. The first step is to separate the input $g_k$'s into those with even $k$ and those with odd $k$, and designate them as

$$f_k = g_{2k} \tag{7.17}$$

$$h_k = g_{2k+1} \quad \text{where} \quad k = 0, 1, 2, 3 \ldots \left(\frac{N}{2} - 1\right). \tag{7.18}$$

The DFT is separated into two parts as follows

$$G_l = \sum_{k=0}^{N/2-1} g_{2k} \, W^{2kl} + \sum_{k=0}^{N/2-1} g_{2k+1} \, W^{(2k+1)l}. \tag{7.19}$$

Inserting (7.17, 18) into (7.19) yields

$$G_l = \sum_{k=0}^{N/2-1} f_k (W^2)^{kl} + W^l \sum_{k=0}^{N/2-1} h_k (W^2)^{kl}. \tag{7.20}$$

The $l$'s are also split into two groups: one is from 0 to $N/2 - 1$ and the other from $N/2$ to $N - 1$, i.e.,

$$G_l = \sum_{k=0}^{N/2-1} f_k (W^2)^{kl} + W^l \sum_{k=0}^{N/2-1} h_k (W^2)^{kl} \quad \text{for} \quad 0 \le l \le N/2 - 1 \tag{7.21}$$

$$G_{N/2+l} = \sum_{k=0}^{N/2-1} f_k (W^2)^{k(N/2+l)} + W^{(N/2+l)} \sum_{k=0}^{N/2-1} h_k (W^2)^{k(N/2+l)}$$
$$\text{for} \quad 0 \le l \le N/2 - 1. \tag{7.22}$$

From the definition of $W$ in (7.6), one obtains

$$W^{Nk} = 1, \quad W^{N/2} = -1.$$

Hence $G_{N/2+l}$ can be further rewritten as

$$G_{N/2+l} = \sum_{k=0}^{N/2-1} f_k (W^2)^{kl} - W^l \sum_{k=0}^{N/2-1} h_k (W^2)^{kl}. \tag{7.23}$$

Therefore the final results of (7.21, 22) are

$$G_l = F_l + W^l H_l \tag{7.24}$$

$$G_{N/2+l} = F_l - W^l H_l \tag{7.25}$$

where $F_l$ and $H_l$ are the $N/2$ point DFT of $f_k$ and $h_k$:

$$F_l = \sum_{k=0}^{N/2-1} f_k (W^2)^{kl} \quad 0 \leq l \leq \frac{N}{2} - 1$$

(7.26)

$$H_l = \sum_{k=0}^{N/2-1} h_k (W^2)^{kl} \quad 0 \leq l \leq \frac{N}{2} - 1.$$

Equation (7.24) shows that the first half of $G_l$ is the sum of the $N/2$-point DFT of the even terms of $g_k$ and the $N/2$-point DFT of the odd terms of $g_k$ multiplied by $W^l$. Equation (7.25) shows that the second half of $G_l$ is the difference of these DFT's. This method is called *method of decimation in time*. The signal flow graph in Fig. 7.10 illustrates the method of calculating DFT using decimation in time for the case $N = 8$. In the same manner as outlined in the previous section, the 8-point DFT is first converted into 4-point DFT's and finally converted into 1-point DFT's. Figure 7.11 shows the completed signal flow graph. When the bent lines in Fig. 7.11 are

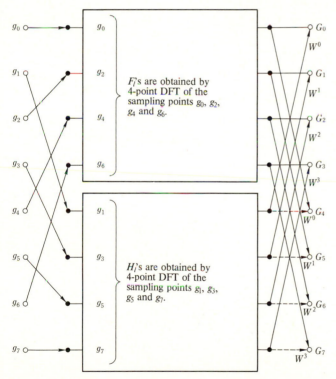

**Fig. 7.10.** Signal flow graph of the first stage of the method of decimation in time; First, DFT $F_l$ and DFT $H_l$ are calculated from the sampled values of $g_{2k}$ and $g_{2k+1}$ and next, $G_l$ is found by combining $F_l$ and $H_l$ in the specified manner

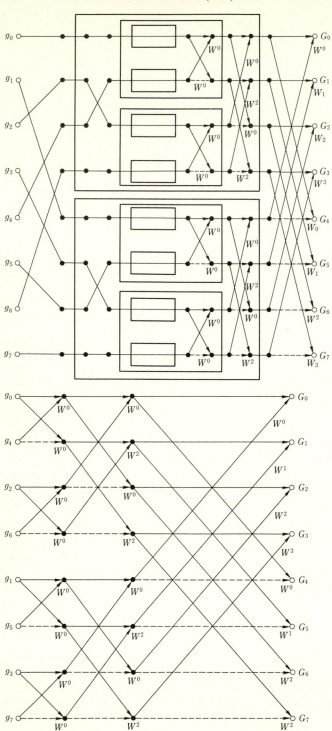

**Fig. 7.11.**
Caption
see opposite
page

**Fig. 7.12.**
Caption
see opposite
page

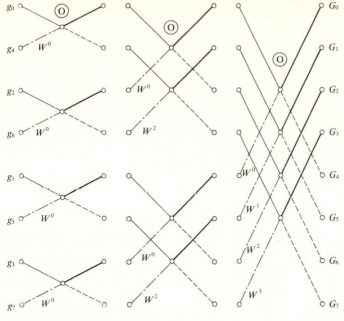

$g_0$  $W^0$

$g_4$  $W^0$

$g_2$  $W^0$

$g_6$  $W^0$  $W^2$

$g_1$  $W^0$

$g_5$  $W^0$

$g_3$  $W^0$

$g_7$  $W^0$  $W^2$

$W^0$

$W^1$

$W^2$

$W^3$

$G_0$
$G_1$
$G_2$
$G_3$
$G_4$
$G_5$
$G_6$
$G_7$

———— Minuend (Number to be subtracted from)
—·—·— Subtrahend (Number to be subtracted)
———— Sum
—————— Difference
  (O)    Operator (Arithmetic and logical unit)

**Fig. 7.13.** Signal flow graph redrawn from that of Fig. 7.12 with the emphasis on using it as an aid to wire the FFT computer. (Method of decimation in time)

straightened, Fig. 7.12 is obtained. Figure 7.13 is the signal flow graph redrawn from Fig. 7.12 to be used as a wiring aid for an FFT special purpose computer. The keys for the lines used in Fig. 7.13 are the same as those used in Fig. 7.9.

In the case of the method of decimiation in time, the inputs are fed into the computer in bit reversed order. All other aspects of the FFT such as the capability of writing over the same memory locations in order to save memory space, and the reduction in the total number of computations, are the same as those in the method of decimation in frequency.

**Fig. 7.11.** The signal flow chart of the method of decimation in time

**Fig. 7.12.** Signal flow graph obtained by removing the frames and straightening the lines of Fig. 7.11. (Method of decimation in time)

## 7.4 Values of $W^k$

Since only a finite number of $W^k$ are needed in an FFT special purpose computer, the values of the $W^k$'s are not recalculated every time they are needed but they are usually stored in the Read Only Memory (ROM). In general, when an $N$ point DFT is performed, $N$ values of $W^k$ are needed. For convenience, the definition of $W$ is once more written here.

$$W^k = \exp\left(-j\frac{2\pi}{N}k\right)$$

$$W^k = \cos\left(\frac{2\pi}{N}k\right) - j\sin\left(\frac{2\pi}{N}k\right). \tag{7.27}$$

The computer needs $2N$ memory locations to store both the real and imaginary parts of (7.27). If the properties of $W^k$ are exploited, the number of memory locations can be substantially reduced.

This will be explained by taking $N = 16$ as an example. $W$ is expressed by a unit circle in the complex plane as in Fig. 7.14. Since the values of $W^k$'s located diametrically opposite each other on the circle are the same in magnitude but different in signs, the values of the $W^k$'s in the range of $180° - 360°$ are obtained from those in $0° - 180°$ by simply changing the sign. If the $W^k$ in $0° - 90°$ are compared with those in $90° - 180°$, the only difference is in the sign of $\cos(2\pi/N)k$, thus the $W^k$'s in $0° - 90°$ can be used to represent both. If the $W^k$ in $0° - 45°$ are further compared with

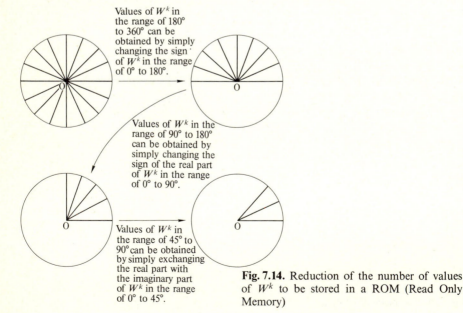

Values of $W^k$ in the range of 180° to 360° can be obtained by simply changing the sign of $W^k$ in the range of 0° to 180°.

Values of $W^k$ in the range of 90° to 180° can be obtained by simply changing the sign of the real part of $W^k$ in the range of 0° to 90°.

Values of $W^k$ in the range of 45° to 90° can be obtained by simply exchanging the real part with the imaginary part of $W^k$ in the range of 0° to 45°.

**Fig. 7.14.** Reduction of the number of values of $W^k$ to be stored in a ROM (Read Only Memory)

those in $45° - 90°$, the values of $\cos(2\pi/N)\,k$ and $\sin(2\pi/N)\,k$ are merely interchanged, thus the $W^k$'s in $0° - 45°$ contain sufficient information to define all the $W^k$. Finally, as shown in Fig. 7.14, *only two complex numbers are to be stored in* ROM *for* $N = 16$, excluding the value of $W^0$ which is 1 and need not be stored.

## Problems

**7.1**  Given input values $g_0$, $g_1$, $g_2$, $g_3$, $g_4$, $g_5$, $g_6$, $g_7$, find the values of $G_1$, $G_5$, $G_2$ and $G_6$ of the DFT using the signal flow graph in Fig. 7.7 and compare the results with those obtained by the signal flow graph in Fig. 7.9.

**7.2**  The DFT $G_{-l}$ in the negative region of $l$ can be obtained by transferring one half of the $G_l$ in the positive region of $l$, as shown in Fig. 7.15, namely,

$$G_{-l} = G_{N-l}.$$

Prove this relationship. Figure 7.15 was obtained by plotting only the real part of $G_l$ with real inputs.

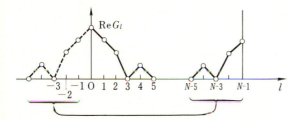

Fig. 7.15. Finding the values of $G_{-l}$ by translating the spectrum near the end

**7.3**  Find the 4-point DFT when all the input $g_k$'s are unity (Fig. 7.16). Perform the Fourier transform of the function $g(x)$ shown in Fig. 7.17. Compare the result of the DFT with that of the Fourier transform.

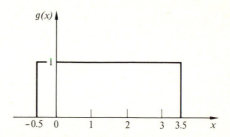

Fig. 7.17.  Curve of

$$g(x) = \begin{cases} 1 & \text{for} \quad -0.5 \le x \le 3.5 \\ 0 & \text{for all others} \end{cases}$$

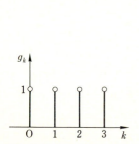

Fig. 7.16. Sampled values of $g_k = 1$

**7.4**  Find the DFT of the input $g_0 = 0$, $g_1 = 1$, $g_2 = 2$, $g_3 = 0$, as shown in Fig. 7.18. Using this result find the DFT of the input $g'_k$'s which were obtained by shifting $g_k$ to the right, as shown in Fig. 7.19. Compare the two results.

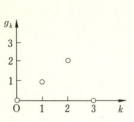

**Fig. 7.18.** Value of $g_k$        **Fig. 7.19.** Value of $g'_k$

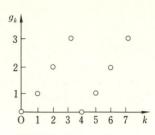

**Fig. 7.20.** Value of $g_k$

**7.5**  As shown in Fig. 7.20, the inputs consist of two periods of the same shape. Using the signal flow graph in Fig. 7.9 prove that one half of the 8-point DFT spectrum of such inputs is zero.

**7.6**  Making use of the chart of decimation in frequency (Fig. 7.9), calculate the DFT's of the following

a)  $g_k = [0, 1, 1, 1, 1, 1, 1, 1]$

b)  $g_k = [0, 0, 1, 1, 0, 0, 1, 1]$

c)  $g_k = [1, A, A^2, A^3, A^4, A^5, A^6, A^7]$

where $A = \exp(j\,\pi/4)$.

# 8. Holography

A photograph records the real image of an object formed by a lens. A hologram, however, records the field distribution that results from the light scattered by an object. Since there is a one-to-one correspondence between the object and its scattered field, it is possible to record information about the object by mapping the scattered field. Actually the *recording of the scattered field provides much more information about the object than that of the real image recorded in a photograph.* For instance, one hologram can readily generate different real images that correspond to different viewing angles.

It would seem possible to record the scattered field by just placing a sheet of film near the scattering object, but unfortunately, phase information about the scattered field cannot be recorded in this way. Nevertheless this approach is used in certain cases, such as the determination of crystal structure from the scattered field of x-rays. The missing information about the phase has to be supplemented by chemical analysis. In holography, in order to record the phase information, a reference wave coming directly from the source to the film is superimposed on the scattered field coming from the object to the film.

## 8.1 Pictorial Illustration of the Principle of Holography

Since any object can be considered as an ensemble of points, the principle of holography will be explained by using a point object. As shown in Fig. 8.1, the incoming plane wave is scattered by a point object O, from which a new diverging spherical wave front emerges. Figure 8.1 is a drawing of the instantaneous distribution of the incident and scattered fields. The contours of the maxima are drawn in solid lines and those of the minima are shown by dashed lines. The points where lines of the same kind intersect become extrema and the points where dissimilar lines intersect correspond to intensity nulls (zero). By connecting the extrema, a spatial fringe pattern between the incident and the scattered field is drawn. These fringes indicate a spatial standing wave as shown by the heavy lines in Fig. 8.1 a.

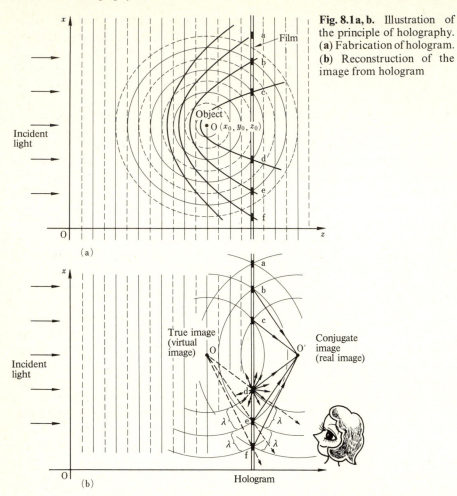

**Fig. 8.1a, b.** Illustration of the principle of holography. (**a**) Fabrication of hologram. (**b**) Reconstruction of the image from hologram

The spatial standing wave pattern can be recorded easily by using a photographic plate. The contours of the extrema will show up as dark lines on the plate. Figure 8.1a is a cross sectional view of the plate placed perpendicular to the $z$ axis. The dark points a, b, c, ..., f, are the cross sections of the extrema lines. The transmittance pattern of such a photographic film is more like that of a Fresnel zone plate. This photographic plate is a hologram of the point object. If the object consists of more than just one point, the hologram of such an object is the superposition of zone plates each of which corresponds to a point on the object.

When the hologram is illuminated by a parallel beam, the points a, b, c, ..., f become scattering centers, as shown in Fig. 8.1b. The difference in path lengths between $d\mathrm{O}$ and $e\mathrm{O}$ is exactly one wavelength, hence the phase of the scattered wave from $d$ is in phase with that from $e$. Similarly, the waves scattered from a, b, c, ..., f are all in phase. Thus, the exact phase distribution that a point object located at O would have generated in the hologram plane is re-established by the scattered field from a, b, c, ..., f. An eye looking into the hologram towards the illuminating source, as shown in Fig. 8.1 b, cannot distinguish whether the light originates just from the surface of the hologram, or indeed from the original object, because the distributions of the light to the eye are the same for both cases, and it would appear as if the original object were present behind the hologram.

Moreover, all the field originally scattered in the direction within the angle subtended by the original object at the photographic plate is simultaneously generated, and slightly different scenes are seen by each eye of the observer, which is necessary for the observer's brain to interpret the scene as three dimensional. Whenever the observer moves his eyes, he sees that portion of the scattered field intercepted by the viewing angle of his eyes, and the scene therefore changes. As a result of these effects, the viewer has the illusion that he is actually seeing a a three-dimensional object in front of him.

Since the light does not actually converge to the location of the image, the image at O is a virtual image. In addition to this virtual image, a real image to which the light converges is formed. Referring to Fig. 8.1 b, at the point O', which is symmetric to point O with respect to the photographic plate, all the scattered waves are in phase and the real image of the point object is created.

One interpretation of holography is that the *field scattered from an object is intercepted by a photographic plate, and the scattered field pattern is frozen onto the plate.* The scattered pattern is stored in the plate as a hologram. Whenever the photographic plate is illuminated by a laser beam, the frozen fringe pattern recreates the original scattered field pattern moving in the original direction.

Because of the zone-plate-like properties of a hologram, the image can still be reconstructed even when only a portion of the hologram is used. The resolution of the image, however, is reduced because the number of available fringe lines on the hologram is reduced, but the location and shape of the image will not be affected.

## 8.2 Analytical Description of the Principle of Holography

An outline of the analysis will be given and then the principle of holography will be discussed in a more quantitative manner [8.1, 2]. Figure 8.2 shows an arrangement for producing a hologram.

A laser beam is split into the object beam which illuminates the object and the reference beam which illuminates the photographic plate directly. The recording of the superposition of the object beam O and the reference

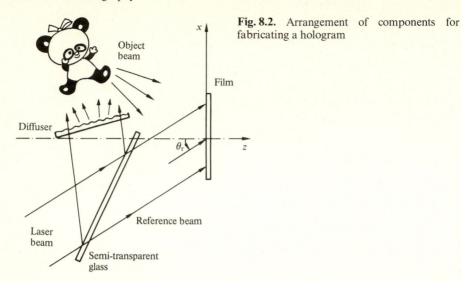

**Fig. 8.2.** Arrangement of components for fabricating a hologram

beam R is a hologram. In order to achieve uniform illumination of the object while avoiding specular reflection from the object, a diffuser such as a sheet of ground glass is inserted between the illuminating source and the object. If one assumes that the amplitude transmittance of the photographic film is proportional to the intensity of the incident light, the distribution of the amplitude transmittance of the hologram is

$$|O + R|^2 = |O|^2 + |R|^2 + OR^* + O^*R. \tag{8.1}$$

When the hologram is illuminated by a laser beam, the image is reconstructed. Assuming that the reconstructing beam is the same as the reference beam ·used to make the hologram, the distribution of light transmitted through the hologram is given by

$$|O + R|^2 R = |O|^2 R + |R|^2 R + O|R|^2 + O^*R^2. \tag{8.2}$$

The 3rd term of (8.2) is of particular interest. If R represents a plane wave, $|R|^2$ is constant across the photographic plate, and the 3rd term of (8.2) is proportional to the distribution O of the object beam. Therefore, the distribution of the transmitted light through the hologram is identical to that of the field scattered from the original object, and it will have the same appearance as the original object to an observer.

This procedure will now be analyzed in a more quantitative manner. The $x$-$y$ coordinates are taken in the plane of the photographic plate. The field distributions of the reference and object beams on the photographic plate are represented by $R(x, y)$ and $O(x, y)$, respectively.

When the photographic film characteristics and contrast reversal are taken into consideration, the expression for the amplitude transmittance

$t(x, y)$ will be given via (8.1) by

$$t(x, y) = t_1(x, y) + t_2(x, y) + t_3(x, y) + t_4(x, y) \tag{8.3}$$

where

$$t_1(x, y) = -\beta|O(x, y)|^2,$$
$$t_2(x, y) = \beta[c - |R(x, y)|^2],$$
$$t_3(x, y) = -\beta R^*(x, y)\, O(x, y),$$
$$t_4(x, y) = -\beta R(x, y)\, O^*(x, y),$$

$\beta$ being the slope of the linear portion of the film's transmittance versus exposure curve, and $c$ a constant.

The fields $O(x, y)$ and $R(x, y)$ will first be calculated by using the Fresnel approximation. Assume that the reference beam is a plane wave incident upon the photographic plate with its direction of propagation in the $x$-$z$ plane as shown in Fig. 8.2. Then, $R(x, y)$ can be expressed as,

$$R(x, y) = R_0\, e^{jkx\sin\theta_r}, \tag{8.4}$$

where $\theta_r$ is the angle between the direction of propagation and the $z$ axis.

The object is considered to be made up of slices, and the field diffracted from only one of the object slices, $\mathcal{O}(x, y)$, located in the plane $z = z_0$, is considered first. When such a field is observed at $z = z$, the field distribution is expressed by

$$O(x, y, z) = \mathcal{O}(x, y) * f_{z-z_0}(x, y), \quad \text{where} \tag{8.5}$$

$$f_{z-z_0}(x, y) = \frac{1}{j\lambda(z - z_0)} \exp\left\{ jk\left[ (z - z_0) + \frac{x^2 + y^2}{2(z - z_0)} \right] \right\}.$$

For a photographic plate placed at $z = 0$, we have

$$O(x, y, 0) = \mathcal{O}(x, y) * f_{-z_0}(x, y), \tag{8.6}$$

where

$$f_{-z_0}(x, y) = \frac{j}{\lambda z_0} \exp\left\{ -jk[z_0 + (x^2 + y^2)/2z_0] \right\}. \tag{8.7}$$

By inserting (8.4, 6, 7) into (8.3), an expression for the amplitude transmittance of the hologram is obtained.

Next, the procedure for obtaining the reconstructed image is described mathematically. The image is reconstructed by illuminating the hologram with a reconstructing beam $P(x, y)$. This beam is assumed to be a plane wave incident upon the hologram tilted again only in the $x$ direction at an angle $\theta_p$, as shown in Fig. 8.3, namely

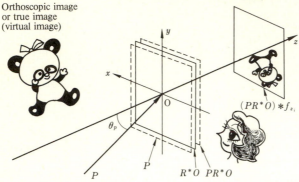

**Fig. 8.3.** Reconstructing of the image from a hologram

$$P(x, y) = P_0 \, e^{jkx\sin\theta_p}. \tag{8.8}$$

The Fresnel diffraction pattern of the illuminated hologram is

$$E(x_i, y_i) = [P(x_i, y_i) \, t(x_i, y_i)] * f_{z_i}(x_i, y_i). \tag{8.9}$$

The expression for $t(x, y)$ consists of four terms. The term $t_2(x, y)$ is related to the uniform beam of light which propagates straight through the hologram. The term $t_1(x, y)$ is associated with noise. Terms $t_3(x, y)$ and $t_4(x, y)$ contain the factor $O(x, y)$ and pertain to the image.

The expression for $E_3(x, y)$ associated with $t_3(x, y)$ can be rewritten by using (8.3, 4, 6–8) as

$$
\begin{aligned}
E_3(x_i, y_i) &= - \beta \, [P(x_i, y_i) \, R^*(x_i, y_i) \, O(x_i, y_i)] * f_{z_i}(x_i, y_i) \\
&= - \beta P_0 R_0 \{\exp[j \, k \, x_i (\sin\theta_p - \sin\theta_r)] \, [\mathscr{O}(x_i, y_i) * f_{-z_0}(x_i, y_i)]\} \\
&\quad * f_{z_i}(x_i, y_i). 
\end{aligned}
\tag{8.10}
$$

In order to simplify (8.10), the double operation of Fourier transform followed by an inverse Fourier transform will be performed:

$$E_3(x_i, y_i) = - \beta P_0 R_0$$

$$\times \mathscr{F}^{-1}\left\{\left[\delta\left(f_x - \frac{\sin\theta_p - \sin\theta_r}{\lambda}\right) * \left(\bar{\mathscr{O}}(f_x, f_y) \, F_{-z_0}(f_x, f_y)\right)\right] F_{z_i}(f_x, f_y)\right\}$$

$$= - \beta P_0 R_0 \mathscr{F}^{-1}\left\{\bar{\mathscr{O}}\left(f_x - \frac{\sin\theta_p - \sin\theta_r}{\lambda}, f_y\right)\right.$$

$$\left. \times F_{-z_0}\left(f_x - \frac{\sin\theta_p - \sin\theta_r}{\lambda}, f_y\right) F_{z_i}(f_x, f_y)\right\}, \tag{8.11}$$

where

$$\mathscr{F}\{\mathscr{O}(x, y)\} = \bar{\mathscr{O}}(f_x, f_y).$$

Using (3.39), one obtains

$$E_3(x_i, y_i) = -\beta P_0 R_0 \exp\left[-jk(z_0 - z_i) + j\pi\lambda z_0\left(\frac{\sin\theta_p - \sin\theta_r}{\lambda}\right)^2\right]$$

$$\times \mathscr{F}^{-1}\left\{\bar{\mathscr{O}}\left(f_x - \frac{\sin\theta_p - \sin\theta_r}{\lambda}, f_y\right) \exp[j\pi\lambda(z_0 - z_i)(f_x^2 + f_y^2)]\right.$$

$$\times \exp[-j2\pi z_0(\sin\theta_p - \sin\theta_r)f_x]\Big\}. \tag{8.12}$$

When $z_i = z_0$, (8.12) becomes

$$E_3(x_i, y_i) = -\beta P_0 R_0 \exp\left[j\pi\lambda z_0\left(\frac{\sin\theta_p - \sin\theta_r}{\lambda}\right)^2\right]$$

$$\times\left[\mathscr{O}(x_i, y_i)\exp\left(j2\pi\frac{\sin\theta_p - \sin\theta_r}{\lambda}x_i\right)\right] * \delta(x_i - z_0(\sin\theta_p - \sin\theta_r)),$$

and finally,

$$E_3(x_i, y_i) = -\beta P_0 R_0 \exp\left[j2\pi\frac{\sin\theta_p - \sin\theta_i}{\lambda}\left(x_i - \frac{z_0(\sin\theta_p - \sin\theta_r)}{2}\right)\right]$$

$$\times\mathscr{O}(x_i - z_0(\sin\theta_p - \sin\theta_r), y_i) \tag{8.13}$$

which means that the image is reconstructed at $z = z_0$. Furthermore, it is located exactly where the object was placed during the fabrication if the condition $\theta_p = \theta_r$ is satisfied. However, if $\theta_p \neq \theta_r$, there is a shift in the $x_i$ direction by $z_0(\sin\theta_p - \sin\theta_r)$. Since the image looks exactly the same as the object, the image is called *orthoscopic* or *true image*.

So far, only the term $t_3(x, y)$ has been considered. The term $t_4(x, y)$ associated with the conjugate image will be considered next. The formula which corresponds to (8.12) is

$$E_4(x_i, y_i) = -\beta P_0 R_0 \exp\left[jk(z_0 + z_i) - j\pi\lambda z_0\left(\frac{\sin\theta_p + \sin\theta_r}{\lambda}\right)^2\right]$$

$$\times \mathscr{F}^{-1}\left\{\bar{\mathscr{O}}^*\left(-f_x + \frac{\sin\theta_p + \sin\theta_r}{\lambda}, -f_y\right)\right.$$

$$\times \exp[-j\pi\lambda(z_0 + z_i)(f_x^2 + f_y^2)]$$

$$\times \exp[j2\pi z_0(\sin\theta_p + \sin\theta_r)f_x]\Big\}. \tag{8.14}$$

In the plane $z_i = -z_0$, the expression simplifies to

$$E_4(x_i, y_i) = -\beta P_0 R_0 \exp\left[j 2\pi \frac{\sin\theta_p + \sin\theta_r}{\lambda}\left(x_i + \frac{z_0(\sin\theta_p + \sin\theta_r)}{2}\right)\right]$$

$$\times \mathscr{O}^*(x_i + z_0(\sin\theta_p + \sin\theta_r), y_i) . \tag{8.15}$$

The interpretation of (8.15) is that a conjugate image is formed at $z_i = -z_0$. This means that the conjugate image is on the other side of the holo- gram from the original object. There is a shift in the $x_i$ direction by $z_0(\sin\theta_p + \sin\theta_r)$ even when $\theta_p = \theta_r$, but this shift disappears when $\theta_p = -\theta_r$ or $\theta_p = \theta_r + \pi$. The conjugate image is a real image; light is actually focused on this location and the projected image can be observed if a sheet of paper is placed at this location.[1]

The conjugate image has a peculiar property and is called a *pseudoscopic* image. The pseudoscopic image looks to an observer as if the image is inside out. Figure 8.4 illustrates why it looks inside out. Since the position of the focused beam is $z_i = -z_0$, a point on the object closer to the holo- gram is focused to a point further away from the observer. For example, referring to Fig. 8.4, point $F$ which is further away from the observer at $E_1$ than $N$, is focused to the point $F'$ which is closer to the observer than $N'$. To the observer, the nose of the doll is seen further than the forehead. Besides, the hatched section (which was not illuminated when the hologram was

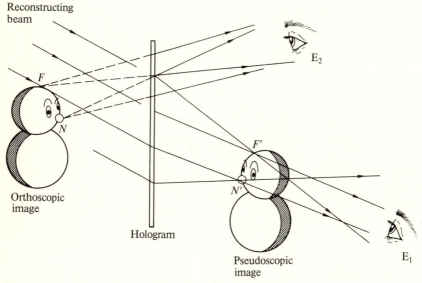

**Fig. 8.4.** Images reconstructed from a hologram

---

1 If $\mathscr{O}$ is a point source, then $\mathscr{O} = [A\exp(jkr)]/r$, and $\mathscr{O}^* = [A\exp(-jkr)]/r$. The expres- sion for $\mathscr{O}^*$ is that of spherical wave converging into one point. It is then clear that an image formed with $\mathscr{O}^*$ is a real image and not a virtual one.

fabricated) is not present in the reconstructed image. Thus, the observer has a strange sensation of seeing the *inside out* face of the doll.

The phase distributions of the reconstructed image and the original object are exactly the same except for a phase shift of 180° indicated by the negative signs in (8.13 and 15). However, since the human eye is only sensitive to the intensity, one cannot recognize this phase shift.

The spatial frequency component of the noise term $t_1(x, y)$ is approximately twice that of the object, and the direction of the diffraction from this term is at twice the diffraction angle of the true image. By properly selecting the values of $\theta_r$ and $\theta_p$, the directions of the true and virtual images can be adjusted so as not to overlap with that of the noise term.

## 8.3 Relationship Between the Incident Angle of the Reconstructing Beam and the Brightness of the Reconstructed Image

Because of the finite thickness of the photographic plate, the brightness of the reconstructed image is influenced by the angle of incidence of the reconstructing beam. Fig. 8.5 shows the cross section taken by an electron microscope of the photographic emulsion of a hologram [8.3]. Platelets are arranged like the fins of a venetian window blind. Each platelet acts like a

**Fig. 8.5.** Photograph of the cross-section of a Kodak 469F holographic plate taken by an electron microscope [8.3]

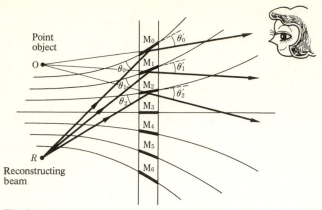

**Fig. 8.6.** The reconstruction of an image from a thick hologram. The same source which was used for the reference beam is used for the reconstructing beam

small mirror. These platelets are oriented in the plane of the bisector of the angle between the object and reference beams (Fig. 8.6). For instance, the surface of the platelet at the point $M_0$ is in a plane which bisects the angle $\angle$ $OM_0R$ made by $\overline{OM_0}$ and $\overline{RM_0}$ .

When a hologram is illuminated by a reconstructing beam, the wave is scattered in all directions by the fringes on the hologram, but the intensity distribution of the scattered wave is very direction sensitive. The intensity of the scattered wave is strongest in the direction in which the reconstructing beam would be reflected if each platelet were a small mirror, namely, when $\theta_0 = \theta_0'$, $\theta_1 = \theta_1'$, $\theta_2 = \theta_2'$, and so on, as shown in Fig. 8.6. The easiest way to have such a condition satisfied is to use the reference beam used for fabricating the hologram as the reconstructing beam. In this case, the orientations of all the platelets satisfy the above condition and the brightest possible reconstructed image is obtained.

## 8.4 Wave Front Classification of Holograms

According to the shape of the wave front of the object beam, holograms can be classified as: (1) Fresnel hologram, (2) Fourier transform hologram, (3) image hologram, and (4) lensless Fourier transform hologram.

### 8.4.1 Fresnel Hologram

When the distribution of the object beam on the hologram is the diffraction pattern in a Fresnel region, the hologram is called a Fresnel hologram. This is the most common type of hologram, and all the holograms described previously are of this type.

### 8.4.2 Fourier Transform Hologram

This hologram is made such that the field distribution of the object beam on the hologram is the Fourier transform of the object. One way to achieve this is to place the object in the far-field region, but in practice the distance required is too great, so that the hologram is fabricated by placing a photographic plate on the back focal plane of a converging lens. Usually a converging lens is used to form the reconstructed image on the back focal plane of the lens.

As seen from Fig. 8.7a, information about the object is concentrated in the vicinity of the focal point and the area needed to record the hologram is small compared to other types of holograms. The Fourier transform hologram is often used for the purpose of high-density recording [8.4, 19].

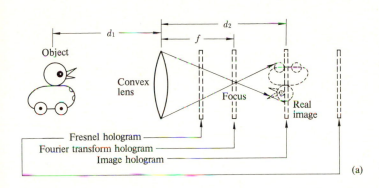

(a)

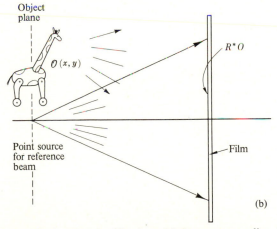

(b)

**Fig. 8.7a, b.** The classification of holograms according to the shapes of the wave front. **(a)** Geometries for Fourier transform hologram, image hologram, Fresnel hologram. **(b)** Lensless Fourier transform hologram

### 8.4.3 Image Hologram

A hologram made by placing a photographic plate near the location of the real image formed by a lens (Fig. 8.7a) is called an *image hologram* [8.5]. Since the position of the reconstructed image is on the hologram itself, the distance from the hologram to the image is very short, and consequently, the coherency of the source is not very critical. A simple explanation is as follows. If a hologram is made by light of one of two different frequencies, and the image is reconstructed using both of these frequencies, two images slightly displaced from each other are formed. The amount of displacement decreases as the distance between the image and the hologram decreases. Therefore, the image hologram, with its short hologram to image distance, is insensitive to the coherency of the reconstructing beam. This principle is used for white-light holograms, to be mentioned later in this chapter.

### 8.4.4 Lensless Fourier Transform Hologram

As shown in Fig. 8.7b, a point source is used as the reference beam and the object is placed in the same plane as that of the point source. In this configuration, the distribution OR* becomes the Fourier transform of the object, thereby allowing a Fourier transform hologram to be fabricated without the use of a lens. Such a hologram is called a *lensless Fourier transform hologram.*

The expression for OR* will be examined in more detail in order to demonstrate that the hologram formed is indeed a Fourier transform hologram. Let the location of the point source used as the reference beam be $(0, 0, -z_r)$ and the location of the film be $z = 0$. Then OR* is given by

$$OR^* = \frac{R_0^*}{j \lambda z_r} [\mathcal{O}(x, y) * f_{-z_0}(x, y)] \cdot \exp \{j \, k \, [z_r + (x^2 + y^2)/2 \, z_r]\} \tag{8.16}$$

or

$$OR^* = \frac{R_0^*}{(\lambda z)^2} \int\limits_{-\infty}^{\infty}\!\!\int \left[ \mathcal{O}(\xi, \eta) \exp\left(-j \, k \, \frac{\xi^2 + \eta^2}{2 z}\right) \right]$$

$$\times \exp\left[\frac{j \, 2 \, \pi}{\lambda \, z} (x \, \xi + y \, \eta)\right] d\xi \, d\eta \tag{8.17}$$

where

$$z_r = z_0 = z \, . \tag{8.18}$$

OR* is further rewritten as

$$OR^* = \frac{R_0^*}{(\lambda z)^2} \, \mathscr{F} \left\{ \mathcal{O}(x, y) \exp\left(-j \, k \, \frac{x^2 + y^2}{2 z}\right)\right\}_{\substack{f_x = -x/\lambda z, \\ f_y = -y/\lambda z}} . \tag{8.19}$$

Thus one can see that $OR^*$ is the Fourier transform of

$$\mathcal{O}(x,y) \exp[-jk(x^2+y^2)/2z]$$

and belongs to the category of Fourier transform holograms.


## 8.5 Holograms Fabricated by a Computer

Once the shape of the object is determined, the distribution of its scattered field can be calculated. In fact, it is possible to obtain the distribution of the scattered field using a computer instead of light. An intensity pattern is drawn according to the intensity calculated by the computer, and this pattern is then photographically reduced to form what is known as a *computed hologram* [8.5−9]. The reconstructed image can be viewed by laser light illumination. In principle, it is possible to view a completed three-dimensional building from an architectural drawing by means of this fabricated hologram.

The methods of fabricating a computed hologram can be categorized as follows:

I)    Recording the amplitude and phase of the scattered field.
II)   Recording the square of the sum of the object and reference beams.
III)  Simulating the phase distribution by modulating the positions of small holes made on a mask.
IV)   Simulating phase and amplitude distributions by modulating both the position and the size of the holes made on the mask.
V)    Using holes with equal spacing.
VI)   Using the computer to make sketches from different viewing angles.

Method I makes use of the calculated result of $t_3(x,y)$ of (8.3), which can be expressed by the complex number

$$t_3(x,y) = A(x,y)\,e^{j\phi(x,y)}\,.$$

The phase and amplitude information are recorded on separate photographic plates. The reconstructed image is obtained by transmitting through the two photographic plates.

The amplitude $A(x,y)$ can be easily recorded by modulating the transmittance of the photographic plate, but the recording of $\phi(x,y)$ is somewhat more difficult. One method of recording $\phi(x,y)$ is through bleaching of a photographic plate. When a photographic plate is bleached, the index of refraction of the plate is changed according to the darkness of the plate, and one can empirically relate the transmittance before the bleaching with the phase after the bleaching. As a matter of fact, by illuminating just one plate modulated by $\phi(x,y)$, the viewer can see the reconstructed image. This phase only hologram has a special name and is called a *kinoform*.

Method II is based on the same principle as that of a normal hologram. The pattern of $|R(x,y)+O(x,y)|^2$ is calculated by a computer, and its pattern is painted on a sheet of paper. The darkness of the paint is modulated according to the computed field intensity. Fabrication of the hologram is completed by photographically reducing the size of the drawn pattern. The image is reconstructed by illumination from a laser beam.

The reconstructed image obtained by the Method II will appear fainter compared with that by the Method I, but Method II is simpler because it uses only one photographic plate. Both Methods I and II use a laser beam to reconstruct the image from the computed holograms. For either microwave or acoustic holography, it is possible to use the computer itself to generate the reconstructed image by coupling the measured field intensity directly to the computer.

Method III will now be explained. In this method, the desired phase distribution is approximated by properly arranging the positions of holes on a mask. It eliminates the need to draw patterns as in Methods I and II. All that is needed is punching holes of equal size at the calculated positions. Figure 8.8 illustrates the principle of this method. As shown in Fig. 8.8 a, an opaque mask is placed at $z = 0$. A light beam illuminates the mask at an angle $\theta$. The distribution of the light field on the surface of the mask is

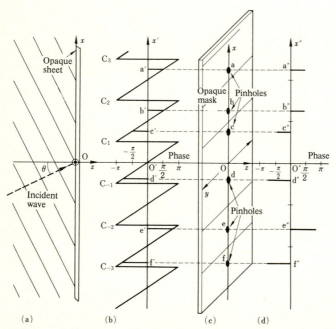

**Fig. 8.8a−d.** Principle of a computer-generated hologram made by modulating the position of the array holes. (**a**) Incident wave. (**b**) Phase distribution of incident light. (**c**) Array of holes. (**d**) Phase of the point sources

$$A \, e^{jkx\sin\theta} \tag{8.20}$$

Hence the phase distribution along the $x$ axis is similar to Fig. 8.8 b. With an increase in $x$, the phase periodically changes between $-\pi$ and $\pi$.

By selecting the locations of the holes, the field with the desired phase can be sampled. If a hole is placed at 1/4 of a period away from the origin, the resulting field has a phase of $\pi/2$ radians with respect to the field at the origin. If one hole is made in each period, labelled as a, b, c, ..., f in Fig. 8.8c, an array of point sources is generated whose phases are indicated by a″, b″, c″, ..., f″ in Fig. 8.8 d. A hologram made in this manner is a kind of kinoform.

Method IV is a revised version of Method III, and records not only the phase but also the amplitude distribution of the object beam. The hologram is made in a manner similar to that shown in Fig. 8.8c except that the size of the holes is made proportional to the field amplitude.

With Methods III and IV the positions of the holes were varied according to the phase, but with Method V the holes are made up of groups of four equally spaced holes. Since the spacing between the holes is fixed, one parameter is reduced. Fig. 8.9 illustrates this method. A plane wave is incident upon a mask at an angle $\theta$ as in the previous case. The phase distribution on the mask is periodic, as shown in Fig. 8.9b. In each period, four

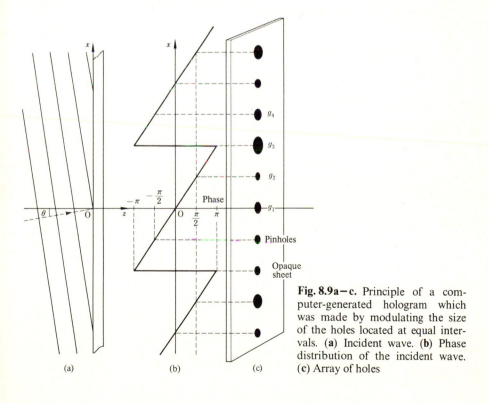

**Fig. 8.9a−c.** Principle of a computer-generated hologram which was made by modulating the size of the holes located at equal intervals. (**a**) Incident wave. (**b**) Phase distribution of the incident wave. (**c**) Array of holes

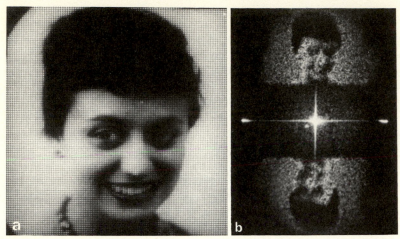

**Fig. 8.10.** (a) Original and (b) reconstructed images from a computer generated hologram of equally spaced holes, modulating the size of the holes [8.7]

holes are made with equal spacing. The sizes of the four holes are chosen so that the amplitude of the transmitted light from these four holes is proportional to certain calculated values, indicated by $g_1$, $g_2$, $g_3$ and $g_4$ in Fig. 8.9. The phase of the light from each hole is 0, $\pi/2$, $\pi$ and $3\pi/2$, respectively, where the phase at the origin is assumed to be zero radian. The resultant light amplitude from the four holes is

$$g = (g_1 - g_3) + j(g_2 - g_4) . \tag{8.21}$$

Thus, by controlling the sizes of the four holes, an arbitrary complex number can be represented. A hologram can be made by recording the object beam $O(x, y)$ in this manner.

Figure 8.10a is the original picture used as the object, while Fig. 8.10b is the reconstructed image of the hologram made by this method. The picture was first divided into $128 \times 128$ picture elements and then the intensity of each element was fed into a computer in order to calculate the object beam $O(x, y)$. From the calculated value of $O(x, y)$, the holes were made in the manner described in Method V.

The brightness of the reconstructed image of a hologram made by such Methods as III, IV and V depends on the number of holes made. In general, the number of holes needed to make a clear reconstructed image is very large and the amount of computation is also quite large. For instance, to obtain the quality of the reconstructed image shown in Fig. 8.10b, a memory of 131,072 words is necessary, which puts a sizable demand on one computer. One way around this is Method VI.

In method VI, the diffraction pattern is obtained by using light, and only a sketch of the object viewed from different angles is made by computer.

The computer time needed for this method is less than 1/1000 of that of the previous method. The principle is based on the fact that *when one sees the reconstructed image only a limited portion of the hologram is being used.* In another words, referring to Fig. 8.11, when the viewer's eyes are at a point $A$, the shape of the reconstructed image is more or less determined by the portion of the hologram indicated by $a-a'$. When the viewer moves his eyes to a new location $B$, the shape of the image is formed only by the portion of the hologram indicated by $b-b'$. This method makes use of the fact that only a small part of the hologram is viewed, once the position of the viewer's eyes is set.

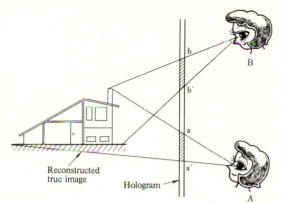

Reconstructed true image

Hologram

**Fig. 8.11.** The portions which contribute to reconstruction of the image

The computer makes up a sketch of the object, as viewed from the point $A$. It is then drawn in one frame of a film. Once a sketch viewed from the point $A$ is made, it is easy to draw a sketch viewed from $B$, because the computation involved is only the rotation of the coordinate system. Similarly, the sketches viewed from slightly different angles $C$, $D$, $E$, ... are made and are sequentially recorded on a roll of film. As shown in Fig. 8.12, the laser beam illuminating from behind each frame of the film is the object beam to a small section of the hologram. Each section of the hologram made corresponds to a particular sequence of the viewing angle. The same reference beam is used for each section of the hologram. When the completed composite hologram is illuminated by a reconstructing beam, and the viewer scans through the reconstructed images with sequentially different viewing angles, the result is an illusion of a three-dimensional object. Figure 8.13 shows a photograph of the reconstructed image from a hologram made in such a manner. The object is a bundle of randomly oriented wires.

In this method, a computer was used for making sketches but this is not always necessary. A series of normal photographs taken from sequentially different viewing angles using white light can substitute for the computer. The rest of the procedure is the same. One short-coming of method VI is that the quality of the reconstructed image is degraded as the location of the viewer's eye deviates from the predetermined position.

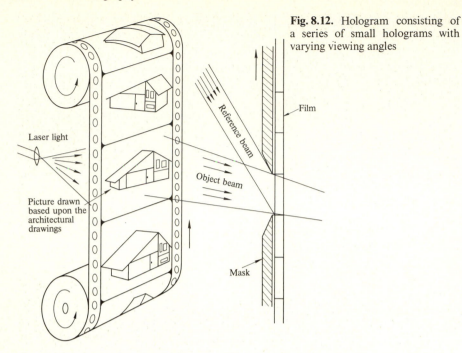

Film

Laser light

Reference beam

Object beam

Picture drawn based upon the architectural drawings

Mask

## 8.6 White-Light Hologram

A hologram whose image can be reconstructed by white light is called a *white-light hologram* [8.10, 11]. The types of white-light holograms are the *reflection hologram,* the *image hologram* and the *rainbow hologram.*

There are two basic mechanisms in forming an image out of a hologram. In one mechanism, the *spacings between the platelets* shown in Fig. 8.6 are modulated such that the interference of their scattered waves forms an image. Such a mechanism depends very critically upon the wavelength of the light. The other mechanism is to make the *surface shape of the platelet* in such a way that the reflected beam forms an image. The platelets serve as mirror surfaces, producing characteristics which are not a sensitive function of the wavelength of the light.

The reflection-type white-light hologram uses the latter mechanism as a means of image formation. The required condition is that the area of each platelet be made large while the number of platelets remains small. Such a condition can be satisfied by making the angle between the object and reference beams large when the hologram is fabricated. In producing the reflection hologram, the object beam illuminates the front of the hologram while the reference beam illuminates the back (Fig. 8.14). In reconstructing the image by white light, the mirror-like platelets form the same image regardless of the wavelength of the light.

**Fig. 8.13.** Reconstructed image from a hologram fabricated at the Bell Telephone Laboratories using the computer only for generating the drawings with varied viewing angles of the same object

Next, the image-type white-light hologram is described. When the principle of holography was first explained using Fig. 8.1, the wavelengths for producing the hologram and for reconstructing the image were the same and the image was formed exactly at the same place as the original object was. If, however, the image were reconstructed using light with a spread in wavelengths, the angle of diffraction from the fringe patterns in the hologram would also have a spread. The angle of diffraction of the shorter wavelengths would be smaller than that of the longer wavelengths. The blur

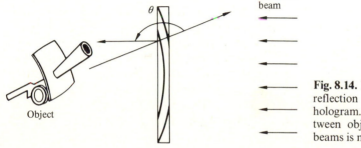

**Fig. 8.14.** Fabrication of the reflection type white light hologram. The angle $\theta$ between object and reference beams is made large

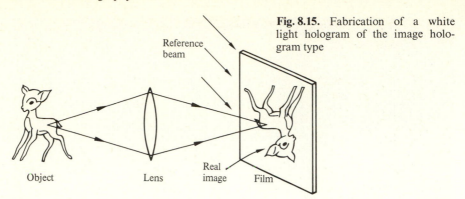

**Fig. 8.15.** Fabrication of a white light hologram of the image hologram type

Reference beam

Object    Lens    Real image    Film

of the reconstructed image resulting from the spread in diffraction angles can be minimized by choosing an image location as close to the hologram as possible. Figure 8.15 shows the arrangement for making a white-light hologram of this type. The real image of the object is formed right onto the film so as to minimize the distance between the image and the hologram.

The last white-light hologram to be mentioned is the rainbow hologram. By projecting a real image of a thin section of a hologram, the overlap of the images due to the spread of wavelengths is removed.

The master hologram is first made in the usual manner as arranged in Fig. 8.16a. The reference beam $R_1$ and object beam O are recorded in the master hologram $H_1$. Next, only a small horizontal strip of the hologram $H_1$ is illuminated by a reconstructing beam so as to form the real image (Fig. 8.16b). The reconstructing beam is actually $R_1^*$ which is the conjugate of $R_1$ in Fig. 8.16a. The conjugate beam can be made by illuminating the hologram strip from the opposite side and converging the beam with lens $L_1$ to the point $S_1$ from which the reference beam initially originated.

The second hologram $H_2$ is made using this reconstructed real image from the horizontal strip of the hologram $H_1$ as if it were an object. A beam converging to point $S_4$ is used as the reference beam in forming the hologram $H_2$. Strickly speaking, the object of the hologram $H_2$ was not the real image of the hologram $H_1$, but the strip of the hologram $H_1$ itself. The hologram $H_2$ is the desired white-light hologram.

In the process of reconstruction of the image, a white-light source is placed at the point $S_4$ so as to illuminate the hologram by the conjugate beam of $R_2$. The reconstructed image from $H_2$ is the real image of the strip of $H_1$. If an observer's eyes are placed in front of this real image of the strip of $H_1$, it appears as if the strip of $H_1$ is physically in front of the eyes, and one can observe the image of the object.

Since the illuminating light is white light, a series of strips of $H_1$ of different colors are formed. The color of the image changes from red to violet as the observer lowers his eye position. This is the reason why this hologram is called a rainbow hologram.

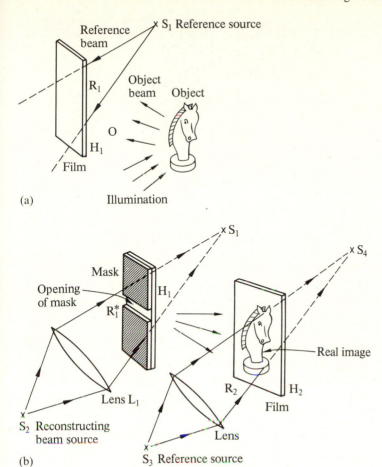

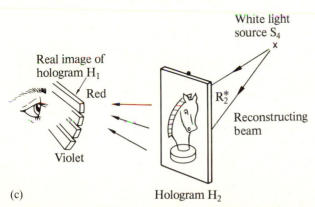

**Fig. 8.16.** Procedure for fabricating a rainbow hologram. **(a)** Fabrication of master hologram. **(b)** Second hologram $H_2$ is made from real image of $H_1$. **(c)** Reconstruction of image using a white light source

**Fig. 8.17.** Reconstructed image from the rainbow hologram. The hologram is displayed in an ordinary laboratory environment, showing the high visibility of the image. The illumination source is a small spotlight above and behind the hologram [8.11]

The width of the strip has to be made narrow enough to avoid interferences between the images of different colors. Making the strip horizontal sacrifices vertical parallax, while the horizontal parallax is least affected. To human eyes, the loss of vertical parallax is not as noticeable as the loss of horizontal parallax.

Figure 8.17 shows the reconstructed image from the rainbow hologram. Since the reconstructing beam illuminates the back of the hologram, and converges to a small area of the real image of hologram $H_1$, a very bright image is reconstructed.

## 8.7 Speckle Pattern

The reconstructed image of a hologram has granular texture while that of a white light hologram does not. As a matter of fact, any image formed using coherent light displays this granular texture or speckle pattern [8.12].

Illumination by two coherent plane waves generates a fringe pattern consisting of an array of bright and dark lines, as shown in Fig. 2.8. If there were two pairs of such coherent wave arranged with each set perpendicular to the other, dark lines in a mesh pattern would be generated. When many sources are present at arbitrary locations, the fringe pattern will contain several dark spots of irregular shapes. Wherever the vectorial sum of the total field is zero, a dark spot is generated. In the case of illumination by incoherent white light the situation is different, and the larger the number of sources, the more uniform in general the illumination becomes. The intensities of an incoherent source such as white light always add and there is no vectorial cancellation.

*Gabor* [8.13] classified the speckle patterns into two categories according to their origins. One is "objective speckle" which originates from inhomogeneity already present in the light illuminating the object. When the object is illuminated through a diffuser, as shown in Figs. 8.2 and 18a, a speckle pattern is projected onto the object. The hologram faithfully reconstructs the shape of the object as well as the projected speckle patterns. One way to cope with objective speckle is to avoid the use of a diffuser.

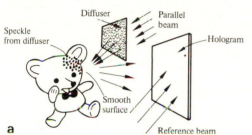

**Fig. 8.18a−c.** Causes of speckle patterns of a hologram. (**a**) Objective speckle due to a diffuser. (**b**) Objective speckle due to rough surface of the object. (**c**) Subjective speckle due to the finiteness of the pupil

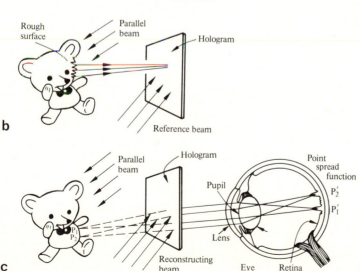

There is another type of objective speckle which is generated by the roughness of the surface of the object. This objective speckle is generated even when a parallel beam is used to illuminate the object. Unless the surface of the object is smooth to the order of the wavelength of light, the reflectivity of a spot on the object critically depends on the relative phases among the wavelets reflected from that point (Fig. 8.18 b).

The other kind of speckle is "subjective speckle" which arises when the detector is diffraction limited. For instance, when a human eye is used as the detector for the reconstructed image, the eye is diffraction limited by the size of the pupil. The images of $P_1$ and $P_2$ on the object projected to the retina have some spread (Fig. 8.18 c). The tails of the point-spread functions interfere with each other. This interference is seen as a speckle pattern to the observer.

A completely incoherent light, like white light, does not generate speckle patterns, but at the same time does not reconstruct an image of a hologram either, unless the hologram is a white-light hologram (Sect. 8.6). A partially coherent light is a compromise; it effectively reduces speckle patterns and yet reconstructs the image. The coherence of any given spectral line from a sodium arc lamp is lower than that of a laser. Such a source is good for observing the hologram with much less subjective speckle. Another method is the use of a laser light whose spatial coherence is reduced by an external means. A closely spaced pair of diffusers which keep moving with respect to each other is used as a means of reducing spatial coherence. A nematic liquid crystal gate whose minute scattering centers are constantly agitated by the applied voltage is also used for this purpose.

Although speckle is an annoying problem when reconstructing hologram images, speckle patterns are not all bad, and they can be put to good use. One such area of study, based on the special properties of speckle, is that of speckle interferometry [8.12, 14]. The basic principle of speckle interferometry is illustrated by referring to Fig. 8.19. The movement of a rough surface M is to be analyzed. As explained in Fig. 8.18 b, an optically rough surface generates objective speckle. The objective speckle pattern is photographed by a imaging lens (Fig. 8.19 a). If the illuminating parallel beam is uniform, then the speckle pattern generated on the surface moves with that surface.

A double exposure is made before and after the movement. In the double exposed photograph, each speckle spot is paired with its identical speckle spot (Fig. 8.19 b). Such a photographic plate $P_1$ is Fourier transformed by a Fourier transforming lens $L_2$ (Fig. 8.19 c) and the distance of the movement is obtained from the Fourier transformed image.

Let the distance of the movement be $2\xi$ and the entire speckle pattern is represented by $g(x, y)$. The Fourier transform of the double exposed plate $P_2$ is

$$F\{g(x - \xi, y) + g(x + \xi, y)\} = G(f_x, f_y)\{e^{-j2\pi f_x \xi} + e^{j2\pi f_x \xi}\}$$

$$= 2G(f_x, f_y)\cos 2\pi f_x \xi. \tag{8.22}$$

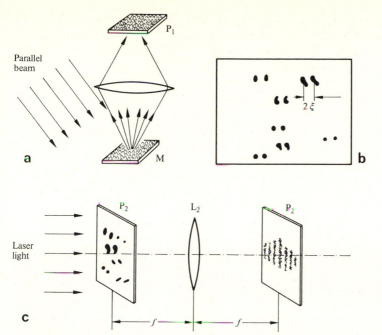

**Fig. 8.19a−c.** Principle of speckle interferometry. (**a**) Recording the speckle patterns. (**b**) Double exposure of the speckle pattern before and after movement. (**c**) Fourier transform analysis of the movement

The Fourier transformed image is modulated by $\cos 2\pi f_x \xi$ and from the pattern, $\xi$ is determined.

## 8.8 Applications of Holography

The principle of holography has been widely applied to various fields. The applications will be explained with an emphasis on the principle of operation. The application of holography to microwave devices and optical signal processing, however, will be treated in separate chapters.

### 8.8.1 Photographs with Enhanced Depth of Field

It is very difficult to take a microscopic photograph of a small moving object such as a swimming microbe or a small particle in Brownian motion, because the depth of field of a microscope is so shallow that it is almost impossible to follow the movement by manipulating its focus. Such a photograph can be taken rather easily by using a hologram [8.17].

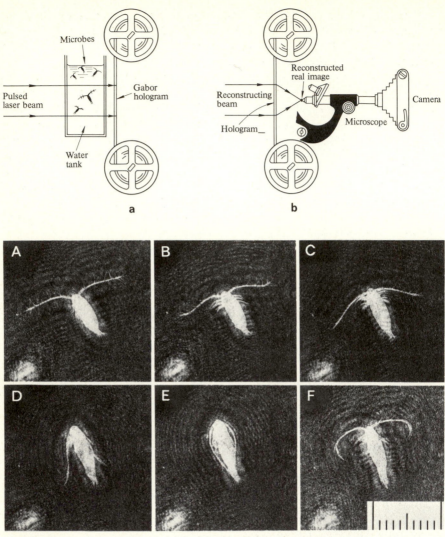

**Fig. 8.20a–c.** A series of reconstructed images to investigate the movement of the microbe. The enhanced field of view of the hologram was applied to make such pictures. (**a**) Fabrication of series of holograms. (**b**) Microscope is focused to a particular object in the series of holograms taken. (**c**) The movement of the microbe was photographed from the reconstructed image of the holograms taken at an interval of 1/70 seconds. The blurred spot on the lower left corner is another microbe and had the microscope been focused to this microbe, the microbe in the center would have been out of focus [8.17]

Holography does not use a lens but records the scattered field from near as well as far objects; once one hologram is made, one can focus the microscope to the reconstructed image of any particular object in a steady state of motion by adjusting the focus of the microscope.

Figure 8.20 illustrates a series of photographs showing the swimming motion of a microbe in a water tank. The procedure for taking such photographs is as follows. As shown in Fig. 8.20a, fine-grain holographic film is installed in a movie camera reel, and a series of holograms are made at short intervals using a pulsed laser. In this particular case, the camera was set at a speed of 70 frames per second. Next (Fig. 8.20b), the reconstructed image is made from one frame of the hologram at a time. In each frame of the reconstructed image, the focus of the microscope is always adjusted to one particular object to follow its movement. Figure 8.20c is a series of microscope pictures of the swimming motion of a microbe taken by this method.

### 8.8.2 High-Density Recording

For conventional high-density recording of real images such as microfiche [8.18], even a small dust particle on the film can create a missing portion on the record, and the missing information can never be recovered. Therefore, one has to be extremely careful about dust or scratches on the film. However, when using holograms for high density recording, *a scratch on the film will not destroy information* but merely causes a slight increase in the noise on the reconstructed image, so that no particular portion of the recording is lost.

The highest achievable density of the recording is theoretically the same whether recorded as a real image or as a hologram. Figure 8.21 is the

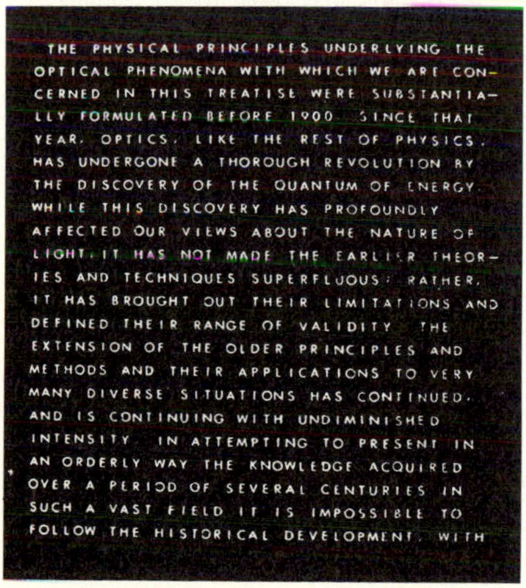

Fig. 8.21. The reconstructed image from a hologram 0.7 mm in diameter. The density of the recording is 2000 characters per mm$^2$ [8.19]

reconstructed image from a small hologram 0.7 mm in diameter containing 800 letters. A density of 2000 letters per square millimeter of hologram is achievable [8.19 – 21].

### 8.8.3 Optical Memory for a Computer

A black and white pattern like the one shown in Fig. 8.22 can be used as a means of recording the binary numbers of 0's and 1's. Figure 8.23 shows an example of an arrangement for using such a pattern as a read-only-memory of a computer. Figure 8.23 a illustrates the mechanism for picking a particular hologram from an array of small holograms. Figure 8.23 b is a photograph of the array of small holograms and Fig. 8.23 c shows an image reconstructed from one of the small holograms. The reconstructed image has $8 \times 8$ small sections and each small section is again divided into $8 \times 8$ picture elements. In this photograph, each section is the same and has one "0" bit represented by a black picture element in the center and 63 white picture elements each representing a "1" bit. Therefore, in this one sheet of hologram, $42 \times 10^6$ bits of information have been stored [8.22].

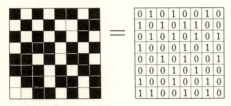

Fig. 8.22. The optical memory; the black and white dots represent "0" and "1", respectively

The selection of a particular hologram from the array of holograms is made by an $X-Y$ beam deflector, as shown in Fig. 8.23 a. The beam deflector used here has been described in Problem 4.2. The real image reconstructed from the selected hologram is projected onto a matrix of photo-transistors. The size of each photo-transistor is $100 \times 100 \ \mu m^2$ and 64 of them are arranged on one wafer (Fig. 8.23 d). In this example, 64 of these wafers are arranged in a square shape. The outputs from the photo-transistors are connected to the electronic circuits of the computer. A high-density recording is achieved through the use of the Fourier transform hologram described in Sect. 8.4.2.

It should be noted that the diffraction pattern of a periodic input like this binary coding mask is characterized by periodic peaks. The darkness of the film at the peak is the limiting factor of the information storage density. One method of raising the limit is to place the hologram slightly away from the focal point when the Fourier transform hologram is used for recording. A better method is to place a sheet of random phase shifter over the bit pattern when the hologram is fabricated. Because of the randomization of the phase, the peaks disappear, and the density limitation of the recording is raised [8.19, 20, 21].

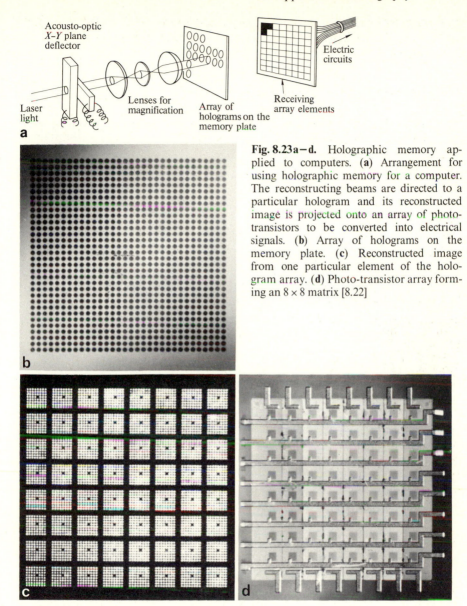

Fig. 8.23a−d. Holographic memory applied to computers. (a) Arrangement for using holographic memory for a computer. The reconstructing beams are directed to a particular hologram and its reconstructed image is projected onto an array of photo-transistors to be converted into electrical signals. (b) Array of holograms on the memory plate. (c) Reconstructed image from one particular element of the hologram array. (d) Photo-transistor array forming an 8 × 8 matrix [8.22]

### 8.8.4 Holographic Disk

The memory system mentioned in the previous subsection used a two-dimensional recording pattern, but the *holographic disk* described here uses a one-dimensional recording pattern with rotational movement of the hologram [8.23]. Figure 8.24 shows the arrangement of the holographic disk. An analogue electrical signal is first converted into a digital signal. The digital

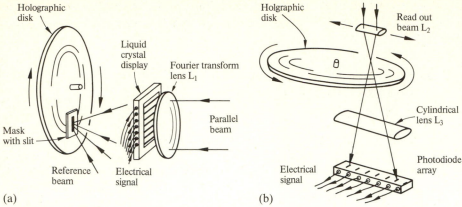

**Fig. 8.24 a, b.** Holographic disk. (**a**) Recording arrangement. (**b**) Playback arrangement

electrical signal is further converted into a one-dimensional dark-and-light pattern on a liquid-crystal display [8.24]. A Fourier transform hologram of the liquid-crystal display is recorded by lens $L_1$ onto the holographic disk (Fig. 8.24a). The mask is for limiting the size of the hologram. A series of holograms are recorded as the holographic disk is rotated.

When the signal is to be retrieved, the read-out beam is scanned over the rotating holographic disk. The cylindrical lens $L_3$ reconstructs the real image of the signal recorded in the holographic disk onto the photodiode array (Fig. 8.24b). The photodiode converts the light pattern back into an electrical signal output. The attractive feature of such a system is that the photodiode array is capable of handling a large number of bits simultaneously (parallel processing) and the recording or replaying rate can be made very high.

### 8.8.5 Laser Machining

A high-intensity laser beam with approximately 10 to 100 MW/cm$^{-2}$ can be easily obtained by focusing a laser beam with a lens. Such a high-intensity laser beam can be used for machining or welding. In many cases micro-machining which is difficult to achieve by mechanical tools can be achieved quite easily by laser beam machining [8.25].

In laser-beam machining, two methods are commonly employed. One method involves the use of a hologram and the other uses a Fresnel-zone plate. The hologram method performs the machining, or cutting, by illuminating the object to be machined with the high-intensity reconstructed real image from a hologram. The hologram is made such that the reconstructed image forms a desired pattern. However there is a limit on the laser intensity because the photographic plate is not heat resistive. The other method of machining uses a metallic Fresnel-zone plate to eliminate this limitation on the intensity of the illuminating laser beam, as already shown in Fig. 4.28.

### 8.8.6 Observation of Deformation by Means of an Interferometric Hologram

Holographic interferometry makes it possible to measure extremely small deformations in an object of the order of a wavelength of light [8.15, 16]. Basically, the method involves *observing the interference patterns resulting from superimposing the reconstructed images of an object in two different states.* This method has a wide range of applications, e.g., it is used to measure such minute changes as, heat expansion, chemical corrosion, plant growth, flow rate of gas or fluid, and detection of damage and fatigue for purposes of quality control.

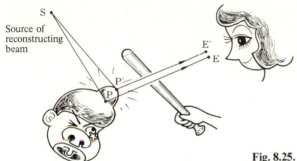

**Fig. 8.25.** Interferometric holography

Figure 8.25 shows the principle of this method. A double-exposure hologram is made of a head, before and after being hit. Not only the amplitude but also the phase information of the object is reproduced in the reconstructed image. If the difference between the optical paths SPE and SP'E' is one half the wavelength of the laser light, the two images add destructively so that the point becomes dark; if the path difference is one wavelength, they add constructively so that the point becomes bright. Thus the contour of the deformation of the object is obtained.

*In the case of normal interferometric measurement, two beams of light have to be present simultaneously at the same spot, but in holographic interferometry the measurement can be made even when there is a lapse of time between the first and the second exposures.*

Because of this property, it is possible to observe the temporal changes of an object such as a growing plant. Another advantage of holographic interferometry is that the three-dimensional pattern of the deformation can be observed. A disadvantage of holographic interferometry, however, is that the analysis becomes complicated if the deformation is a combination

**Fig. 8.26a, b.** A method of finding a repair area of a valuable art work by means of interferometric holography [8.26]. (**a**) Picture of Santa Caterina by Pier Francesco Fiorentino. (**b**) Interference of the reconstructed image of the holograms taken at room temperature and that taken at 40°C is used for detecting the area of the lifted paint

of rotation and translation because both of these create a similar interference pattern.

Real-time holographic interferometry can be produced by using the object itself to interfere with the reconstructed image of the hologram.

Figure 8.26 shows an example of using the interferometric hologram for the repair of old art work. If the art work is warmed by blowing warm air over the surface, the place where the paint is lifted expands more and the spots in need of repair can therefore be detected using this holographic method [8.26].

Interferometric holography is also used in the field of medicine. Deformation by force that takes place in such locations as foot and arm bones, teeth, or skulls has been studied [8.27]. The displacement contours measured by holographic interferometry are used to study pressure distributions.

Interferometric holograms are also widely used for product quality control. An interferometric pattern is made between the product and the reconstructed image from the hologram. Figure 8.27 shows the configuration for detecting a high spot on a mirror during surface polishing, by means of the interference between the reconstructed image of a computer generated hologram and the surface of the mirror [8.28].

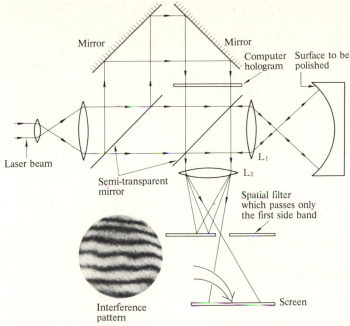

Mirror

Mirror

Computer hologram

Surface to be polished

$L_1$

$L_2$

Laser beam

Semi-transparent mirror

Spatial filter which passes only the first side band

Interference pattern

Screen

**Fig. 8.27.** A method of checking the degree of polish from the interference pattern between the reconstructed image from a computer hologram and that of the surface to be polished [8.28]

Holographic interferometry is very sensitive and can detect a displacement of a fraction of a light wavelength, but there is an upper limit to the amount of displacement that can be measured by this means [8.15, 16]. If the displacement becomes more than a few micrometers, the interference pattern becomes too fine for detection. The upper limit of speckle interferometry (Sect. 8.7), however, is much larger and is of the order of submillimeters.

### 8.8.7 Detection of the Difference Between Two Pictures

As mentioned in Sect. 8.2, the phase distribution of the reconstructed image is 180° out of phase with that of the object. This fact can be used to detect the difference between an object and the reconstructed image of a similar object. If the reconstructed image of the hologram of one picture is laid over another picture, the difference between the two pictures is obtained. Figure 8.28 shows an example of the difference obtained by this method. The difference between a picture with a jeep present and a picture without it was obtained. The presence of the jeep was detected [8.29].

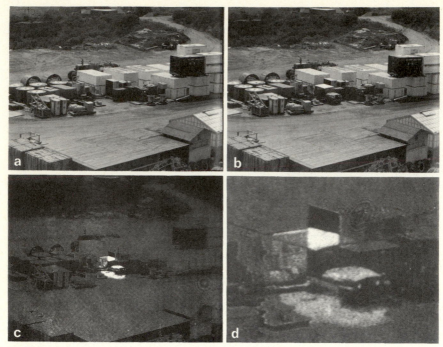

**Fig. 8.28a–d.** Example showing the difference between two similar pictures [8.29]. (**a**) Picture without a jeep. (**b**) Picture with a jeep. (**c**) Only the portion which is different between (a) and (b) is detected; a picture of a jeep. (**d**) Expanded picture of the jeep

### 8.8.8 Observation of a Vibrating Object

It is mandatory to minimize the movement of the object during the fabrication of a hologram. Since the spatial frequency of the fringe pattern to be recorded on a holographic plate is of the order of a wavelength, a movement of the hologram of the order of a wavelength of light will prevent the reconstructed image from being produced. This fact can be utilized for detecting minute vibrations of the surface of the object. The vibrating part becomes dark in the reconstructed image because no image is formed. In this manner, the standing-wave pattern of the vibration can be observed. Figure 8.29 is a pattern of the vibration of a guitar obtained by this method.

In addition, a variation of this method enables one to observe the object during the vibration. If the laser light is strobbed at the same frequency as that of the vibration, a hologram of the vibrating object with a specified phase angle between the vibration and the strobe light can be obtained. This method is used for measuring the dynamic characteristics of a structure [8.30].

**Fig. 8.29.** Picture of a vibration pattern of a guitar taken by holographic means [8.30]

### 8.8.9 Generation of Contour Lines of an Object

By using coherent-source interference patterns, contour lines of equal heigths can be made in several ways. The first method uses a spatial standing-wave pattern generated by two point sources. One way to attain this is to use a point source and its image made by a mirror. An alternative approach to creating a spatial standing wave pattern is through a double exposure hologram. The point source is then used as an object. One exposure made with the point source at one location is followed by another exposure with the point source slightly displaced. This makes use of a special property of a hologram, namely *the images of two exposures taken at different times can form an interference pattern.*

The second method utilizes the fact that the location of the reconstructed image shifts from the original location if the hologram is made by one laser wavelength and the image is reconstructed by another wavelength [8.31, 32]. The amount of the shift increases with the distance from the source. One exposure made by one laser wavelength is followed by another exposure made by a different laser wavelength. The image reconstructed by one of the laser wavelengths is a slightly displaced doubled image. The place where the displacement between the two images is an odd multiple of a half wavelength becomes dark, and the place where the displacement is an even multiple of a half wavelength becomes bright. Figure 8.30 shows the contour lines made by this method.

The third method which uses a single laser is a modification of the second method which requires the use of two lasers. In order to change the

**Fig. 8.30.** The contour lines generated by making a hologram with two wavelengths and reconstructing it with one of the two [8.32]

wavelength, the object is submerged in two gasses or fluids of different refractive indices. A double exposure is made first by placing the object in one medium, and then in the other. The reconstructed image has fringe lines indicating equidistance from the source.

# Problems

**8.1** In this chapter, the size of the hologram was considered infinite; find the expression for the reconstructed image when the hologram dimension in the $x$ direction is reduced to $a$. Assume that the object is one dimensional and is a function of only $x$, and assume that $\theta_r = \theta_p$.

**8.2** The reconstructing beam is incident along the $z$ axis. What is the change in the position of the reconstructed image when the plane of the hologram is rotated by an angle $\alpha$ from a position perpendicular to the $z$ axis?

**8.3** A hologram was made with a reference beam whose incident angle to the surface of the hologram was $\theta$. This hologram is illuminated by a reconstructing beam whose incident angle to the surface of the hologram is $-\theta$. Find the location of the conjugate image of the hologram.

**8.4**  Find how the reconstructed image changes when the orientation of the hologram is changed as shown in Fig. 8.31. The reconstructing beam is always incident along the $z$ axis.

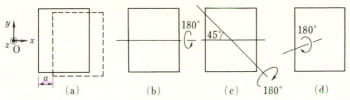

**Fig. 8.31.** Variety of orientations of a hologram

**8.5**  As shown in Fig. 8.32, a Fourier transform hologram is made by placing an object with dimensions $l_1 \times l_2$ at the front focal plane of the lens $L_1$, and a photographic plate at the back focal plane of the lens. The focal length of the lens $L_1$ is $f_1$. In order to reconstruct the image, the hologram is placed at the front focal plane of a lens $L_2$ with a focal length of $f_2$.

Find the location of the reconstructed image. The incident angle of the reference beam is $\theta$.

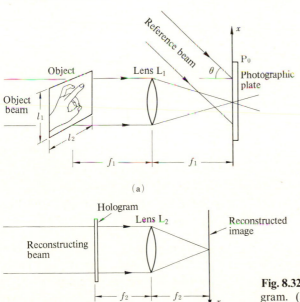

**Fig. 8.32a, b.** Fourier transform hologram. (**a**) Fabrication of hologram. (**b**) Construction of the image from the Fourier transform hologram

# 9. Laboratory Procedures for Fabricating Holograms

What may seem simple and easy to understand in theory can often be difficult to put into practice. While space does not permit a description of all the interesting experiments that could be performed to illustrate the theory of this book, the fabrication of a hologram has been chosen as a representative example. This section is intended as a step-by-step guide to the experiment, complete with several illustrative photographs.

## 9.1 Isolating the Work Area from Environmental Noise

A specially prepared bench has to be used to fabricate a hologram. Since details in the fringe pattern which is being recorded are of the order of an optical wavelength, extreme caution must be used to avoid vibration of the optical bench during the time of exposure.

The electrical equivalent circuit of the optical bench is a series resonance circuit. The noise reaches the surface of the bench through this resonance filter. The environmental noise spectrum is spread over a very wide range but the predominant component of the noise which reaches the bench top is at this resonance frequency. It is therefore desirable to make this resonance frequency as low as possible, preferably below the inverse of the photographic exposure time.

Figure 9.1 shows a drawing of the bench and its electrical equivalent circuit. Let the weight of the bench top be $M$, and the compliance of an air cushion between the stand and the bench top be $K$. The resonant frequency of the combination of the bench top and the air cushion is

$$f_0 = \frac{1}{2\pi \sqrt{MK}} . \tag{9.1}$$

In order to make $f_0$ small, the product of $M$ and $K$ has to be made large.

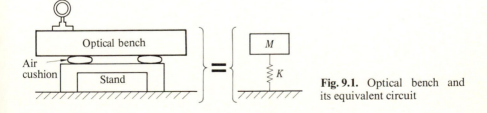

**Fig. 9.1.** Optical bench and its equivalent circuit

**Fig. 9.2.** Air is being pumped into air cushions which support the stainless steel optical bench

Speaking in qualitative terms, it is as if trying to shake a heavy mass through a soft air cushion. Likewise, environmental noise from the floor is not easily transmitted to the bench top. Figure 9.2 shows a photograph of a steel optical bench with four cushions at the corners. The air cushion is pumped up with a bicycle pump. Figure 9.3 illustrates various improvised optical benches. Figure 9.3a shows an ordinary bench consisting of braces across the legs of a table with a metal sheet on the top. The entire bench is placed over the inner tube of a tire. When the output power of the laser is large and hence the exposure time is short, even this simple arrangement is practical. When ease of transportation is an important consideration, the bench in Fig. 9.3b is often used. A metal sheet is placed over an air-cushioned sand box. The weight of the bench would, of course, depend on

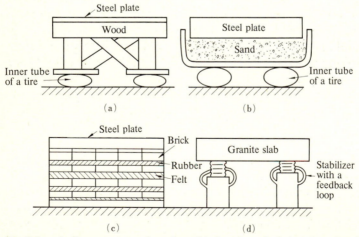

**Fig. 9.3a–d.** Construction of various kinds of optical benches for holography experiments

the amount of sand. The bench in Fig. 9.3c consists of a metal sheet, bricks and sheets of felt or rubber. This again is designed for easy transportation into the laboratory. Figure 9.3d shows a bench with special legs. The leg has a servo-loop which automatically absorbs any external vibration. A metal sheet or granite slab is placed over the legs.

It is necessary for all the bench tops to be smooth. Roughness in the surface could cause optical elements to rattle, or could result in a change of height as the elements are slid across the bench top, making fine adjustment very difficult.

## 9.2 Necessary Optical Elements for Fabricating Holograms

Requirements vary depending upon the purpose and the desired quality of the hologram, but essential optical elements are:

1) optical bench
2) laser
3) beam director
4) spatial filter

5) beam splitter
6) plate holder
7) film
8) front surface mirror

9) variable attenuator
10) light meter
11) lens
12) lens holder.

In the subsections to follow the individual items are examined closely.

### 9.2.1 Optical Bench

As mentioned in Sect. 9.1, select a bench with a resonant frequency as low as possible. The basement of the building is the best place to install optical benches but basements which are too close to subways, street car tracks or roads with heavy traffic should be avoided. One also must be aware of vibration which may come from water pumps or air ventilation ducts.

The bench should be carefully adjusted to a comfortable height in order to avoid undue strain on the experimenter. It is desirable to place the bench in the center of the room, rather than a corner, so that the bench is accessible from all sides.

### 9.2.2 Laser

If the output power of the laser is less than a few milliwatts, the exposure time becomes excessively long and very inconvenient.

Some types of lasers have ventilation holes in the external casing where light tends to leak out. A suitable cover or screen should be used to prevent this extraneous light from reaching the film. Those lasers in which ventilation takes place by heat conduction have no holes and do not need covers.

### 9.2.3 Beam Director

In some experimental arrangements, the output beam of the laser is not at the same height as the other optical elements. In these cases, a beam director is used to control the height of the beam. As shown in Fig. 9.4, the beam director consists of two small mirrors and is of similar structure to a periscope. Both the upper mirror $M_1$ and the lower mirror $M_2$ can be rotated with respect to the vertical support and consequently the direction and height of the light beam can be adjusted.

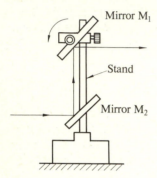

**Fig. 9.4.** Beam director

### 9.2.4 Spatial Filter

The spatial filter is comprised of a microscope objective and a pinhole. This is used for the purpose of filtering light whose wave front is disturbed due to scattering, reflection or some other cause. Figure 9.5 shows the structure of the spatial filter. The light beam passing through the microscope objective is focused onto the pinhole. If the size of the pinhole is selected such that only the main lobe of the Fourier transform of the beam can go through, then only the light ray with an angle of incidence parallel to the optical axis can pass through, and all other rays will be eliminated.

When a $\times N$ microscope objective is used, a pinhole with a diameter given by

$$D_p = 300 \, \frac{\lambda}{N} \, [\mu m] \tag{9.2}$$

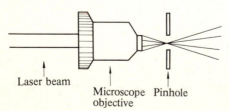

Laser beam     Microscope   Pinhole
               objective                    **Fig. 9.5.** Spatial filter

should be selected, where the unit of light wavelength $\lambda$ is micrometers. For instance, when a helium neon laser ($\lambda = 0.6328$ µm) is combined with × 10 lens, a hole 20 µm in diameter should be chosen.

### 9.2.5  Beam Splitter

A beam splitter is a sheet of glass which is partially transparent and partially reflecting, and is used for splitting the beam into two, the reference and object beams. Most convenient for hologram experiments is the beam splitter whose transmittance varies linearly along the glass length. This beam splitter allows easy adjustment of the ratio of the intensities of object and reference beams.

### 9.2.6  Photographic-Plate Holder

Since the photographic-plate holder must be manipulated in the dark, it should be simple and be able to hold the film firmly in the same position every time. The reason being that when a real time interferometric hologram is made, it is necessary to put the developed film back exactly in the original position.

### 9.2.7  Film

Since the details in the fringe pattern are of the order of the wavelength of light, high-resolution film is needed to record the hologram [9.1]. Specially made holographic film is available commercially with typical resolutions of 2000 to 2500 lines/mm, which should be sufficient for most experiments. Care should be taken to select film with a good spectral response at the wavelength of the laser light.

To minimize deformation of the film during the developing process, the following method may be used. The unexposed film is immersed in a water bath in a transparent developing tank. An exposure is taken, after which developer is added.

## 9.3  Photographic Illustration of the Experimental Procedures for Hologram Fabrication

To start with, the allocation of the optical elements is drawn on paper with the following points in mind [9.2, 3]:

1) The optical paths of the object and reference beams should be equal in length.

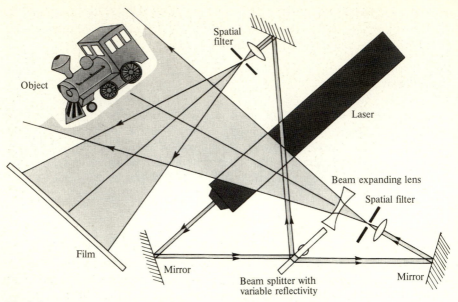

**Fig. 9.6.** Arrangement of optical components for hologram fabrication

2) The angle between the object and reference beams should be neither too large nor too small. Angles between 30° and 60° are adequate.
3) It is easier to manipulate the photographic plate in the dark when it is placed near the edge of the optical bench.
4) The viewing angle of the reconstructed image would be the same as that through a window of the same size and location as the particular hologram. Make sure that the scene viewed by looking through the film holder frame is the same as the scene desired to be recorded.
5) The laser does not necessarily have to be placed on the optical bench. It can be placed on a reasonably steady normal bench.

A possible arrangement of the optical elements is shown in Fig. 9.6.

After making sure the optical bench is level, start putting the optical components, biggest first, onto the optical bench, as shown in Fig. 9.7. Figure 9.8 is a photograph showing the initial stage of the arrangement of the optical components.

A photographic emulsion cross-section in thickness of the fringe pattern of a hologram looks like a venetian blind. When the distortion of the fringe pattern is critical, as in the case of an interferometric hologram, the incident angle $\theta_r$ of the reference beam to the film is made equal to the incident angle $\theta_o$ of the object beam to the film. Then the surface of each element of the "venetian blind" fringe pattern becomes perpendicular to the surface of the photographic emulsion.

**Fig. 9.7.** The largest item, usually the laser, is placed first on the optical bench

**Fig. 9.8.** Installation in progress; items from left to right are: photographic plate holder, beam director, laser, and mirror

A procedure for making $\theta_r$ equal to $\theta_o$ is shown in Fig. 9.9. First, any used photographic plate is attached temporarily onto the plate holder to act as a mirror. The plate is then rotated so that the reference beam reflected from the plate meets the object beam on the surface of a sheet of paper located where the object will be placed. When the two beams meet, the shaft of the plate holder is tightened, completing the adjustment. Figure 9.10 illustrates this procedure, showing the two beam spots on the sheet of paper. In Fig. 9.11, the sheet of paper is replaced by the object.

The distances travelled by the object and reference beams from the laser to the film should be made equal. As shown in Fig. 9.12, the distances are checked by a measuring tape. This is done so that the interference patterns on the photographic plate are formed with two light beams, each of which left the laser at the same time. By so doing, the requirement for temporal stability of the laser becomes less critical. If, however, the distance traversed by either reference or object beam is longer than the other, the beam travelling the longer distance interferes with the other beam which has left the laser later and travelled the shorter distance. The allowable difference in the distances depends upon the stability of the laser used, but

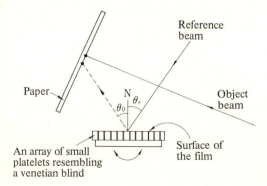

**Fig. 9.9.** Orientation which minimizes the distortion of the reconstructed image due to shrinkage during the development.

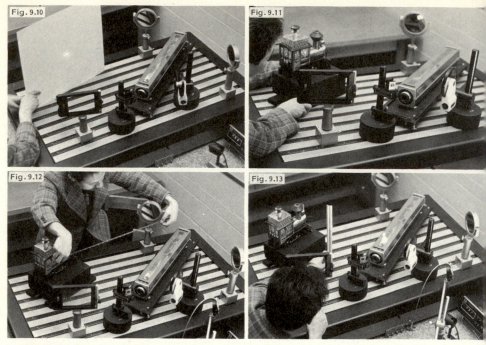

**Fig. 9.10.** The film holder containing a blank plate is rotated until the reflection of the reference beam hits the paper at the location where the object beam hits

**Fig. 9.11.** Place the object at the location determined by the method illustrated in Figs. 9.9, 10

**Fig. 9.12.** The total distance from the source to the photographic plate of the reference beam should be the same as that of the object beam

**Fig. 9.13.** The height of the beam from the bench should be made as constant as possible everywhere

it is desirable to be within a few centimeters and at most not more than a couple of tens of centimeters.

Next, it is necessary to keep the beam at the same height from the surface of the optical bench everywhere (Fig. 9.13). Almost all optical elements are designed under the assumption that the optical axis is parallel to the surface of the bench. Thus, if the beam is not perfectly horizontal, adjusting the elements becomes troublesome.

For optimum illumination of object and film plate, the reference and object beams are expanded by placing spatial filters in the beam paths (Fig. 9.14). If the beam is expanded more than necessary, stray light scattered either from the surface of the bench or supports for the optical elements becomes intolerable and the quality of the hologram suffers. The size of the expanded beam can be adjusted by properly locating the spatial filter.

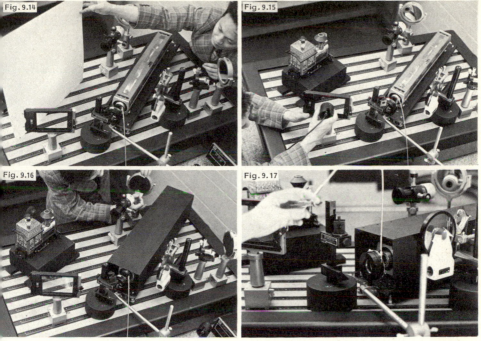

**Fig. 9.14.** The spatial filter is used not only for producing a uniform wavefront but also for expanding the diameter of the beam

**Fig. 9.15.** Measurement of the ratio of the intensity of the reference beam to the object beam

**Fig. 9.16.** Unused parts of mirrors are covered with black paper to minimize undesirable reflections. If the laser has air ventilation holes, they should be covered to eliminate stray light

**Fig. 9.17.** The shutter is installed so that there is no direct physical contact with the optical bench during the plate exposure

Since the wave front of the reference wave must be undisturbed, the spatial filter for the reference beam is inserted in the path just before the photographic plate. Optical elements such as mirrors should not be inserted between the spatial filter and the photographic plate because they will most certainly disturb the wave front. This, however, is not applicable to the object beam since the object beam is to be scattered by the object before it reaches the photographic plate. If the illumination is too strong and sharp shadows form behind the object, illuminate the object through a piece of ground glass. Moreover, the ground glass can be an alternative to a spatial filter.

The next step is to measure the ratio of the intensities of the reference beam to the object beam on the surface of the photographic plate. When the path of the incident beam is not perpendicular to the surface of the photographic plate, the power at normal incidence is first measured and its result is multiplied by the cosine of the incident angle. This must be done because most photometers give correct values only when the beam is at normal incidence. In Fig. 9.15, the light power is being measured at its normal incidence.

The standing wave fringe pattern peak-to-valley ratio of the light intensity becomes maximum, and infinity, when the amplitudes of the object and reference waves are made equal. In this case, the light intensity at the minimum becomes zero, but the characteristic curve expressing the relationship between exposure time and transmittance of the film is very nonlinear near zero exposure. Therefore the ratio of the intensities of the reference and object beam should be something other than unity. The intensity of the reference beam is normally taken 4 to 8 times that of the object beam.

Stray light must be avoided as much as possible. A means of doing so is shown in Fig. 9.16. A black sheet of paper is pasted over the unused portion of the mirrors and a black optical fence is set up to intercept the stray light. If there is any leakage of stray light from the laser, such as on those types with ventilation holes, a cover should be made.

The shutter should not be installed directly onto the optical bench. As shown in Fig. 9.17, its support should be outside the bench so that no vibration propagates into the bench when it clicks. It is important to ensure that the laser beam reflected from the shutter diaphragm in its closed state does not illuminate unexpected areas.

## 9.4 Exposure Time

The types of safety lamps that can be used in a dark room depend upon the type of film which is used. Safety lamps are not recommended when Kodak 649F or Agfa 8E75 are used, because these films are sensitive over a wide range of the visible light spectrum.

When a photographic plate is installed onto a plate holder, the emulsion side should be directed toward the incoming laser light. The emulsion side of a glass plate can be determined by rubbing the surface with a moistened finger. The side which is more viscous is the emulsion side. In films, the convention is to indicate the emulsion side by knotches. When the knotches are positioned at the bottom right-hand edge, the emulsion side faces the experimenter, as illustrated in Fig. 9.18.

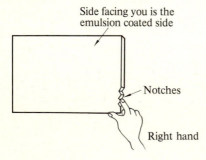

Side facing you is the emulsion coated side

Notches

Right hand

**Fig. 9.18.** Method for determining the emulsion side

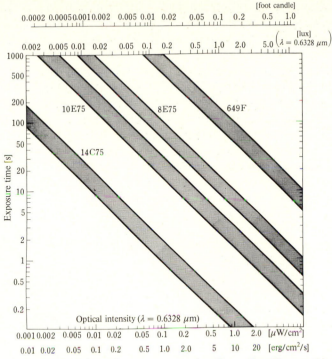

**Fig. 9.19.** Chart for optimum exposure time for various films used in holography

In the process of installing either a photographic plate of a film onto the plate holder, vibrations may be induced in the optical bench and optical components. It is best to wait a short while before making an exposure so that these vibrations can die down.

Figure 9.19 can be used to estimate the exposure time for a number of commonly used films. Once a rough estimate of the exposure time is made, the method shown in Fig. 9.20 can be used to refine this estimate. In Fig. 9.20, the exposure time is determined by using a single photographic

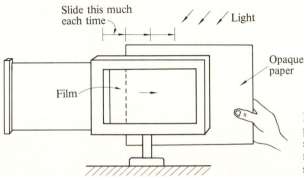

**Fig. 9.20.** By sliding the black paper after each exposure, several test exposures can be made using one photographic sheet

plate and sequentially intercepting portions of the photographic plate by a black sheet of paper. The amount of interception is decreased stepwise after each trial exposure. Three to five test exposures can be made with one photographic plate. At no time during these exposures should the sheet of paper, or the hand holding the sheet of paper, touch any part of the optical bench.

A rule of thumb governing the darkness of the developed photographic plate is that the fine print of a newspaper shoud be legible through it. Care should be taken that the hologram is not too dark.

## 9.5  Dark-Room Procedures

The procedures for developing a holographic plate are similar to those for fine-grain film [9.4]. The photographic plate is developed by immersing it into chemicals in the following order:

1) Kodak D-165 developer for 3 ~ 5 min
2) stop bath            30 s ~ 1 min
3) fixer bath           4 ~ 7 min
4) hypo eliminator      2 min
5) rinsing by water     30 min ~ 1 h
6) Kodak Photo-Flo      30 s ~ 1 min
7) drying               30 min.

### 9.5.1  Developing

The photo-sensitive material used for film is a silver halide compound. When the exposed film is put into the developer, only the silver halide in the exposed portion is hydrated and changed into silver grains, forming the image. The developer does not react with the silver halide in the unexposed portion.

A fine-grain developer is used for developing holographic films. There are various kinds of developers but it is safer to use the kind recommended by the film manufacturer. When preparing a developer, first warm half of the required quantity of water to around 50 °C so that it will be able to dissolve the chemicals more effectively. The rest of the water is added cold to bring the final temperature to as close to 20 °C as possible. One should be aware that some chemicals do not dissolve if the order of mixing is not adhered to.

The developing time is influenced by temperature. The time is shortened by 30 seconds with a 1 °C increase in the developer temperature. With an increase in the developing time, the value of the slope (called $\gamma$) of the photographic density versus exposure time curve is increased, as shown in Fig. 9.21.

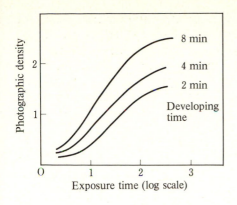

**Fig. 9.21.** The $\gamma$ value of the film decreases with an increase in the developing time

Photographic density

8 min

4 min

2 min

Developing time

Exposure time (log scale)

In order to prevent uneven development it is necessary to agitate the developer. For the first few second it is necessary to agitate it as thoroughly as possible.

The developer is soon oxidized by the oxygen in the air and becomes ineffective. It is, therefore, desirable to use new developer every time a new plate is developed.

### 9.5.2 Stop Bath

The developer is alkali and chemical reaction stops when it becomes even slightly acid. The stop bath, as its name implies, stops the development of the film, and consists primarily of acetic acid. Rinsing with water can be substituted for the stop bath, but the stop bath halts the developing action more evenly. In addition to putting an end to the developing action, the stop bath can prevent the gelatin of the film from swelling up, as well as prolonging the life of the fixer, which is also acid.

### 9.5.3 Fixer

The silver halide in the unexposed portion is not affected by the developer. The fixer converts the unaffected silver halides into soluble acid. It does not react with the developed silver grain. The major ingredient of fixer is sodium thiosulfate $Na_2S_2O_3 \cdot 5\,(H_2O)$, or ammonium thiosulfate $((NH_4)_2S_2O_3)$. It is sometimes called hypo.

When the film is transferred into the fixer, the unexposed silver portion first changes into a transparent substance, and then into a soluble substance. The fixing process should be continued until the silver halide has been transformed completely into soluble substance. The total fixing time is approximately twice the time required for the film to become transparent.

The time required to become transparent can be used as a criterion to determine the degree of exhaustion of the fixer. When the time required to reach the transparent stage becomes twice as long as the fresh fixer, the fixer is considered to have reached the exhaustion point and should be replaced by fresh fixer.

### 9.5.4 Water Rinsing

The fixer should be washed off by immersing it in running water at $18-20\,°C$ for 30 minutes. The rate of water flow should be such that the water is completely replaced every 5 minutes. If the film is immersed in hypo eliminator for two minutes, the water rinsing time can be cut in half.

### 9.5.5 Drying

After removing streaks of water from the photographic plate with a soft squeegee, it is suspended by a corner and dried. It is not good to attempt to accelerate the drying by directly blowing hot air because the gelatin may distort its shape. If the film is immersed in Kodak Photo-Flo for 30 seconds to 1 minute it is easier to squeeze out the water.

### 9.5.6 Bleaching

A phase hologram can be fabricated by bleaching after rinsing the fixed hologram [9.5]. The bleach converts the transmittance variation of the fringe pattern into phase variation. The brightness of the reconstructed image from the phase hologram is much greater than that of the unbleached hologram.

When using Kodak chromium intensifier bleach, immerse the hologram in $A$ solution and $B$ solution, wash for 5 to 10 minutes, and then immerse in a copper chloride $(CuCl_2)$ solution for about 5 minutes. The copper chloride solution is prepared by dissolving 5 grams of copper chloride in 1 litre of water. The fabrication of the hologram is complete after drip drying.

## 9.6 Viewing the Hologram

Searching for the location of the reconstructed image can be facilitated by placing the developed hologram back into the original film holder, and illuminating it with the same reference beam used during the fabrication.

Look through the hologram, and the reconstructed image will appear right where the object was originally situated. It is such a thrilling experience to take a glimpse of your first hologram. Optics is indeed light work!

# 10. Analysis of the Optical System in the Spatial Frequency Domain

In the field of electrical engineering, analysis can be divided into two broad categories that are related by Fourier transforms, namely, analysis in the time domain and analysis in the frequency domain. If phenomena are known in one domain, those in the other domain can be found by means of an appropriate transformation.

Similarly, in the field of optics, *analysis can be carried out either in the space domain or in the spatial frequency domain.* By knowing one, the other can be found by using a Fourier transform relationship. Previous chapters have dealt with problem solving in optics from the space domain viewpoint. In this chapter, treatment in the spatial frequency domain is introduced.

## 10.1 Transfer Function for Coherent Light

Frequency response is widely used to characterize the quality of electrical products such as video recorders, microphones and audio amplifiers. Similarly, spatial frequency response is used to characterize imaging systems such as cameras, microscopes, telescopes, copying machines, and facsimiles.

### 10.1.1 Impulse Response Function

*The expression for the output from a system to which an impulse (delta function) signal is applied is called the impulse response function of that system.* The impulse response function not only specifies the quality of the system, but also provides an expression for the output of the system for an arbitrary input signal.

Only an ideal imaging system can reproduce an infinitesimally small point as an infinitesimally small point. Hence, the impulse response function is a delta function only if the system is ideal. The impulse response functions of all non-ideal systems will always have finite widths. Generally speaking, the system quality degrades with increasing response width. As a matter of fact, the width of the impulse response function is used as a system quality criterion.

The Fourier-transform spectrum of the delta function spreads across the entire Fourier transform domain with a constant magnitude. If the system is

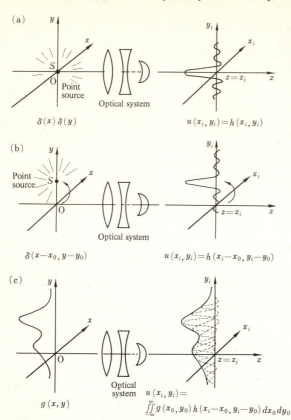

**Fig. 10.1 a – c.** Illustration of the impulse response function $h(x, y)$. **(a)** Point source at origin. **(b)** Point source at $(x_0, y_0)$. **(c)** Spatially spread source $g(x_0, y_0)$

incapable of responding to the entire range of spatial frequencies, the delta function can never be reproduced by the system.

It is therefore pertinent to consider the response of an imaging system to a point source input. First, a point source delta function described by $\delta(x_0) \delta(y_0)$ is placed at the origin of the input coordinates, as shown in Fig. 10.1 a. Let the impulse response function of this imaging system be $h(x_i, y_i)$. Then, the delta function source is moved from the origin to a new point $(x_0, y_0)$. Assuming a system magnification of unity, the position of the output image should move by the same amount. For simplicity, the sense of the output coordinates is taken such that the magnification is always plus one. The moved image is then expressed by

$$u(x_i, y_i) = h(x_i - x_0, y_i - y_0) . \tag{10.1}$$

As a further step, the delta function source is now replaced by a distributed source $g$, as shown in Fig. 10.1 c. If the distributed source is considered as an ensemble of point sources, the amplitude of the constituent point source at a particular point $(x_0, y_0)$ is naturally $g(x_0, y_0)$. The output due to this constituent point source is then

$$g(x_0, y_0) \, h(x_i - x_0, y_i - y_0) \tag{10.2}$$

and the output image $u(x_i, y_i)$ due to all point sources comprising $g$ is

$$u(x_i, y_i) = \int\!\!\int_{-\infty}^{\infty} g(x_0, y_0) \, h(x_i - x_0, y_i - y_0) \, dx_0 \, dy_0 \,. \tag{10.3}$$

Since (10.3) is exactly in the form of a convolution, it can be rewritten as

$$u(x_i, y_i) = g(x_i, y_i) * h(x_i, y_i) \,. \tag{10.4}$$

*Once the impulse response function $h(x_i, y_i)$ is known, the output image for any arbitrary input function can be calculated by (10.4).* In this respect, the impulse response function is one of the most important functions used to characterize the system. However, it should be noted that (10.3) is true only when the ratio of the translation of the input source to that of the output image is always the same regardless of the location of the source, or in other words, only when the system is shift invariant. In actual optical systems, this condition is satisfied only in very limited regions of the input coordinates.

## 10.1.2 Coherent Transfer Function (CTF)

In the case of an audio amplifier, its frequency characteristics can be easily determined by finding the frequency spectrum distribution of the output by means of a spectrum analyzer (or Fourier transform) when a delta function signal is applied to the amplifier input. This is possible because the input delta function signal has a constant magnitude throughout the entire frequency spectrum. The same analogy is applicable to optical systems and the Fourier transform $H(f_x, f_y)$ of the impulse response function $h(x_0, y_0)$ specifies the system. In optics, $H(f_x, f_y)$ is called the coherent transfer function or CTF [10.1].

Because (10.4) expresses the output image in the form of a convolution, its Fourier transform is the product of the Fourier transforms of each of the functions being convolved. This fact is one of the advantages of analysis in the spatial frequency domain as opposed to that in the space domain. The Fourier transform of (10.4) is

$$U(f_x, f_y) = G(f_x, f_y) \, H(f_x, f_y) \qquad \text{or} \tag{10.5}$$

$$H(f_x, f_y) = \frac{U(f_x, f_y)}{G(f_x, f_y)} \,. \tag{10.6}$$

**Exercise 10.1** Using coherent light, a picture is taken by a camera according to the geometry shown in Fig. 10.2. Find the CTF of the camera

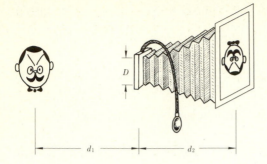

**Fig. 10.2.** Calculation of the coherent transfer function (CTF) of a camera

and determine the highest spatial frequency resolved by this camera when the diameter of its lens is $D$. For simplicity, treat the question as a one-dimensional problem. It is assumed that the camera is diffraction limited (Sect. 6.5.2).

*Solution*    The results obtained in Sect. 6.5.2 can be applied directly. Let the one-dimensional pupil function be $P(x) = \Pi(x/D)$. From (6.44) the light distribution $u(x_i)$ on the film is

$$u(x_i) = K g\left(-\frac{x_i}{M}\right) * \mathrm{sinc}\left(\frac{D}{\lambda \, d_2} x_i\right) \tag{10.7}$$

where the first factor of (6.44) was assumed constant and represented by $K$. Comparing the form of this formula with (10.4), the impulse response function $h(x_i)$ is

$$h(x_i) = K \, \mathrm{sinc}\left(\frac{D}{\lambda \, d_2} x_i\right).$$

Therefore, by Fourier transforming, the coherent transfer function $H(f_x)$ is obtained as

$$H(f_x) = K \frac{\lambda \, d_2}{D} \Pi\left(\frac{\lambda \, d_2 \, f_x}{D}\right). \tag{10.9}$$

The value of this equation is

$$H(f_x) = \begin{cases} K \dfrac{\lambda \, d_2}{D} & \text{for} \quad |f_x| \leq \dfrac{D}{2 \, \lambda \, d_2} \\[3mm] 0 & \text{for} \quad |f_x| > \dfrac{D}{2 \, \lambda \, d_2}. \end{cases} \tag{10.10}$$

From (10.10), the CTF has its cut-off at

$$f_{i_{\max}} = \frac{D}{2 \, \lambda \, d_2} \tag{10.11}$$

and $f_{i_{\max}}$ is the maximum resolvable spatial frequency of the output image.

The corresponding maximum resolvable spatial frequency $f_{0_{max}}$ of the input image is obtained by multiplying $f_{i_{max}}$ by the magnification ratio $d_1/d_2$, i.e.,

$$f_{0_{max}} = D/2\lambda\, d_1\ .$$

For the more general case of the two-dimensional pupil function $P(x, y)$, the coherent transfer function is obtained by Fourier transforming both sides of (6.45), so that

$$\bar{E}(f_x, f_y) = KG(-Mf_x, -Mf_y)\cdot P(\lambda\, d_2 f_x, \lambda\, d_2 f_y) \tag{10.12}$$

where the first factor of (6.45) was assumed constant. Aside from a constant factor, comparing (10.12) and (10.5) gives

$$H(f_x, f_y) = P(\lambda\, d_2 f_x, \lambda\, d_2 f_y)\ . \tag{10.13}$$

An interesting result has been obtained. *The coherent transfer function of an imaging system by a lens of finite aperture is obtained simply by replacing the variables of the pupil function by $\lambda\, d_2 f_x$ and $\lambda\, d_2 f_y$, aside from a constant factor.* This fact makes the calculation of the coherent transfer function of a diffraction-limited imaging system very simple.

## 10.2  Spatial Coherence and Temporal Coherence

Consider a fringe pattern made by Young's double pinhole, as shown in Figure 10.3a. If the phase of the field at one of the pinholes is perfectly coherent with that of the other, a fringe pattern of 100% modulation is observed on the screen. As the coherency between the two pinholes deteriorates, the contrast of the fringe pattern decreases. When the coherency is totally lost, the fringe pattern disappears. Thus Young's fringe pattern is very often used as a means to measure the coherency between two points in space [10.2].

To make a quantitative measurement of spatial coherency, the spacing between the pinholes is varied. When the spacing is increased, two things happen. One is an increase in the spatial frequency of the fringe pattern and the other is a decrease in the fringe contrast. The limit of the pinhole spacing beyond which the fringe pattern can no longer be observed is called the *spatial coherence width* and is used as a measure of the spatial coherence between the locations of the pinholes.

Next, Fig. 10.3b shows an arrangement for measuring the temporal coherence of the source. The light from the source is split into two rays. One ray goes straight to the photo-detector while the other ray goes through a delay path made of a movable prism before it reaches the photo-detector. As the prism is moved, the amount of delay in the prism path with respect to the straight path is varied. The output from the photo-detector goes through a series of maxima and minima with the movement of the prism, as shown in the graph in Fig. 10.3b.

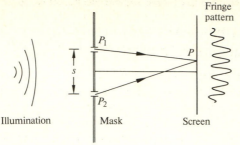

**Fig. 10.3 a, b.** Methods of measuring spatial and temporal coherence. (**a**) Young's fringe pattern for measuring spatial coherency. (**b**) Arrangements to measure temporal coherence

(a)

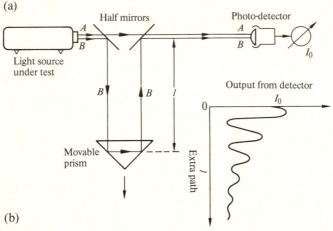

(b)

At any particular moment, these two rays incident upon the photo-detector will have left the light source at two different times. Consequently, the stability of the source is essential for the two rays to interfere with each other at the location of the photo-detector, thereby producing a modulation of the detector output with the movement of the prism. If the source is not stable i.e., not perfectly temporally coherent, the degree of the modulation of the output decreases as the prism is moved away until finally no modulation with the movement of the prism can be observed.

The furthest distance $l$ where the modulation with respect to the movement just disappears is used as a measure of the coherency of the source. This length $l$ is known as the *coherence length*. The difference $\Delta t$ in the departure times of the two rays at this limit of coherency is $\Delta t = l/c$, where $c$ is the speed of light, and $\Delta t$ is called the *coherence time*.

The arrangement just described for measuring temporal coherence is often used to measure the stability of a monochromatic source such as a laser. The coherence length of the laser source is an important consideration when fabricating a hologram. The difference between the total path of the reference beam from the laser to the film and the total path of the object beam from the laser to the film must be much less than the source coherence length in order to avoid a washing out of the fringe pattern on the hologram. In fact, the size of the field of view of the hologram is determined by $l$ obtained above.

## 10.3 Differences Between the Uses of Coherent and Incoherent Light

One often wonders what are the differences in the resolution of an imaging system when a coherent light source and incoherent light source is used. We will make a comparison in this section.

Let the output intensity distribution be $I(x_i, y_i)$ when a spatially distributed source $g(x_0, y_0, t)$ is applied as an input to the imaging system. The case of the spatially coherent source is precisely what we have so far dealt with and there is no problem. The output intensity distribution is

$$I(x_i, y_i) = |g(x_i, y_i) * h(x_i, y_i)|^2. \tag{10.14}$$

When the distributed source is spatially incoherent it is quite a different matter. The rigorous analysis is quite complicated. A simplified version of the analysis will be presented here. The output intensity is given by

$$I(x_i, y_i) = \langle u u^* \rangle \tag{10.15}$$

where $u$ is the light amplitude on the output plane, and $u^*$ is the complex conjugate of $u$ [10.1, 3]. The brackets $\langle \rangle$ represent the operation of averaging in time. The source is time dependent and is expressed by $g(x_0, y_0, t)$. Inserting (10.4) into (10.14), the intensity of the output image from the system due to the input field $g(x_0, y_0, t)$ is obtained as[1]

---

1 To be more exact, in the process of obtaining $u(x_i, y_i, t)$, one more convolution with respect to $t$ is necessary

$$u(x_i, y_i, t) = \int_{-\infty}^{t} \left[ \iint_A g(x_0, y_0, \tau) h(x_i - x_0, y_i - y_0, t - \tau) dx_0 dy_0 \right] d\tau. \tag{10.16}$$

In the case of a monochromatic (coherent) source, the time delay due to the system is simply expressed by one phase constant. In the incoherent case, the source is a arbitrary function of $t$, and the input is normally treated as a chain of impulses whose amplitudes follow the shape of the source function. The response of the system to each one of the impulses has its own trail (impulse response). In order to determine the output, the contributions of all the trails have to be summed up from $-\infty$ to the present time $t$.

Now, (10.16) is expressed in terms of the frequency spectrum:

$$u(x_i, y_i, t) = \int_{-\infty}^{\infty} \tilde{u}(x_i, y_i, f) e^{j 2\pi f t} df \tag{10.17}$$

where the symbol "~" means the Fourier transform with respect to time. If (10.17) is used in (10.16), one obtains

$$u(x_i, y_i, t) = e^{j 2\pi f_0 t} \int_{-\infty}^{\infty} \iint_A \tilde{g}(x_0, y_0, f) \tilde{h}(x_i - x_0, y_i - y_0, f) e^{j 2\pi (f - f_0) t} dx_0 dy_0 df \tag{10.18}$$

where $f_0$ is the center frequency of the spectrum.

It should be recognized that $\tilde{h}(x_i - x_0, y_i - y_0, f)$ is not necessarily frequency independent. (Continued to the footnote in the next page.)

$$I(x_i, y_i) = \left\langle \iint_A g(x_0, y_0, t) \, h(x_i - x_0, y_i - y_0) \, dx_0, dy_0 \right.$$

$$\left. \times \iint_A g^*(x_0', y_0', t) \, h^*(x_i - x_0', y_i - y_0') \, dx_0' \, dy_0' \right\rangle.$$

Since the choice of a small element area $dx_0 \, dy_0$ for integration is not related to the choice of $dx_0' \, dy_0'$, the area segments $dx_0 \, dy_0$ and $dx_0' \, dy_0'$ are independent, and the above integration can be rewritten as

$$I(x_i, y_i) = \left\langle \iint_A \iint_A h(x_i - x_0, y_i - y_0) \, h^*(x_i - x_0', y_i - y_0') \right.$$

$$\left. \times g(x_0, y_0, t) \, g^*(x_0', y_0', t) \, dx_0 \, dy_0 \, dx_0' \, dy_0' \right\rangle. \tag{10.20}$$

If it is assumed that the propagation medium is not subject to disturbances such as air turbulance or haze, and does not change with time, (10.20) can be rewritten as

$$I(x_i, y_i) = \iint_A \iint_A h(x_i - x_0, y_i - y_0) \, h^*(x_i - x_0', y_i - y_0')$$

$$\times \left\langle g(x_0, y_0, t) \, g^*(x_0', y_0', t) \right\rangle dx_0 \, dy_0 \, dx_0' \, dy_0'. \tag{10.21}$$

Note that $t$ designates the departure time from the source rather than the arrival time at the output plane.

In the incoherent case, the phase at one point of the input field has no relationship with that of another point. The time average of the integrand in (10.21) becomes zero, unless the two points are identically the same point, and the radiating source power per unit area is thus expressed by

$$\left\langle g(x_0, y_0, t) \, g^*(x_0', y_0', t) \right\rangle = |g(x_0, y_0)|^2 \, \delta(x_0 - x_0') \, \delta(y_0 - y_0'). \tag{10.22}$$

The value of $I(x_i, y_i)$ for the spatially incoherent case is calculated by inserting (10.22) into (10.21)

$$I(x_i, y_i) = \iint_A \iint_A h(x_i - x_0, y_i - y_0) \, h^*(x_i - x_0', y_i - y_0')$$

$$\times |g(x_0, y_0)|^2 \, \delta(x_0 - x_0') \, \delta(y_0 - y_0') \, dx_0' \, dy_0' \, dx_0 \, dy_0.$$

Integrating with respect to $x_0'$ and $y_0'$ gives

---

A spectrometer separates the output spectrum by the action of a grating or a prism whose $\tilde{h}(x_i - x_0, y_i - y_0, f)$ strongly depends upon the frequency. Even with a simple aperture diffraction, the transfer function depends upon the wavelength.

If a severe restriction is imposed of a narrow frequency spectrum and a narrow spread of the phase angle $(2\pi \Delta f/c) \, d \ll 1$, where $d$ is the maximum distance involved in the geometry, then (10.18) can be rewritten as

$$u(x_i, y_i, t) = \iint_A g(x_0, y_0, t) \, \tilde{h}(x_i - x_0, y_i - y_0) \, dx_0 \, dy_0 \tag{10.19}$$

where $f_0$ in $\tilde{h}(x_i - x_0, y_i - y_0, f_0)$ was suppressed.

$$I(x_i, y_i) = \iint_A h(x_i - x_0, y_i - y_0) \, h^*(x_i - x_0, y_i - y_0) \, |g(x_0, y_0)|^2 \, dx_0 \, dy_0$$

$$= \iint_A |h(x_i - x_0, y_i - y_0)|^2 \, |g(x_0, y_0)|^2 \, dx_0 \, dy_0$$

or using the convolution notation,

$$I(x_i, y_i) = I_0(x_i, y_i) * |h(x_i, y_i)|^2 \quad \text{where} \tag{10.23}$$

$$I_0(x_0, y_0) = |g(x_0, y_0)|^2.$$

*The difference between (10.14) for coherent light and (10.23) for incoherent light is the order in which the square modulus and the convolution occur.* For coherent light, the intensity is the square modulus of the convolution between the input function and the impulse response function, whereas for incoherent light, the intensity is the convolution between the square modulus of the input function and the square modulus of the impulse response function.

## 10.4  Transfer Function for Incoherent Light

Similar to the case of spatially coherent light, a transfer function is defined for spatially incoherent light and is used for characterizing the optical system. However, for incoherent light, the transfer function is defined in terms of the Fourier transform of the intensity rather than the Fourier transform of the amplitude. Fourier-transforming both sides of (10.23) gives

$$\mathscr{F}\{I(x_i, y_i)\} = \mathscr{F}\{|g(x_i, y_i)|^2\} \, \mathscr{F}\{|h(x_i, y_i)|^2\}. \tag{10.24}$$

Equation (10.24) corresponds to (10.5) of the coherent case.

Equation (10.24) is usually normalized with respect to the field intensity of the background light which is the component of light with $f_x = f_y = 0$. Incorporating this normalization, (10.24) is rewritten as

$$\mathscr{I}(f_x, f_y) = \mathscr{I}_0(f_x, f_y) \, \mathscr{H}(f_x, f_y) \quad \text{where} \tag{10.25}$$

$$\mathscr{I}(f_x, f_y) = \frac{\mathscr{F}\{I(x_i, y_i)\}}{\mathscr{F}\{I(x_i, y_i)\}_{f_x = f_y = 0}} \tag{10.26}$$

$$\mathscr{I}_0(f_x, f_y) = \frac{\mathscr{F}\{|g(x_i, y_i)|^2\}}{\mathscr{F}\{|g(x_i, y_i)|^2\}_{f_x = f_y = 0}} \tag{10.27}$$

$$\mathscr{H}(f_x, f_y) = \frac{\mathscr{F}\{|h(x_i, y_i)|^2\}}{\mathscr{F}\{|h(x_i, y_i)|^2\}_{f_x = f_y = 0}}. \tag{10.28}$$

The expression $\mathscr{H}(f_x, f_y)$ is called the *Optical Transfer Function* (OTF).

When the intensity $|g(x_i, y_i)|^2$ of the input light is a delta function, then $\mathscr{I}_0(f_x, f_y)$ becomes unity and $\mathscr{I}(f_x, f_y)$ becomes $\mathscr{H}(f_x, f_y)$. The optical transfer function is therefore the Fourier transform of the output intensity when an impulse intensity input is applied to the system.

Next, it will be demonstrated that the auto-correlation of the CTF is the OTF. The correlation operation, denoted by the symbol $\bigstar$, is defined as

$$X(f_x, f_y) \bigstar Y(f_x, f_y) = \iint\limits_{-\infty}^{\infty} X(x, y) \, Y^*(x - f_x, y - f_y) \, dx \, dy. \tag{10.29}$$

For the special case when $X$ and $Y$ are equal, this operation is called auto-correlation. The numerator of the OTF defined in (10.28) can be rewritten as

$$\mathscr{F}\{|h(x_i, y_i)|^2\} = \mathscr{F}\{h(x_i, y_i) \, h^*(x_i, y_i)\}$$
$$= H(f_x, f_y) * H^*(-f_x, -f_y). \tag{10.30}$$

Equation (10.30) can be further rewritten using the notation of the correlation operator as

$$\mathscr{F}\{|h(x_i, y_i)|^2\} = H(f_x, f_y) \bigstar H(f_x, f_y). \tag{10.31}$$

The denominator of (10.28) is

$$\mathscr{F}\{|h(x_i, y_i)|^2\}_{f_x = f_y = 0} = \iint\limits_{-\infty}^{\infty} H(x, y) \, H^*(x - f_x, y - f_y) \, dx \, dy \,\big|_{f_x = f_y = 0}$$
$$= \iint\limits_{-\infty}^{\infty} |H(x, y)|^2 \, dx \, dy. \tag{10.32}$$

Combining (10.28, 31, 32) gives

$$\mathscr{H}(f_x, f_y) = \frac{H(f_x, f_y) \bigstar H(f_x, f_y)}{\iint\limits_{-\infty}^{\infty} |H(x, y)|^2 \, dx \, dy}. \tag{10.33}$$

An example of how to use (10.33) follows.

**Exercise 10.2**   Find the OTF of the system described in Exercise 10.1 when spatially incoherent light is used in place of coherent light. Also, find the maximum resolvable spatial frequency that this camera can resolve, and compare this resolution with that of coherent light.

*Solution*   Inserting (10.9) into (10.33) gives

$$\mathscr{H}(f_x) = \frac{\left(K \dfrac{\lambda d_2}{D}\right)^2 \Pi\left(\dfrac{\lambda d_2 f_x}{D}\right) \bigstar \Pi\left(\dfrac{\lambda d_2 f_x}{D}\right)}{\displaystyle\int\limits_{-\infty}^{\infty} \left(K \dfrac{\lambda d_2}{D}\right)^2 \left\{\Pi\left(\dfrac{\lambda d_2 x}{D}\right)\right\}^2 dx}$$

and therefore,

$$\mathcal{H}(f_x) = \Lambda\left(\frac{\lambda \, d_2}{D} f_x\right).$$

A comparison between the result of this exercise and that of the previous one is made in Fig. 10.4. The cut-off spatial frequency of the incoherent light shown in Fig. 10.4b is twice as large as that of coherent light shown in Fig. 10.4a. Glancing at these figures for the CTF and OTF, it would appear that the resolution of the camera becomes twice as good when incoherent light is used, but this is not necessarily true. It must be kept in mind that, in order to reproduce an output image which resembles the input, the transfer function should have a bandwidth wide enough to cover the entire bandwidth of the spectrum of the input. The spectral bandwidth of the intensity of the input, in most cases, is twice as wide as that of the amplitude. Hence for the same resolution of the output, the OTF has to be twice as wide as the CTF. Unless the bandwidth of $\mathcal{H}(f_x, f_y)$ is more than twice that of $H(f_x, f_y)$, the resolution of the output image is more or less the same for either light.

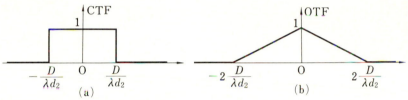

**Fig. 10.4 a, b.** Comparison between (**a**) the coherent transfer function and (**b**) optical transfer function for Exercises 10.1, 2 respectively

Next, the optical transfer function is obtained by a graphical method. Equation (10.13) which defines the coherent transfer function of a diffraction limited imaging system with pupil function $P(x, y)$ is combined with (10.29, 33) to express the optical transfer function. The result is

$$\mathcal{H}(f_x, f_y) = \frac{\int\limits_{-\infty}^{\infty} P(\lambda \, d_2 x, \lambda \, d_2 y) \, P(\lambda \, d_2 (x - f_x), \lambda \, d_2 (y - f_y)) \, dx \, dy}{\int\limits_{-\infty}^{\infty} [P(\lambda \, d_2 x, \lambda \, d_2 y)]^2 \, dx \, dy} \qquad (10.34)$$

where $P(\lambda \, d_2 x, \lambda \, d_2 y)$ is real valued. Putting $\lambda \, d_2 x = \xi$, $\lambda \, d_2 y = \eta$, (10.34) becomes

$$\mathcal{H}(f_x, f_y) = \frac{\int\limits_{-\infty}^{\infty} P(\xi, \eta) \, P(\xi - \lambda \, d_2 f_x, \eta - \lambda \, d_2 f_y) \, d\xi \, d\eta}{\int\limits_{-\infty}^{\infty} [P(\xi, \eta)]^2 \, d\xi \, d\eta}. \qquad (10.35)$$

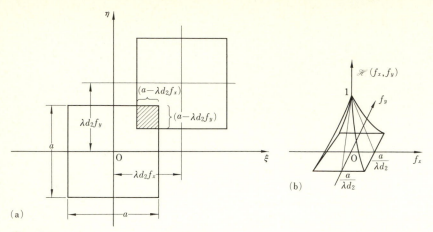

**Fig. 10.5 a, b.** Graphical method of finding the optical transfer function (OTF). (a) Geometry to assist in the computation of the numerator of $\mathcal{H}(f_x, f_y)$. (b) Illustration for the case of a square pupil

Figure 10.5 shows an example of graphically obtaining the OTF of an imaging system with a square pupil. Since the pupil function is unity throughout the aperture, the denominator represents the area of the aperture, and the *numerator represents the area of overlap of two pupil functions, one of which is displaced by $\lambda d_2 f_x$ in the $\xi$ direction and by $\lambda d_2 f_y$ is the $\eta$ direction.* Equation (10.35) can be generalized to the statement of

$$\mathrm{OTF} = \frac{\text{area of overlap of displaced pupil functions}}{\text{complete area of pupil function}}. \tag{10.36}$$

**Exercise 10.3**   The image of an input mask whose distribution function is $g(x) = |\cos 2\pi f_0 x|$ was formed by using an imaging system whose cut-off spatial frequency is $1.2 f_0$. Would it be better to use coherent light illumination or incoherent light illumination?

*Solution*   $g(x)$, and $I_0(x) = |g(x)|^2$ can be rewritten as

$$|\cos 2\pi f_0 x| = \sum_{n=-\infty}^{\infty} \frac{2}{\pi} \frac{(-1)^n}{1-(2n)^2} e^{j2\pi n \cdot 2 f_0 x}$$

$$|\cos 2\pi f_0 x|^2 = \cos^2 2\pi f_0 x = \tfrac{1}{4}(2 + e^{-j4\pi f_0 x} + e^{j4\pi f_0 x}).$$

The spatial frequency spectra obtained by Fourier transforming these functions are indicated in the top diagrams of Fig. 10.6a and b. The CTF and OTF, respectively, are indicated by the middle diagram of the same figure. The diagrams at the bottom of the figure are the spatial frequency spectra of the output amplitude or intensity, as the case may be, which were formed by taking the product of the upper and middle diagrams.

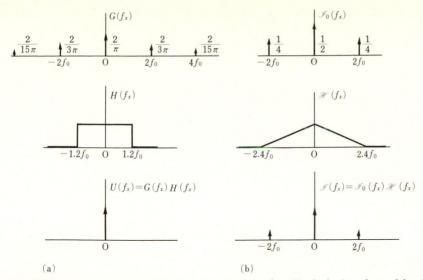

(a)                                              (b)

**Fig. 10.6 a, b.** Comparison of the images of $g(x) = |\cos 2\pi f_0 x|$ when formed by (a) coherent light and (b) incoherent light

In the coherent case, $U(f_x) = G(f_x) H(f_x)$ winds up with only one spectral component located at $f_x = 0$. This means that the output becomes nothing but a plain background. In the incoherent case, on the other hand, there are two spectral components at $f_x = \pm 2 f_0$ besides the one at $f_x = 0$, and the image can be recovered even though its contrast is not as good as the original.

This example is somewhat unusual in that the fundamental spatial frequency of the amplitude is identical to that of the intensity. Usually the latter is twice the former.

## 10.5 Modulation Transfer Function (MTF)

It is difficult to establish an exact measure of image quality since quality is an entity which usually finds itself subject to human judgement. Nevertheless, certain criteria have been set forth in an attempt to quantify image quality, one of these being the cut-off spatial frequency. This, however, gives no information about the shape of the spatial frequency curve. The MTF (or OTF), which is about to be described, is a better criterion since it applies to the shape of the entire spatial frequency range.

The contrast $m$ is defined as

$$m = \frac{I_{\max} - I_{\min}}{I_{\max} + I_{\min}} \tag{10.37}$$

where $I_{max}$ and $I_{min}$ are the maximum and the minimum of the light intensity.

Suppose the input light intensity $I(x)$ is modulated as

$$I(x) = 1 + A \cos 2\pi f x . \tag{10.38}$$

Then when $A$ is positive and real, the maximum and minimum intensities are

$$I_{max} = 1 + A \qquad I_{min} = 1 - A .$$

Inserting these into (10.37) the relationship

$$m = A \tag{10.39}$$

is obtained. This means that, when the amplitude of the background is taken as unity, $m$ immediately indicates the amplitude of modulation. This fact makes such a definition of $m$ more useful.

In general, the contrast of any image which has gone through a normal imaging system is worse than that of the input image. The ratio of the contrast of the output image to that of the input image is defined as the *modulation transfer function* (MTF) [10.4]

$$\text{MTF} = \frac{\text{contrast of output image}}{\text{contrast of input image}} . \tag{10.40}$$

The change in contrast by passing through an optical system is expected to have a lot to do with the optical transfer function that specifies the quality of the system. In the next section, the exact relationship between the MTF and OTF is explored.

## 10.6 Relationship Between MTF and OTF

When such an image as is expressed by (10.38) is put through an imaging system, the output image can be calculated by (10.23),

$$I(x_i) = (1 + m \cos 2\pi f x_i) * |h(x_i)|^2 \tag{10.41}$$

$$I(x_i) = \int_{-\infty}^{\infty} |h(x)|^2 \, dx + \frac{m}{2} e^{j2\pi f x_i} \int_{-\infty}^{\infty} |h(x)|^2 e^{-j2\pi f x} \, dx$$

$$+ \frac{m}{2} e^{-j2\pi f x_i} \int_{-\mu}^{\infty} |h(x)|^2 e^{j2\pi f x} \, dx$$

$$= \int_{-\infty}^{\infty} |h(x)|^2 \, dx + \frac{m}{2} [e^{j2\pi f x_i} H \star H + (e^{j2\pi f x_i} H \star H)^*] . \tag{10.42}$$

Since $H \star H$ is in general a complex number, it is written as

$$H \star H = |H \star H| \, e^{j\phi} . \tag{10.43}$$

By inserting (10.43) into (10.42)

$$I(x_i) = \int_{-\infty}^{\infty} |h(x)|^2 \, dx + m \, |H \star H| \, \cos(2\pi f \, x_i + \phi) \tag{10.44}$$

is obtained. The maximum and minimum of (10.44) are

$$I_{\text{max}} = \int_{-\infty}^{\infty} |h(x)|^2 \, dx + m \, |H \star H| \tag{10.45}$$

$$I_{\text{min}} = \int_{-\infty}^{\infty} |h(x)|^2 \, dx - m \, |H \star H| . \tag{10.46}$$

Inserting (10.45, 46) into (10.37), the contrast of the output image is

$$m_{\text{out}} = \frac{m \, |H \star H|}{\int_{-\infty}^{\infty} |h(x)|^2 \, dx} . \tag{10.47}$$

Using the definition of (10.40), the modulation transfer function is written as

$$\text{MTF} = \frac{m_{\text{out}}}{m} = \frac{|H \star H|}{\int_{-\infty}^{\infty} |h(x)|^2 \, dx} . \tag{10.48}$$

By applying Parseval's theorem

$$\int_{-\infty}^{\infty} |h(x)|^2 \, dx = \int_{-\infty}^{\infty} |H(f_x)|^2 \, df_x$$

to (10.48) and comparing with (10.33), it is easy to see that

$$\text{MTF} = |\text{OTF}| . \tag{10.49}$$

In conclusion, the *modulation transfer function has been found to be identical to the absolute value of the optical transfer function.* From (10.43) it is also noticed that there is a phase shift of $\phi$, which is called the phase transfer function (PTF). The PTF is sometimes an important quantity to consider. For instance, when $\phi = \pi$, the contrast completely reverses. If $h(x)$ is symmetric with respect to the origin, then $H(f_x)$ is a real quantity. (Problem 10.2) and $\phi = 0$ so that no phase shift takes place.

## Problems

**10.1**   Prove that the auto-correlation $A \star A$ is symmetric with respect to the origin when the function $A (f_x, f_y)$ is a real function.

**10.2**   Prove that the optical transfer function (OTF) is real if the impulse response function $h (x, y)$ is symmetric with respect to the origin.

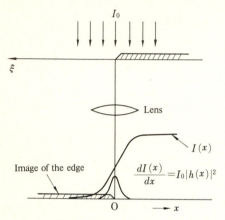

**Fig. 10.7.** A method of determining $|h (x)|^2$ from the image of a knife edge

**10.3**   Figure 10.7 shows an arrangement to determine the impulse response function. The image of a knife edge placed at $\xi = 0$ is formed by the imaging system under test. Show that intensity $I (x)$ along $x$ is related to the square of impulse response $|h (x)|^2$ by

$$|h (x)|^2 = \frac{1}{I_0} \frac{dI (x)}{dx} ,$$
(10.50)

where $I_0$ is the intensity of the incident light.

**10.4**   Find the optical transfer function of a diffraction-limited camera whose pupil function consists of two square apertures side by side, as shown in Fig. 10.8. The squares have 1 cm sides and the spacing between the centers of the squares is 3 cm. The wavelength used is centered at 0.5 μm ($5 \times 10^{-4}$ mm). The distance from the lens to the film is 10 cm.

**10.5**   Figure 10.9 shows a camera shutter in the shape of an octagon with edge dimension $a$. The distance from the lens to the film is $d_2$ and the wavelength used is $\lambda$. Sketch the curve of the optical transfer function along the $f_x$ axis by a graphical method.

**10.6**   The object of this problem is to determine the optical transfer function of an imaging system with a circular aperture.

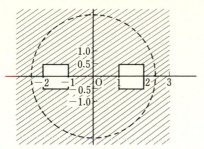

**Fig. 10.8.** Pupil made up of two squares side by side

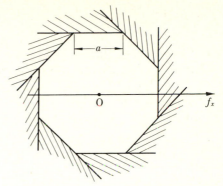

**Fig. 10.9.** Octagonal shutter with edge dimension $a$

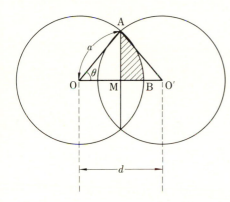

**Fig. 10.10.** Calculation of the overlapped area of two circles for the purpose of determining the optical transfer function of the system with a circular pupil

(a) Show that the shaded portion of the overlapped area of the two circles in the Fig. 10.10 is expressed by

$$S(d) = \frac{a^2}{2} \cos^{-1}\left(\frac{d}{2a}\right) - \frac{ad}{4}\sqrt{1 - \left(\frac{d}{2a}\right)^2} . \tag{10.51}$$

(b) Making use of (10.51), find the expression for the optical transfer function of the system and draw its curve. The distance from the lens to the screen is $d_2$ and the wavelength is $\lambda$.

# 11. Optical Signal Processing

Optical signal processing has become a useful tool in a wide variety of fields including pattern recognition, computer aided vision, computed tomography, aerial photography, meteorology, image storage, and image improvement, to name a few. As various techniques for optical processing are available that range from the simple to the sophisticated, examples representative of this wide range will be described.

## 11.1 Characteristics of a Photographic Film

Photographic film is one of the most common means for optical recording, so that a study of film characteristics is beneficial for the understanding of optical signal processing. Film is composed of a base material such as glass, acetate, or paper, which has been coated with a photographic emulsion consisting of a suspension of silver halides in gelatin. The silver halides used are silver chloride AgCl, silver bromide AgBr, and silver iodide AgI, where silver iodide is the most sensitive to light and silver chloride is the least sensitive. The silver halides found in print paper are usually silver chloride and silver bromide, whereas those in the negative film are silver bromide and silver iodide [11.1, 2].

When film is subjected to light and developed, the exposed grains of silver halides are changed into metallic silver which creates a change in the light transmittance of the film. The intensity transmittance $\tau$ which is defined as the ratio of the transmitted light intensity to the incident light intensity decays exponentially with respect to the increase in $d$ which is a quantity proportional to the amount of metallic silver.

$$\tau = A\,e^{-d} \quad \text{or} \quad d = \log(1/\tau) + \log A\,.$$

In order to define the degree of opaqueness of the film, a quantity somewhat proportional to the metallic silver content is used

$$D = \log_{10}(1/\tau) \tag{11.1}$$

where $D$ is called optical density.

F. Hurter and V. C. Driffield represented the characteristics of a film by plotting the optical density as a function of the logarithm of the exposure.

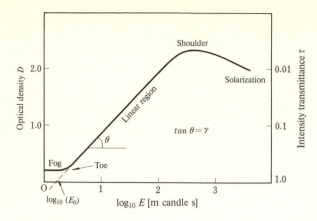

**Fig. 11.1.** Typical H-D curve used to characterize a photographic film

This characteristic curve is known as the *H-D curve* and has been widely used. Figure 11.1 shows a typical H-D curve. The important features of this curve will be examined more closely. Even when unexposed, the film is not completely transparent. Consequently, the H-D curve never crosses the origin, but the curve becomes parallel to the horizontal axis near the origin. This region of the curve is called fog, and the turning point is called the toe. As the exposure is increased beyond the toe, the curve becomes linear and it is this linear region that is preferred for most applications. The slope of the linear region is denoted by the symbol $\gamma$. The point at the top of the linear region is called the shoulder. Beyond the shoulder there is a region where the optical density decreases (rather than increases) with the exposure time. This phenomenon is called solarization.

The formula relating the exposure to the transmittance can be obtained from Fig. 11.1, which yields

$$D = \gamma \left[ \log_{10}(E) - \log_{10}(E_0) \right] \tag{11.2}$$

so that from (11.1)

$$\tau = \left( \frac{E}{E_0} \right)^{-\gamma}. \tag{11.3}$$

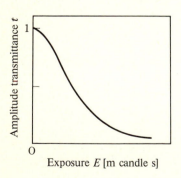

**Fig. 11.2.** Typical $t-E$ curve used to characterize film

The amplitude transmittance $t$ is defined in terms of the intensity transmittance $\tau$ as $t = \sqrt{\tau}$, and (11.3) can also be expressed as

$$t = \left(\frac{E}{E_0}\right)^{-(\gamma/2)}. \tag{11.4}$$

To be more exact, a phase factor which accounts for the relief and the change in the refractive index in the film should be included, but was omitted for simplicity. The $t - E$ curve demonstrates the relation between transmittance and exposure and is an alternate way of representing film characteristics. Figure 11.2 shows an example of a typical $t - E$ curve.

## 11.2 Basic Operations of Computation by Light

Optical signal processing can be described as a combination of certain basic operations which are summarized in this section [11.3, 4].

### 11.2.1 Operation of Addition and Subtraction

Let $g_1(x, y)$ and $g_2(x, y)$ be the input functions for which the operations of addition and subtraction are to be performed. For convenience, these functions will be abbreviated as $g_1$ and $g_2$ provided there is no cause for ambiguity. The input from two perpendicular collimated light beams can be added by means of a beam splitter oriented at 45° with respect to the beams, as shown in Fig. 11.3. The lens forms the images of both $g_1$ and $g_2$ on the screen, and the sum $g_1 + g_2$ is obtained. It should be noted that if the lens is eliminated, the distribution on the screen becomes the addition of the diffraction patterns of $g_1$ and $g_2$ rather than the addition of the images of $g_1$ and $g_2$.

If the optical path difference of $g_1$ and $g_2$ is made an odd multiple of half wavelengths, the operation becomes a subtraction whereas if the path difference is made an even multiple, the operation becomes an addition. If the glass plate in the left part of Fig. 11.3 is rotated, the path length of the light in the glass changes and the phase of the transmitted light is altered.

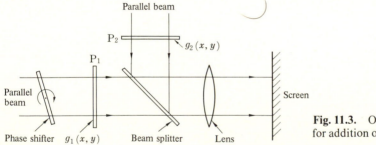

**Fig. 11.3.** Optical system for addition of $g_1$ and $g_2$

In applications where performing real-time operations is not essential, the hologram provides another method for adding input functions. The photographic plate is first exposed to the sum of the reference $R$ and the input $g_1$, and is then reexposed to the sum of the reference $R$ and the input $g_2$. The transmittance of the hologram, or strictly speaking, the opacity of the hologram, is

$$|R + g_1|^2 + |R + g_2|^2 = 2|R|^2 + |g_1|^2 + |g_2|^2$$
$$+ R^*[g_1 + g_2] + R[g_1^* + g_2^*]. \qquad (11.5)$$

The reconstructed image from the fourth term in (11.5) represents the sum of $g_1$ and $g_2$.

## 11.2.2 Operation of Multiplication

Real-time multiplication of the inputs $g_1$ and $g_2$ can be performed easily by laying one input on top of the other. However, if one input plate is to be moved with respect to the other, damage could result from close contact and a set of lenses is used to avoid scratching the plates, as shown in Fig. 11.4. Lenses $L_1$ and $L_2$ form an inverted magnified, real image of $g_1$ on the plane $P_2$ where the magnification is determined by $f_2/f_1$. (Recall from Sect. 6.4 that an image formed by only one lens introduces a phase error.) The product of $g_1$ and $g_2$ can be obtained by placing $g_2$ on the plane $P_2$. The lens $L_3$ forms the real image of the product on the screen.

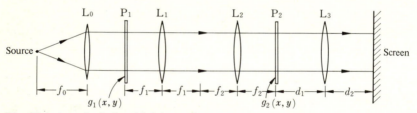

**Fig. 11.4.** A method of multiplying $g_1$ by $g_2$

A holographic approach may also be used to perform the product operation. In this method, $g_1(x, y)$ and $g_2(x, y)$ are incident on a photographic plate at two different angles as shown in Fig. 11.5. The transmittance of the plate is

$$|g_1(x \cos \alpha, y) e^{-jkx\sin\alpha} + g_2(x, y)|^2 = |g_1(x \cos \alpha, y)|^2 + |g_2(x, y)|^2$$
$$+ g_1(x \cos \alpha, y) e^{-jkx\sin\alpha} g_2^*(x, y) + g_1^*(x \cos \alpha, y) e^{jkx\sin\alpha} g_2(x, y)$$
$$(11.6)$$

where $\alpha$ is the angle of incidence of the input $g_1$. When the hologram is illuminated by a reconstructing beam incident perpendicular to the surface of the hologram, the distribution of $g_1(x \cos \alpha, y) g_2^*(x, y) \exp(-jkx \sin \alpha)$

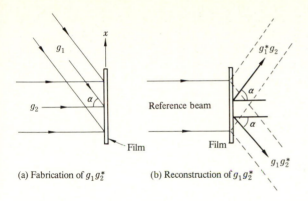

**Fig. 11.5 a, b.** A method of multiplication by means of holography. (**a**) Fabrication of $g_1 g_2^*$. (**b**) Reconstruction of $g_1 g_2^*$

(a) Fabrication of $g_1 g_2^*$     (b) Reconstruction of $g_1 g_2^*$

is obtained in the direction defined by the angle $\alpha$. It should be noted that this product is made by the complex conjugate $g_2^*$ rather than by $g_2$. The value of the factor $\exp(j k x \sin \alpha)$ can be made unity by using a prism.

### 11.2.3 Operation of Division

The operation of division $g_1/g_2$ is equivalent to the product operation of $g_1$ and $1/g_2$. The problem of division is thus reduced to obtaining the transmittance of the inverse. For the operation of the inverse, two plates are needed corresponding to the two factors in the right hand side of (11.7), namely

$$\frac{1}{g_2} = \frac{1}{|g_2|^2} g_2^*. \tag{11.7}$$

The first factor, $1/|g_2|^2$ can be realized by referring to (11.4) and using a photographic film with $\gamma = 2$. The second factor, $g_2^*$ can be realized by making a hologram with $g_1 = 1$ so that the third term in (11.6) yields the desired $g_2^*$.

### 11.2.4 Operation of Averaging

Obviously, averaging can be performed by using the operation of addition mentioned in Sect. 11.2.1 [11.5]. However, in that method there is a limit on the number of inputs. The method described below is a real-time operation and has the merit that any unwanted input can be attenuated by observing the output as the inputs are being added.

Figure 11.6 illustrates the principle of this operation. As shown in Fig. 11.6a, the input "$A$" is placed in the front focal plane of a lens $L_1$ and a mask of $\mathrm{III}(x/a)$ made by an array of pinholes is placed in the back focal plane of the lens. The output from the mask is

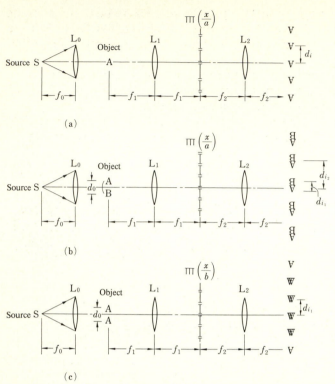

**Fig. 11.6 a – c.** Arrangement for taking an average of several inputs. (a) Formation of many "$A$" images. (b) Formation of many "$AB$" images. (c) Adjustment of $d_0$ such that "$A$" and "$A$" of "$AA$" image overlap

$$E_M = K\bar{A}\left(\frac{x}{\lambda f_1}\right) \, \mathrm{III}\left(\frac{x}{a}\right). \tag{11.8}$$

If this output is further Fourier transformed by the lens $L_2$, the result is

$$\bar{E}\left(\frac{x}{\lambda f_2}\right) = K'A\left(-\frac{f_1}{f_2}x\right) * \mathrm{III}\left(\frac{a x_1}{\lambda f_2}\right), \tag{11.9}$$

where the bar represents Fourier transform. In this way, a large number of images of the input "$A$" can be created on the screen. Now suppose that the input "$A$" is replaced by "$AB$". Naturally the output on the screen becomes a series of "$AB$" images. As the spacing between "$A$" and "$B$" is increased, the "$A$" starts to overlap the adjacent "$B$", as shown in Fig. 11.6b. After the spacing between "$A$" and "$B$" has been adjusted so that the image of "$A$" exactly overlaps that of "$B$", the input letter "$B$" is replaced by "$A$", then the output becomes a series of "$A$" overlapped with "$A$" images, as shown in Fig. 11.6c.

Extending this concept to an input made up of $N$ number of "$A$"'s, the output on the screen will consist of an array of images, and, except for the outermost elements of the array, each image is the addition of $N$ number of "$A$"'s.

The photographs in Fig. 11.7 illustrate the two-dimensional analog of the operation in Fig. 11.6a. In the case of Fig. 11.7, one lens $L_1$ plays the dual role of Fourier transforming and image forming.

Figure 11.8 shows the average of 100 faces, each of which is incomplete in the input stage. As a result of the averaging, all faces become complete.

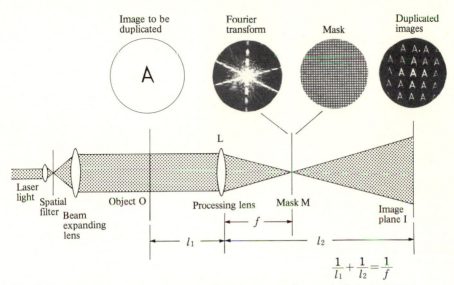

Fig. 11.7. A method for generating several output images from a single input image

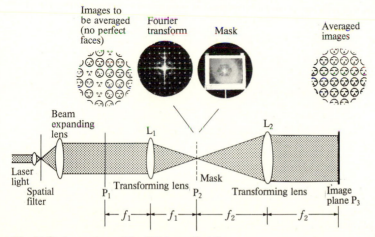

**Fig. 11.8.** Averaging of the 100 imperfect faces

## 11.2.5 Operation of Differentiation

At first glance, forming the derivative of an input $g(x)$ according to (11.10) seems quite a formidable task, i.e.

$$g'(x) = \frac{\partial}{\partial x} g(x).$$

(11.10)

The method described here makes use of the fact that

$$g'(x) = \mathscr{F}^{-1} \{\mathscr{F} [g'(x)]\}.$$

(11.11)

It turns out that taking the Fourier transform of the derivative is much easier than performing the differentiation directly. Once $\mathscr{F}[g'(x)]$ is obtained, the derivative $g'(x)$ can be obtained through the inverse transform according to (11.11).

The Fourier Transform of (11.10) is

$$\mathscr{F} \{g'(x)\} = j\, 2\pi f_x\, G(f_x)$$

(11.12)

so that $2\pi f_x\, G(f_x)$ is obtained by laying the mask of $2\pi f_x$ over that of $G(f_x)$. Since the value $2\pi f_x$ ranges from negative to positive, it has to be realized by two masks as shown in Fig. 11.9. The transmittance of one of the masks is proportional to $|2\pi f|$ and that of the other is a phase plate step function distribution for attaining the correct sign.

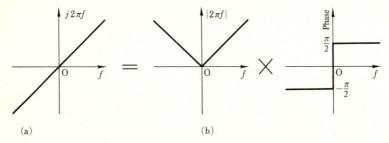

**Fig. 11.9 a, b.** A composite mask realizing "$2\pi f_x$" consisting of a mask for $|2\pi f_x|$ overlapped with that of the step function for the sign. (a) $2\pi f$ filter. (b) Means of achieving $2\pi f$ filter

Figure 11.10 shows the lens arrangement used to perform the differentiation. The lens $L_1$ performs the Fourier transform of the input on the $2\pi f_x$ mask, and the lens $L_2$ further Fourier transforms $2\pi f_x\, G(f_x)$ onto the screen. The image on the screen is a $f_2/f_1$ magnified, inverted image of the derivative, where the inversion results from applying $\mathscr{F}\mathscr{F}$ instead of $\mathscr{F}^{-1}\mathscr{F}$ in (11.11). Figure 11.11 shows an example of the results of the derivative operation [11.6].

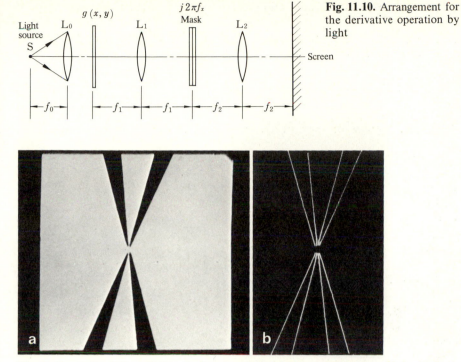

**Fig. 11.10.** Arrangement for the derivative operation by light

**Fig. 11.11 a, b.** By differentiating the input pattern (**a**), the output pattern (**b**) is obtained. The derivative was taken only in the horizontal direction [11.6]

Differentiation concludes the analysis of the basic operations. In the following sections, signal processing by combinations of these basic operations will be considered.

## 11.3 Optical Signal Processing Using Coherent Light

A filter is a device used for optical signal processing. Filters used in the spatial frequency domain are called spatial frequency filters or simply spatial filters. This section will be devoted to spatial filtering techniques using coherent light. In the next section, similar techniques using incoherent light will be presented.

### 11.3.1 Decoding by Fourier Transform

Decoding by Fourier transform is a technique based on periodic features in a given input [11.7]. As an example, consider the meteorologist who wishes to find out oceanic wind conditions. The direction and the magnitude of the

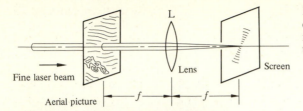

**Fig. 11.12.** A method to decode the wind speed direction from the Fourier transform of the ocean wave pattern

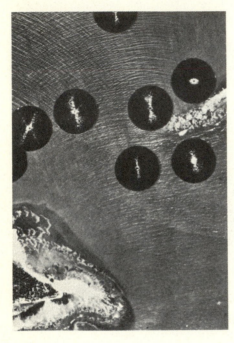

**Fig. 11.13.** From the Fourier transform of a local region of an aerial picture, the wind direction and magnitude can be decoded [11.7]

wind on the surface of the ocean can be determined from the shape of the ocean waves. The period and the direction of the waves are decoded from an aerial picture as follows. The Fourier transform of a portion of an aerial photograph is made, as shown in Fig. 11.12. The transform pattern has a spread in a direction perpendicular to the ocean wave pattern and the amount of the spread is used to determine the spatial frequency distribution of the ocean wave. Figure 11.13 shows the aerial picture amalgamated with the Fourier transform of its local regions obtained in the manner described above. From the figure, the direction of the surface wind and period of the ocean waves is interpreted.

### 11.3.2 Inverse Filters

The Fourier transform of an image using a coherent source is given by (10.5), namely

$$U(f_x, f_y) = G(f_x, f_y) \, H(f_x, f_y) \tag{11.13}$$

where $U(f_x, f_y)$ is the Fourier transform of the output, $G(f_x, f_y)$ is the Fourier transform of the input, and $H(f_x, f_y)$ is the coherent transfer function.

For an ideal optical system, the value of $H(f_x, f_y)$ is constant and unity for all spatial frequencies, but in real systems there is a cut-off at some finite spatial frequency and the output is not a faithful reconstruction of the input. The inverse filter is one means of improving the quality of the output image.

When the spectrum of the output is multiplied by $1/H(f_x, f_y)$, we get

$$U'(f_x, f_y) = U(f_x, f_y) \frac{1}{H(f_x, f_y)} = G(f_x, f_y) \tag{11.14}$$

so that the output can be made identical to the input. Such a filter with a transmittance $T(f_x, f_y) = 1/H(f_x, f_y)$ is called an inverse filter. The fabrication of the inverse filter is similar to that of the two-mask filter made for the division operation in Sect. 11.2.3. Thus

$$T(f_x, f_y) = \frac{1}{H(f_x, f_y)} = \frac{1}{|H(f_x, f_y)|^2} H^*(f_x, f_y).$$

Exercise 11.1 investigates the problem of designing an inverse filter to suit a particular purpose.

**Exercise 11.1**  Suppose that a certain doubly exposed picture consists of identical images separated from each other by $2a$. Design an inverse filter to reconstruct a single image from the doubly exposed images.

*Solution*  The impulse response function of the double exposure $h(x, y)$ is

$$h(x, y) = \delta(x + a) + \delta(x - a). \tag{11.15}$$

The Fourier transform of (11.15) is

$$H(f_x) = e^{j2\pi f_x a} + e^{-j2\pi f_x a} = 2 \cos 2\pi f_x a. \tag{11.16}$$

The graph of $H(f_x)$ appears in Fig. 11.14a. The transmittance $T(f_x, f_y)$ of the inverse filter is

$$T(f_x, f_y) = \frac{1}{2 \cos 2\pi f_x a}. \tag{11.17}$$

Equation (11.17) is shown in Fig. 11.14b. Since there is a section that becomes negative, it has to be realized by a combination of an amplitude mask and a phase mask as shown in Fig. 11.14c.

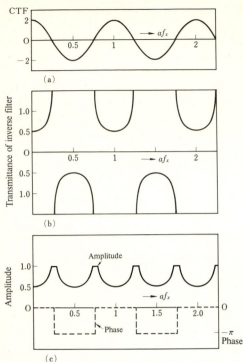

(a)

(b)

(c)

**Fig. 11.14 a – c.** The design of an inverse filter to reconstruct a single image from a picture with double images. (**a**) Coherent transfer function. (**b**) Transmittance of inverse filter. (**c**) Amplitude and phase masks making up inverse filter

Even though the principle of the inverse filter is simple, it has the following drawbacks.

1) Sometimes the transmittance of the mask should be infinity, as indicated by Fig. 11.14b, but as long as passive elements are used, the transmittance can never be larger than unity. It has to be approximated by a distribution such as that of Fig. 11.14c.
2) The inverse filter has to be designed for individual configurations.
3) It is necessary to match the amplitude and the phase plates properly.
4) The image quality becomes poor near the region where $T(f_x, f_y)$ should be infinite.
5) This method is impractical when the transfer function is time varying in an unpredictable manner, such as the case of propagation through turbulent media.

The last two drawbacks in the above list can be compensated for by the Wiener filter which is explained in the next section.

### 11.3.3 Wiener Filter

Undesirable characteristics that cause a masking of the true image are denoted as noise. When taking aerial photographs, for example, the atmosphere in the form of clouds, mists, and turbulences represents a source of noise which tends to mask the true picture that would be obtained in the absence of atmospheric disturbances. *The Wiener filter uses information about the statistical distribution of the noise spectrum in order to improve the noisy image* [11.8, 9].

In almost every case, there are some peculiar characteristics in the distribution of the noise spectrum that can be used for processing. The transmittance of the Wiener filter is modulated according to the noise-spectrum distribution by attenuating where the probability of noise is high and by accentuating where the noise probability is low. The *transmittance is designed in such a way that the mean-square deviation of the corrected image from the true image is minimized.*

Let the amplitude distribution of the image with noise be represented by

$$u(x, y) = g(x, y) * h(x, y) + n(x, y) \tag{11.18}$$

where $g(x, y)$ is the true image, $h(x, y)$ is the impulse response function in the absence of noise, and $n(x, y)$ is the noise. Let the amplitude of the corrected image be represented by

$$g'(x, y) = u(x, y) * t(x, y) \tag{11.19}$$

where $t(x, y)$ is a function whose Fourier transform gives the desired transfer function. The filter is designed such that the mean square deviation of $g'(x, y)$ from $g(x, y)$ is minimized, i.e., the quantity

$$E = \int\int |g'(x, y) - g(x, y)|^2 \, dx \, dy \tag{11.20}$$

is minimized.

The transmittance $T(f_x, f_y)$ that makes the value of $E$ minimum is

$$T(f_x, f_y) = \frac{1}{H(f_x, f_y)} \frac{\dfrac{\Phi_{gg}(f_x, f_y)}{\Phi_{nn}(f_x, f_y)}}{\dfrac{\Phi_{gg}(f_x, f_y)}{\Phi_{nn}(f_x, f_y)} + |H(f_x, f_y)|^{-2}} \tag{11.21}$$

where $H(f_x, f_y)$ is the coherent transfer function of the noiseless system, and $\Phi_{gg}(f_x, f_y)$ and $\Phi_{nn}(f_x, f_y)$ are the power spectral densities of the signal and the noise, respectively. Their values are the Fourier transforms of the autocorrelations $R_{gg}(\xi, \eta)$ and $R_{nn}(\xi, \eta)$, i.e.

$$\left.\begin{aligned}
\Phi_{gg}(f_x, f_y) &= \mathcal{F}\{R_{gg}(\xi, \eta)\} \\
\Phi_{nn}(f_x, f_y) &= \mathcal{F}\{R_{nn}(\xi, \eta)\}, \quad \text{where} \\
R_{gg}(\xi, \eta) &= g \star g = \int g^*(x, y)\, g(x + \xi, y + \eta)\, dx\, dy \\
R_{nn}(\xi, \eta) &= n \star n = \int n^*(x, y)\, n(x + \xi, y + \eta)\, dx\, dy\,.
\end{aligned}\right\} \tag{11.22}$$

Equation (11.21) is the transmittance of the Wiener filter and its derivation and an excellent description are found in [11.8]. The first factor of (11.21) is identical to the inverse filter and the second factor is a consequence of the statistical treatment of the design.

The experimentally measured value of $H(f_x, f_y)$ can be combined with the computed value of the second factor of (11.21) to determine the transmittance distribution of the filter. Since the filter has rotational symmetry, a light source shaped according to (11.21) is imaged onto an unexposed filter film, and the film is rotated in order to achieve the desired two-dimensional transmittance.

Figure 11.15 illustrates image processing for a picture taken through a perturbing medium. The result using the Wiener filter is compared with that using the inverse filter. It can be seen that the Wiener filter gives a much better picture [11.9].

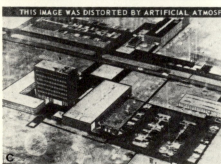

Fig. 11.15 a – c. Comparison of the results with an inverse filter and Wiener filter [11.9]. (a) A normal picture taken through perturbing medium. (b) Picture processed by an inverse filter. (c) Picture processed by the Wiener filter

### 11.3.4 A Filter for Recovering the Image from a Periodically Sampled Picture

In addition to explaining image recovery from sampled pictures, this section provides a good opportunity to illustrate the principles of sampling theory [11.10–12]. An example of a periodically sampled picture is the screened photograph used for half-tone printing in newspapers. The sampled picture is expressed by

$$g_s(x, y) = \text{III}\left(\frac{x}{a}\right) \text{III}\left(\frac{y}{b}\right) g(x, y). \tag{11.23}$$

If this is Fourier transformed by a lens, it becomes

$$G_s(f_x, f_y) = a\, b\, [\text{III}\,(a\, f_x)\,\text{III}\,(b\, f_y)] * G(f_x, f_y)|_{f_x = x_i/\lambda f, f_y = y_i/\lambda f} \tag{11.24}$$

where a constant phase factor was ignored.

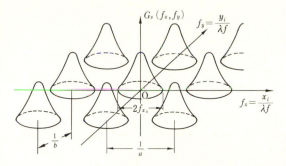

**Fig. 11.16.** Fourier transform of $g_s$ consisting of a grid of identical functions $G(f_x, f_y)$

**Fig. 11.17.** An optical system used to recover the original image from a periodically sampled image

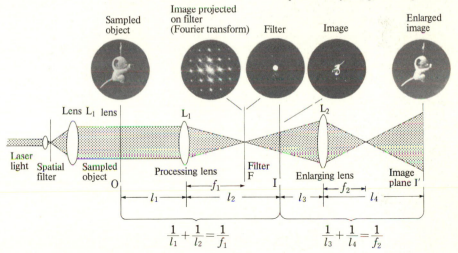

$$\frac{1}{l_1} + \frac{1}{l_2} = \frac{1}{f_1} \qquad\qquad \frac{1}{l_3} + \frac{1}{l_4} = \frac{1}{f_2}$$

The general shape of (11.24) is shown in Fig. 11.16. In the $f_x, f_y$ plane, the function $G(f_x, f_y)$ appears in the form of a two-dimensional array. Each one of the array elements is the Fourier transform of the original picture. A filter made up of an opaque mask with a pinhole in its center is placed in the Fourier transform plane so that only one of the array elements is transmitted. The original image is recovered by inverse Fourier transforming the beam transmitted through the mask.

Figure 11.17 shows the optical system for recovering $g(x, y)$. The required condition for the feasibility of such a processing is that the spectral width of $G(f_x, f_y)$ be finite. The highest spatial frequency $f_{x_B}$ in the x-directon and the highest spatial frequency $f_{y_B}$ in the y-direction have to satisfy the conditions

$$\frac{1}{a} > 2 f_{x_B} \quad \text{and} \quad \frac{1}{b} > 2 f_{y_B}. \tag{11.25}$$

If (11.25) is not satisfied, the array spectra overlap each other, and $G(f_x, f_y)$ can not be retrieved. In other words, the spacing of the sampling in the x-direction has to be shorter than $1/2 f_{x_B}$ and that in the y-direction has to be shorter than $1/2 f_{y_B}$. These are precisely the *Nyquist's criteria* for the sampling spacing in the sampling theorem. As long as these conditions are satisfied, a perfect image can be recovered. This is a special characteristic of this method.

### 11.3.5 Matched Filter

This filter is used for locating a particular picture $g(x, y)$ in the midst of several other similar pictures. Figure 11.18 demonstrates how to use this filter [11.7]. The particular picture $g(x, y)$ is put at the front focal plane of a lens $L_1$ to produce its Fourier transform $G(f_x, f_y)$ in the back focal plane $P_2$. In that plane $P_2$, various trial masks of Fourier transforms $G'^*(f_x, f_y)$ are inserted one by one for examination. When the mask $G'^*(f_x, f_y)$ is matched with $G(f_x, f_y)$, the reading from a meter located in the back focal plane of the lens $L_2$ will be maximum. When a mask of any other distribution is used, the reading from the light meter will be less than this value. Thus the searched for shape can be identified.

In better quantitative terms, when the transmittance of the mask is matched and is $G^*(f_x, f_y)$, the light transmitted through the mask in Fig. 11.18 is

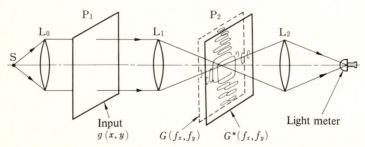

**Fig. 11.18.** Illustration of how a matched filter works. The reading from the light meter becomes maximum when the mask transmittance distribution matches the light distribution immediately in front of the mask

$$\underbrace{|G(f_x,f_y)|\,e^{j\phi(f_x,f_y)}}_{\substack{\text{Fourier transform}\\\text{of the input}}}\,\underbrace{|G(f_x,f_y)|\,e^{-j\phi(f_x,f_y)}}_{\substack{\text{Transmittance of}\\\text{the mask}}}\,. \tag{11.26}$$

The amplitude distribution of the input matches up with that of the transmittance and the output becomes maximum when both distributions are identical. The phase of (11.26) becomes zero throughout the mask and the distribution beyond the mask becomes a parallel beam so that all the output from the lens $L_2$ is converged into the light meter. This system is phase matched as well as amplitude matched, thus providing a great degree of sensitivity.

The matched filter can be also used with incoherent light as described in Sect. 11.6. Phase-matching cannot be used in the case of incoherent light.

The above method is also applicable to searching for a particular figure hidden inside one large picture. As a specific example, a search for the letter "$a$" in a picture containing the letters "$a$" and "$b$" will be undertaken. As shown in Fig. 11.19, both letters are located on the $y$ axis with "$a$" being located at $(0, l)$ and "$b$" at $(0, -l)$. Let the transmittance distribution of the letters "$a$" and "$b$" be represented by $a(x, y)$ and $b(x, y)$, respectively. Then the transmittance of the input mask is $f(x, y) = a(x, y - l) + b(x, y + l)$. This input mask is placed in the front focal plane of a lens $L_1$ to obtain its Fourier transform in the back focal plane. The Fourier transform is

$$F(f_x,f_y) = e^{-j2\pi f_y l}\,A(f_x,f_y) + e^{j2\pi f_y l}\,B(f_x,f_y)$$

where $\mathscr{F}\{a(x, y)\} = A(f_x,f_y),\quad \mathscr{F}\{b(x, y)\} = B(f_x,f_y).$

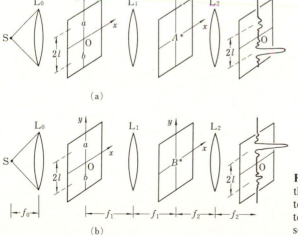

(a)

(b)

Fig. 11.19 a, b. A search for the location of a particular letter by means of a matched filter: (a) search for "$a$", (b) search for "$b$"

As shown in Fig. 11.19a, a mask with the transmittance distribution $A^*(f_x, f_y)$ is placed at the back focal plane of the lens $L_1$. The lens $L_2$ Fourier transforms the product $F(f_x, f_y) A^*(f_x, f_y)$ and projects it onto the output plane. The light distribution on the output plane is

$$a^*(x, y) * a(-x, -y-l) + a^*(x, y) * b(-x, -y+l)$$
$$= [a(x, y) \star a(x, y-l) + a(x, y) \star b(x, y+l)]^* \qquad (11.27)$$

where the definition of the correlation (10.29) was used. The first term in (11.27) is the auto-correlation of $a(x, y)$ as defined by

$$a(x, y) \star a(x, y-l) = \int\limits_{-\infty}^{\infty}\!\!\int a(\xi, \eta)\, a^*(\xi - x, \eta - y - l)\, d\xi\, d\eta. \qquad (11.28)$$

The value of $a \star a$ in (11.28) reaches its maximum

$$\int\limits_{-\infty}^{\infty}\!\!\int |a(\xi, \eta)|^2\, d\xi\, d\eta$$

at $x = 0$, $a = -l$, and a sharp peak is generated at $(0, -l)$. The second term in (11.27) is the cross-correlation of $a(x, y)$ and $b(x, y)$ as defined by

$$a(x, y) \star b(x, y+l) = \int\limits_{-\infty}^{\infty}\!\! \int a(\xi, \eta)\, b^*(\xi - x, \eta - y + l)\, d\xi\, d\eta.$$

The value of the cross-correlation has a small peak at $x = 0$, $y = +l$. In conclusion, a large peak is formed at a point $(0, -l)$ and a small peak at $(0, l)$, as shown in Fig. 11.19a.

If the mask is replaced by another mask whose transmittance distribution is $B^*(f_x, f_y)$, a similar result is obtained, namely

$$[b(x, y) \star a(x, y-l) + b(x, y) \star b(x, y+l)]^*. \qquad (11.29)$$

The first term of (11.29) generates a small peak at $(-l, 0)$ and the second term a large peak at $(l, 0)$, as shown in Fig. 11.19b. The position of the two peaks are the reverse of the previous case. By means of the location of the larger peak, one can find the location of the sought after letter.

Figure 11.20 illustrates a method for obtaining the wind velocity from the correlation peaks of cloud pictures taken from the satellite ATS III. The vector diagrams of the wind velocities obtained by this method [11.7] are shown in Fig. 11.21.

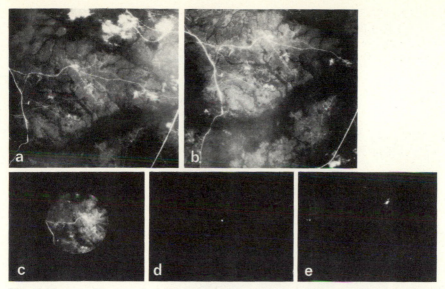

**Fig. 11.20 a − e.** The matched filter is used to trace cloud movement [11.7]. (**a**) Original picture of the cloud. (**b**) Another cloud picture a short time after (**a**) was taken. (**c**) A small section of the cloud picture (**a**). (**d**) The correlation between (**a**) and (**c**). (**e**) The correlation between (**b**) and (**c**). By using the difference in the positions of the peaks in (**d**) and in (**e**) the wind velocity can be calculated

## 11.4 Convolution Filter

Since the convolution filter is placed in the Fourier transform plane, it belongs to the category of spatial frequency filters [11.4, 11]. However, filter design is sometimes simpler if it is considered from the point of view of the space domain rather than the spatial frequency domain. The convolution filter is one such filter [11.13].

The design procedure for the filter will be explained by using a specific example. One wants to recover a good photograph from a picture which was mistakenly exposed four times. The four identical images are displaced from one another by the length $\Delta$. The impulse response of such a system consists of 4 delta functions spaced by $\Delta$, as shown in the top portion of Fig. 11.22. The mathematical expression for the impulse response $h_w(x, y)$ is

$$h_w(x, y) = \sum_{n=0}^{3} a_n \delta(x - n\Delta). \tag{11.30}$$

The image with quadruple exposure is given by

$$u_w(x, y) = g(x, y) * h_w(x, y) \tag{11.31}$$

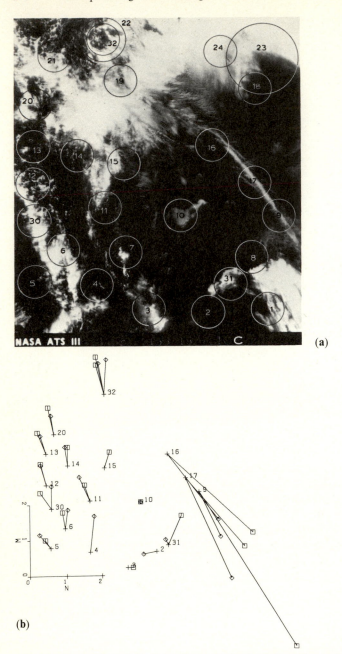

**Fig. 11.21 a, b.** The wind vector diagram obtained from the movement of the correlation peaks [11.7]. (a) In order to determine the cloud movements, the circled sections were first correlated with the picture they belong to and then with a picture taken at a short time later. (b) The vector diagram of the wind obtained from the movement of the correlation peaks

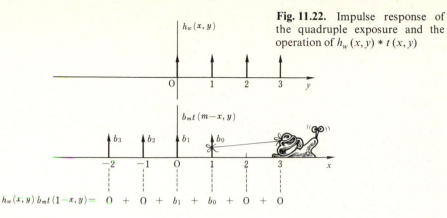

**Fig. 11.22.** Impulse response of the quadruple exposure and the operation of $h_w(x,y) * t(x,y)$

$$h_w(x,y)\, b_{mt}(1-x,y) = \quad 0 \quad + \quad 0 \quad + \quad b_1 \quad + \quad b_0 \quad + \quad 0 \quad + \quad 0$$

where $g(x, y)$ is the image that would have been obtained with a single exposure. The way to extract $g(x, y)$ from (11.31) is to find a function $t(x, y)$ whose convolution with $u_w(x, y)$ becomes $g(x, y)$, namely

$$u_w * t = g * h_w * t = g \tag{11.32}$$

where all functions are understood to be functions of $x$ and $y$. The condition for (11.32) to be satisfied is

$$h_w * t = \delta(x) . \tag{11.33}$$

Suppose that $t(x, y)$ also consists of 4 delta functions with the same spacing as $h_w(x, y)$ and is expressed by

$$t(x, y) = \sum_{m=0}^{3} b_m \delta(x - m\varDelta) \tag{11.34}$$

where the $b_m$ are unknown and are to be determined such that (11.33) is satisfied. To perform the operation of the convolution $h * t$, as shown in Fig. 11.22, $t$ is first flipped with respect to the vertical axis, and then $t$ and $h$ are multiplied together as $t$ is slid in the horizontal direction. If it is assumed that the four exposures are identical and $a_n = 1$, (11.33) becomes

$$\begin{aligned} b_0 &= 1 \\ b_0 + b_1 &= 0 \\ b_0 + b_1 + b_2 &= 0 \\ b_0 + b_1 + b_2 + b_3 &= 0 \end{aligned} \quad \text{or}$$

$$\begin{bmatrix} 1 & 0 & 0 & 0 \\ 1 & 1 & 0 & 0 \\ 1 & 1 & 1 & 0 \\ 1 & 1 & 1 & 1 \end{bmatrix} \begin{bmatrix} b_0 \\ b_1 \\ b_2 \\ b_3 \end{bmatrix} = \begin{bmatrix} 1 \\ 0 \\ 0 \\ 0 \end{bmatrix} . \tag{11.35}$$

The solution of this set of simultaneous equations is

$$b_0 = \quad 1$$
$$b_1 = -1$$
$$b_2 = \quad 0$$
$$b_3 = \quad 0 .$$
(11.36)

By inserting (11.36) into (11.34) the transmittance distribution of the filter is found to be like that shown in Fig. 11.23a.

The expression $h_w(x, y) * t(x, y)$ now has one extra delta function at $x = 4$ besides the one at $x = 0$, as shown in Fig. 11.23b, so that the output of the filter has two images. The two output images will be separated from each other if the image $g(x, y)$ is not too wide, as shown in Fig. 11.23c. It will be left as a problem to show that, by increasing the number of terms in $t(x, y)$, the separation between the two delta functions in Fig. 11.23b is increased.

Figure 11.24 shows an arrangement for picture processing using this filter. The input image $u_w(x, y)$ is placed in the front focal plane of the lens $L_1$. The Fourier transform $U_w(f_x, f_y)$ of the input image is made at the focal plane and illuminates the convolution filter $T(f_x, f_y)$ placed in the same plane, where $T(f_x, f_y) = \mathscr{F}\{t(x, y)\}$. The output from the filter which is the product of $U_w(f_x, f_y) \cdot T(f_x, f_y)$ is then Fourier transformed by the lens $L_2$ to give the final image.

An arrangement for fabricating the convolution filter $T(f_x, f_y)$ is shown in Fig. 11.25. The Fourier transform of $t(x, y)$ can be obtained by means of the Fourier transform hologram described in Sect. 8.4.2 where the ob-

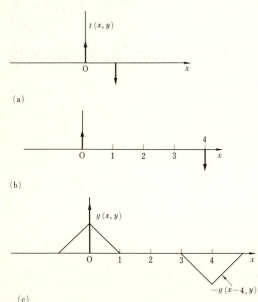

(a)

(b)

(c)

Fig. 11.23 a – c. Design of a filter to recover the original image from a photograph with four exposures. (a) The transmittance $t(x, y)$ of the convolution filter. (b) Graph of $h_w(x, y) * t(x, y)$. (c) Graph of $u_w(x, y) * t(x, y)$

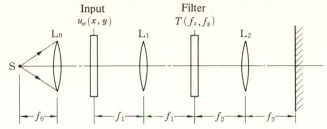

Fig. 11.24. Optical system to recover the original image from a picture with quadruple exposure

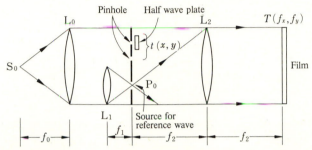

Fig. 11.25. Arrangement for fabricating the convolution filter with transmittance $T(f_x, f_y)$. The filter is a Fourier transform hologram of the two pinholes, one of which undergoes a phase reversal

**(a)**

**(b)**

Fig. 11.26 a, b. The recovered image from a picture of quadruple exposure. (a) The picture of the quadruple exposure. (b) The processed image of (a) by means of a convolution filter [11.13]

ject in this case corresponds to two pinholes, one of which is reversed in phase by means of a half-wavelength plate. The reference beam is formed by the lens combination of $L_0$ and $L_1$. Figure 11.26 shows the results of processing a picture of quadruple exposure.

A picture blurred by camera motion during the exposure can be treated in the same manner. The blurred picture can be considered to be made up of several closely spaced multiple exposures. The design of the filter is exactly the same as for the quadruple exposure, except that the spacing $\Delta$ has to be made as small as practically possible.

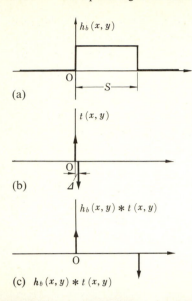

(a)

(b)

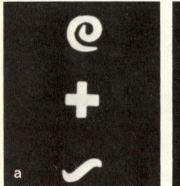

(c)  $h_b(x, y) * t(x, y)$

**Fig. 11.27 a − c.** A method for designing a convolution filter to process a blurred image. (**a**) Impulse response function of a moving camera. (**b**) Transmittance distribution of the filter for processing a blurred image. (**c**) Graph of $h_b(x, y) * t(x, y)$

**Fig. 11.28 a − c.** The processed image obtained from the blurred image by means of a convolution filter [11.13]. (**a**) The original image. (**b**) The processed image obtained by using a convolution filter

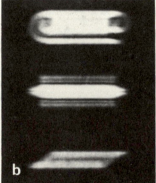
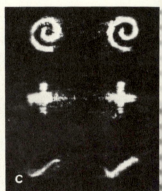

The impulse response function $h_b(x, y)$ and the transmittance distribution $t(x, y)$ of the convolution filter for a blurred image are presented in Fig. 11.27 a and b. Figure 11.27 c shows the result of the convolution $t(x, y) * h_b(x, y)$. The expression for the blurred image is

$$u_b = g * h_b \tag{11.37}$$

and when this is convolved with $t$, the image $g$ is recovered, namely

$$u_b * t = g * h_b * t = g * \delta(x) = g. \tag{11.38}$$

The rest of the procedure is exactly the same as the convolution filter for the quadruple exposure.

Figure 11.28 shows an example of the result of processing a blurred image by means of the convolution filter. The two recovered images are located in positions corresponding to the beginning and the end of the blurred image.

## 11.5 Optical Signal Processing Using Incoherent Light

Optical signal processing with incoherent light cannot utilize the phase information of the image and the sensitivity is then generally smaller than that achieved using coherent light, but the input requirements are less stringent. With incoherent light, it is not necessary to ensure optical flatness of the input and even images printed on such optically irregular surfaces as paper can be used directly as an input.

### 11.5.1 The Multiple Pinhole Camera

*A pinhole camera is advantageous because of its almost infinite depth of focus and its usefulness over a broad spectrum of sources*, which range from visible light to x-rays and $\gamma$-rays [11.14]. However, a major disadvantage of the single-pinhole camera is its low sensitivity, which is a consequence of the fact that only a very small fraction of the incident rays are transmitted through the pinhole. The multiple pinhole camera alleviates this disadvantage by letting in more rays. This is particularly important for x-ray and $\gamma$-ray sources since it reduces the amount of time necessary for irradiation. The picture taken by this camera is similar to the image of a multiple exposure. The peculiarity of this camera is that the positions of the pinholes are randomly distributed, and this randomness is taken advantage of in processing the image. Figure 11.29 shows the impulse response function of a multiple pinhole camera.

Figure 11.30a shows an arrangement used for taking a picture of the object "F" illuminated by incoherent light using a multiple pinhole camera. If the light intensity distribution of the object is designated by $F_0(x, y)$, the intensity distribution $F_m$ on the camera film is then

$$F_m(x, y) = F_0(x, y) * h_m(x, y) \tag{11.39}$$

where $|h(x, y)|^2 = h_m(x, y)$ is called the impulse response function[1]. Notice that each pinhole produces an image of the input. Because of a curious

---

1 In coherent systems, the term "impulse response function" refers to the output amplitude from a system to which an amplitude delta function has been applied. In incoherent systems, impulse response function usually refers to the intensity output from a system to which an intensity delta function has been applied.

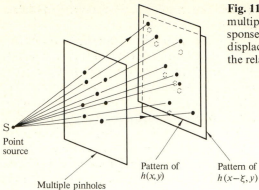

**Fig. 11.29.** Impulse response $h_m(x, y)$ of a multiple pinhole camera. The impulse response function $h_m(x - \xi, y)$ which is slightly displaced from $h_m(x, y)$ is included to show the relationship $h_m(x, y) \bigstar h_m(x, y) = \delta(x)$

S
Point source

Pattern of $h(x, y)$

Pattern of $h(x - \xi, y)$

Multiple pinholes

property in the geometry, the image can be processed by using the same pinholes that were used for taking the picture. Figure 11.30 b shows this arrangement. The intensity $F_c(x, y)$ of the reconstructed image is

$$F_c(x, y) = F_0(x, y) * h_m(x, y) * h_m(-x, -y). \tag{11.40}$$

When $h_m(x, y)$ is real, we get

$$F_c(x, y) = F_0(x, y) * [h_m(x, y) \bigstar h_m(x, y)]. \tag{11.41}$$

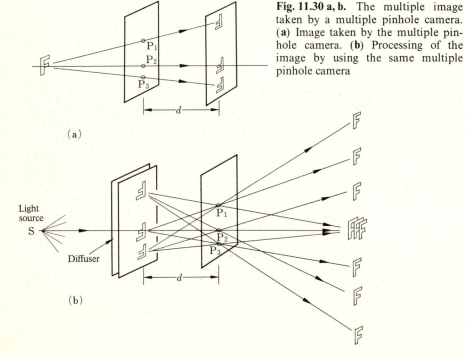

**Fig. 11.30 a, b.** The multiple image taken by a multiple pinhole camera. (**a**) Image taken by the multiple pinhole camera. (**b**) Processing of the image by using the same multiple pinhole camera

If it can be shown that $h_m(x, y) \star h_m(x, y)$ in (11.41) is a delta function, then $F_c(x, y)$ will be identical to $F_0(x, y)$.

Recall that the meaning of the $h_m(x, y)$ is just the light-intensity distribution on the film when a point source is used as the object, as shown in Fig. 11.29. The distribution $h(x - \xi, y)$ which is shifted from $h(x, y)$ by $x$ in the $x$ direction, is included in Fig. 11.29 as an aid to understanding the convolution operation of (11.41). When $\xi = 0$, the two distribution functions precisely overlap each other and the product of the function becomes maximum. As $\xi$ is gradually increased from $\xi = 0$, the product of the two functions sharply decreases and becomes zero as soon as the value of $\xi$ exceeds the diameter of the pinhole. The fact that the product becomes zero is a consequence of the random distribution of the pinholes, which ensures negligible probability of two holes overlapping for $\xi$ greater than the pinhole diameter. In short, the correlation $h_m(x, y) \star h_m(x, y)$ behaves like a delta function $\delta(\xi)\,\delta(\eta)$.

Another explanation of the operation of the multiple pinhole camera is given in Fig. 11.30. The 3-pinhole camera in Fig. 11.30a first produces a potograph that has three identical images of the input letter "F". The same camera is then put to use once again, only this time the photograph taken in Fig. 11.30a becomes the new input. A second photograph is produced which has a total of nine images, as shown in Fig. 11.30b. Moreover, when the distance in Fig. 11.30a is made equal to that in Fig. 11.30b, three of the nine images will overlap and produce one bright image, while the six remaining images will be randomly distributed. The ratio of the intensity of the brightest image to that of any of the other images increases with the number of pinholes. Fig. 11.31a shows the image taken by a 7-pinhole camera, and Fig. 11.31b shows its processed image.

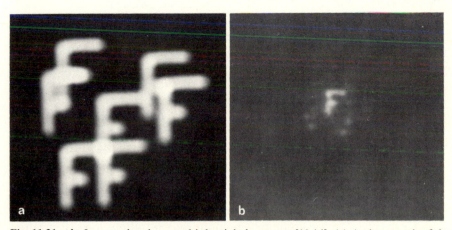

**Fig. 11.31 a, b.** Image taken by a multiple pinhole camera [11.14]. (**a**) A photograph of the letter "$F$" taken by a 7-pinhole camera. (**b**) The reconstruction image obtained by using the photograph in (**a**) as the camera input

### 11.5.2 Time Modulated Multiple Pinhole Camera

Next, a similar multiple pinhole camera but with a time modulated aperture
will be explained. The transmission of each pinhole is varied with respect to
time. A specific example of the time modulated pinhole camera is the
gamma ray camera illustrated in Fig. 11.32. The principle of the modulation is described as follows. Instead of plugging or unplugging each pinhole,
a lead plate with a particular pattern of pinholes is slid over the aperture.
As the plate is slid, the transmission of a particular point in the aperture
changes between 0 and 1. The change of the transmission at points *A*, *B*, and
*C* as a function of time is at the left-hand side of the figure. The mode of
the temporal modulation is either random, pseudorandom or sinusoidal.

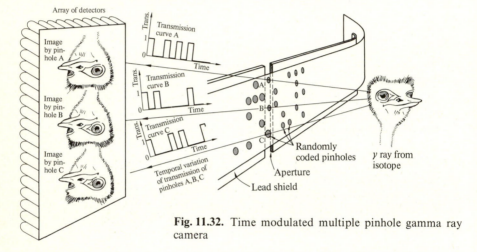

**Fig. 11.32.** Time modulated multiple pinhole gamma ray
camera

The output image at any specific moment is a superposition of the
images that are formed by all pinholes located within the aperture at that
moment. The image thus formed is projected onto an array of detectors
whose output is connected to the memory of a processing computer. The
images are stored preserving the sequence of events. The correlation technique is used to generate a clear image from the superimposed images. The
image that would have been formed had only one pinhole been opened,
say *A* alone, is generated by performing the correlation operation between the stored images and the temporal transmission curve for pinhole *A*.
Only the component of the stored images that has the same temporal variation as the transmission curve for *A* predominates. The contributions of
the other components of the stored images to the correlation operation
become very small and would eventually reach zero if the temporal
modulation curve were absolutely random and infinitely long. Repeating
the same with transmission curve *B*, the image that would have been
formed had the pinhole *B* alone been opened and all others closed is

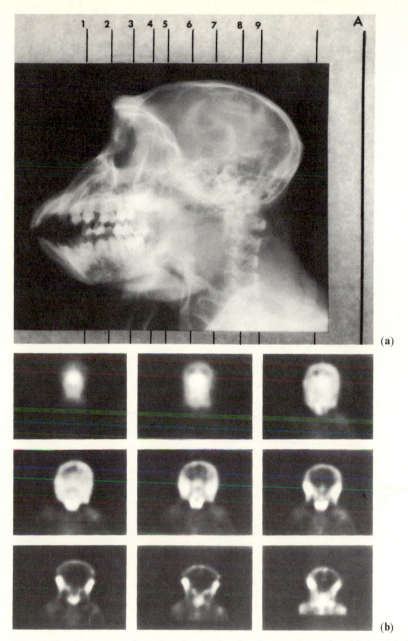

**Fig. 11.33 a, b.** Images of monkey head taken by time modulated multiple pinhole gamma ray camera [11.15]. (**a**) *X*-ray image of a monkey head to show the approximate location of the reconstructed image. (**b**) Slices of the reconstructed image

obtained. Similarly with all other transmission curves, clear images from the respective pinholes are generated.

Now that the images that would have been formed with various locations of the pinholes are available, the depth information of the image can be extracted by computer processing. The depth information cannot be obtained by a single-pinhole camera, because it is a similar situation as that when trying to acquire depth information with only one eye open. This is one of the attractive features of the time-modulated multiple-pinhole camera. Another feature of this method is that the correlation is performed with respect to time rather than with respect to space, and there are no limitations imposed by the finite number of sampling points. The temporal correlation can be continued as long as the dose of $\gamma$-ray exposure to the patient or object permits, and as long as the capacity of the computer permits. Figure 11.33 shows the image of a monkey head obtained by this method. A 15 millicurie dose of isotope $_{43}Tc^{99}$ was administered as the $\gamma$-ray emitting source. Figure 11.33a shows the approximate depthwise location of the slices of the image, while Fig. 11.33b shows the corresponding image slices. This figure demonstrates the image slicing capability of the multiple pinhole $\gamma$-ray camera [11.15].

### 11.5.3 Low-Pass Filter Made of Randomly Distributed Small Pupils

An aperture made up of randomly distributed tiny pinholes behaves as a low-pass filter. The cut off frequency of such a filter is calculated from the optical transfer function of each pinhole. The OTF is expressed by the autocorrelation of the pupil function as given by (10.35). Here, the graphical method demonstrated in Fig. 10.5 is used to obtain the OTF. Fig. 11.34a examines the total overlap area of the two identical pinhole distributions as a function of the relative displacement of the two pupil functions. As the displacement increases, the overlap area decreases and finally becomes zero

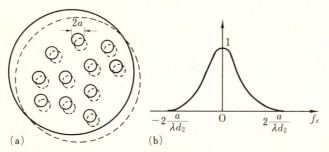

(a)                    (b)

**Fig. 11.34 a, b.** The pupil function and the $OTF$ of a mask with randomly located tiny pinholes. (a) Two identical pupil functions are slid past each other to calculate the auto-correlation. (b) The $OTF$ of the pupil function shown in (a)

**Fig. 11.35 a, b.** Filtering of the vertical lines using an incoherent light low pass filter. The filter is made out of randomly located tiny pinholes as shown in Fig. 11.34a. (**a**) The original lined photograph. (**b**) Image whose vertical lines are filtered out by the low pass filter (Courtesy of Le Chateau Champlain, Montreal)

when the displacement becomes larger than the diameter $2\,a$. That is to say, the integrand of (10.35) becomes zero when

$$\lambda\, d_2 f_x \gtreqless 2\, a$$

and the cut-off spatial frequency $f_i$ of the output image is

$$f_i = \frac{2\,a}{\lambda\, d_2}\,. \tag{11.42}$$

The OTF of the filter is shown in Fig. 11.34 b.

Since the ratio of the sizes of the input to the output images is $d_1/d_2$, the cut-off spatial frequency $f_0$ of the input image is

$$f_0 = f_i \frac{d_2}{d_1} = \frac{2\,a}{\lambda\, d_1}$$

where $d_1$ is the distance between the input image and the lens, and $d_2$ is the distance between the output image and the lens.

The procedure for using this filter is simple. The lens is first focussed so as to form the image and then the filter is inserted immediately in front of the lens. Fig. 11.35 shows an example of filtering out the vertical lines from a lined photograph using the low pass filter.

## 11.6 Incoherent Light Matched Filter

It has been already mentioned in Sect. 11.3.5 that the matched filter designed for coherent light can be readily used with incoherent light at the expense of sensitivity. *The incoherent matched filter makes use of the intensity matching condition in (11.26), but cannot utilize its phase matching condition.*

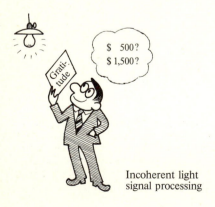

Incoherent light
signal processing

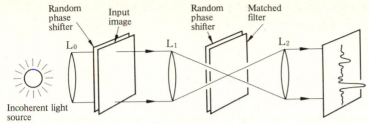

Fig. 11.36. Optical system for incoherent light matched filtering. The system is quite similar to the case of coherent light filtering shown in Fig. 11.18. The additional plates appearing in this figure are random phase filters

As pointed out earlier, the advantage of using incoherent light is that optical flatness in the input is no longer necessary. When using coherent light, an input image on a rough sheet of paper such as the page of a book would have to be copied onto a flat photographic plate, but when using incoherent light, this step is eliminated. Another advantage of incoherent light signal processing is that it is not as sensitive to the orientation of the input image as coherent light processing. The disadvantages are relatively poor sensitivity and, in some cases, the difficulty of coupling the output image with other optical systems.

Figure 11.36 shows an optical system for processing with an incoherent-light matched filter. This system is quite similar to the coherent-light signal processing system shown earlier in Fig. 11.18. In fact, the matched filter for incoherent light is fabricated by the same method as the coherent matched filter except that the phase information is destroyed by adding a random phase filter. If the phase distribution of the input image is not quite random, a random phase shifter is also placed over the input image.

The result obtained using an optical system similar to Fig. 11.36 is shown in Fig. 11.37. The incoherent light matched filter was used to find

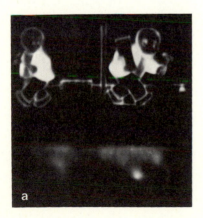

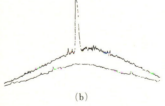

(b)

Fig. 11.37 a, b. The location of a cartoon character was found by using an incoherent light matched filter [11.16]. (a) The bright spot below the character indicates the location of the character. (b) The output from a light meter scanned over the bright spot

the location of a cartoon character. A peak in the output of a light meter is obtained as the light meter is scanned across the location of the desired character [11.16].

## 11.7 Logarithmic Filtering

In this section, a filter which uses photographic film with a logarithmic sensitivity curve is described [11.17]. The logarithmic filter is useful when the signal is contaminated with noise in the form of the product $S(x, y)$ $N(x, y)$ where $S(x, y)$ represents the signal and $N(x, y)$ represents the noise. A picture taken with uneven illumination is one example of noise belonging to this category.

In this case, the light distribution $g(x, y)$ of the noisy input is

$$g(x, y) = S(x, y) N(x, y). \tag{11.43}$$

The signal is to be extracted using a logarithmic filter. The noisy input image is photographed by a film with logarithmic characteristics. The transmittance $t(x, y)$ of the imaged film is

$$t(x, y) = \log S(x, y) + \log N(x, y). \tag{11.44}$$

The noise term now becomes additive rather than multiplicative. The Fourier transform of (11.44) is formed with a lens, i.e.,

$$\mathcal{F}\{t(x, y)\} = \mathcal{F}\{\log S(x, y)\} + \mathcal{F}\{\log N(x, y)\}. \tag{11.45}$$

The term $\mathcal{F}\{\log N(x, y)\}$ is removed by means of an opaque absorber painted in the shape of $\mathcal{F}\{\log N(x, y)\}$ so that only $\mathcal{F}\{\log S(x, y)\}$ is transmitted. A second lens is used to perform another Fourier transform on the transmitted signal so that $S(x, y)$ may be obtained by using film with exponential characteristics.

On the other hand, if (11.43) is Fourier transformed, it becomes

$$\mathcal{F}\{S(x, y)\} * \mathcal{F}\{N(x, y)\}. \tag{11.46}$$

It is hard to generalize whether it is easier to remove the noise by using the additive form of (11.45) or the convolution form of (11.46). It all depends on the nature of the signal and noise. Quite often, however, the additive form proves more advantageous.

Figure 11.38 shows an arrangement for removing the fringe pattern on a picture. The system input consists of a fringe pattern mask superimposed on a transparency made from a photograph of a face. Let the transmittance function of the face be $S(x, y)$ and that of the fringe pattern be $N(x, y)$.

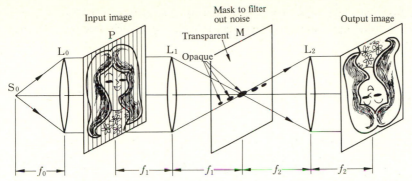

**Fig. 11.38.** Logarithmic filtering. The noise component is filtered out by means of a logarithmic filter $\mathscr{F}\{\log N(x, y)\}$

The input function is then $S(x, y)\, N(x, y)$. As preparation for processing, transparencies of $\log N(x, y)$ and of $\log\{S(x, y)\, N(x, y)\}$ are made by photographing the fringe pattern and the input image with logarithmic film.

First, the transparency of $\log N(x, y)$ is placed in the front focal plane of the lens $L_1$ to obtain the Fourier spectrum $\mathscr{F}\{\log N(x, y)\}$ in the back focal plane. In order to fabricate the mask M, a transparent sheet of glass is placed in the back focal plane, and the areas where the Fourier spectrum exists are painted with an opaque absorber.

As the next step, the input transparency of $\log\{S(x, y)\, N(x, y)\}$ is placed in the front focal plane of the lens $L_1$ while the mask M is left in the

(a)                                                                                (b)

**Fig. 11.39 a, b.** The result of logarithmic filtering [11.17]. (a) Input with a grid pattern. (b) The output image with the grid pattern removed by the logarithmic filter

back focal plane. In this arrangement, the noise spectrum $\mathscr{F}\{\log N(x, y)\}$ is blocked by the mask M and only the signal spectrum is transmitted into the lens $L_2$. In the back focal plane of the lens $L_2$, a photographic plate with exponential characteristics is placed to recover the output image.

Figure 11.39a shows an input image containing vertical lines and Fig. 11.39b the output image with the lines removed through logarithmic filtering.

This method requires the use of films with logarithmic and exponential characteristics. One way of achieving these characteristics is to record the image in two steps; first, it is photographed into a screened photo consisting of tiny dots similar to those used for half-tone photos in a newspaper and next, it is recopied by a regular photographic film. The $t-E$ characteristics of the screened film is quite nonlinear and it happens to approximate the required distribution. By empirically adjusting the two exposure times, the desired characteristics of the film can be achieved.

## 11.8 Tomography

The Greek word "tomos" means slice or section, and tomography is the technique of producing an x-ray photograph of a selected cross section within the body [11.18]. The picture normally taken by an X-ray camera is a projected shadow of several internal organs, bones, and tissues. For diagnostic purposes, it is often desirable to image a single plane or a single structure without obstructions from structures in other planes. This is accomplished by tomography. This technique was probably first used when the French doctor G. Pélissier (1931) told a patient to keep moving his jaw during the exposure of an x-ray camera so as to eliminate the shadow of the jaw and thus take a clear picture of the patient's cervical spine located behind the jaw.

Tomography can be classified into two broad categories. *One is planigraphic tomography or focused tomography, and the other, computed tomography (CT).* The former uses blurring of the shadow from the obstructing organ by moving the source and the film in opposite directions. The latter uses a computer to mathematically combine the x-ray images taken from numerous angles and electronically process them to produce a picture of the cross-section or a tomogram.

### 11.8.1 Planigraphic Tomography

Consider the relative geometry between an x-ray source, an object and a film, as shown in Fig. 11.40. With the movement of the x-ray source from right to left, the shadows of the objects move from left to right. The

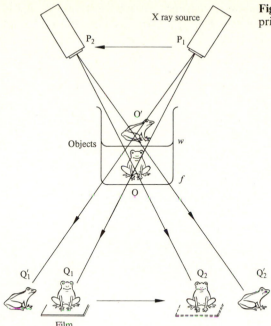

important point is that the amount of movement of the shadow of the object O is different from that of O′. The shadow of O moves from $Q_1$ to $Q_2$, whereas that of O′ moves from $Q_1'$ to $Q_2'$. The distance $\overline{Q_1' \, Q_2'}$ is longer than $\overline{Q_1 \, Q_2}$. The speed of the movement of the shadow of O′ is faster than that of O. If the film is moved to the right at the same speed as that of the shadow of O, a clear picture of O can be formed and that of O′ is blurred. With movement, images of any points on the surface of the plane $f$ are all sharp and those on all other planes are blurred to varying degrees. Thus, only the objects of the plane $f$ are selectively focused. This is the principle of planigraphic tomography [11.19]. The plane $f$ is called the focal plane. The difference $Q_1' \, Q_2' - Q_1 \, Q_2$ is known as is the distance of blurring.

The principle can be explained more quantitatively from Fig. 11.41. Because of symmetry, only one half of the movement is considered. The source located at $X_0$ projects the shadow of the object O to $X_i$. The same source casts the shadow of the object O′, which is $\Delta t$ away from O, to $X_i'$. The thickness $2\Delta t$ (which gives the distance of blurring $B_m$) is calculated. Using the similarity between the triangles $\Delta C_0 X_0 O$ and $\Delta C_i X_i O$ and between $\Delta C_0 X_0 O'$ and $\Delta C_i X_i' O'$ one obtains

$$\overline{C_i X_i} = \frac{b}{a} \, \overline{C_0 X_0}$$

$$\overline{C_i X_i'} = \frac{b + \Delta t}{a - \Delta t} \, \overline{C_0 X_0}$$

(11.47)

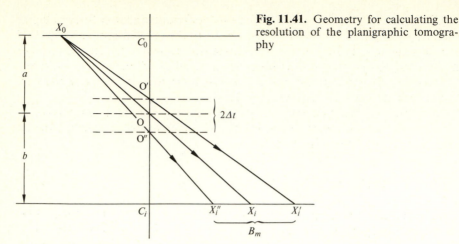

Fig. 11.41. Geometry for calculating the resolution of the planigraphic tomography

where $a$ is the vertical distance from the source to the object O and $b$ is the same from O to the film. The difference between the two equations is

$$\overline{X'_i X_i} \doteq \frac{B_m}{2} \doteq \frac{b}{a} \left( \frac{1}{a} + \frac{1}{b} \right) \overline{C_0 X_0} \, \Delta t$$

$$2\Delta t = \frac{B_m}{\dfrac{b}{a} \left( \dfrac{1}{a} + \dfrac{1}{b} \right) \overline{C_0 X_0}} . \tag{11.48}$$

The larger the movement $2 \, \overline{C_0 X_0}$ , the finer the thickness resolution is.

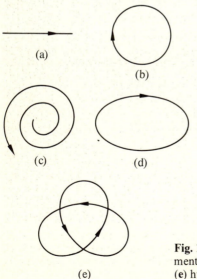

Fig. 11.42 a − e. Various types of tomographic movement. (a) Linear, (b) circular, (c) spiral, (d) elliptical, (e) hypercycloidal

If a structure located above or below the focal plane is a linear object, such as a straight bone, and if it is oriented in the same direction as that of the movement, the shadow of such a bone overlaps with the focused image despite the movement and deteriorates the quality of the image. This can be avoided by moving the source in either circular, spiral, elliptical or hypocycloidal paths, as shown in Fig. 11.42.

Planigraphic tomography has the advantage that no computation is necessary to see the image, but it has the disadvantage of not being able to effectively remove the shadow effect.

### 11.8.2 Computed Tomography (CT)

Computed tomography is able to remove shadow effects more effectively. As shown in Fig. 11.43, amazingly clear details in the cross-section of a human body can be reconstructed by computer tomography, but they are obtained at the expense of rather involved computation.

Consider the case when an x-ray penetrates through a composite material whose layers are characterized by the attenuation coefficients

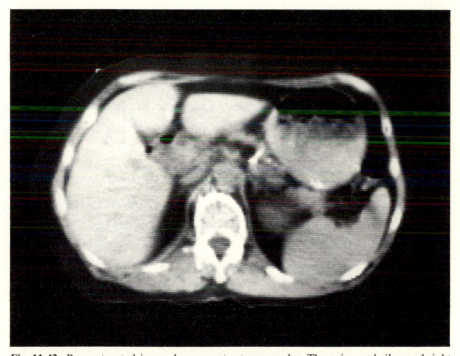

**Fig. 11.43.** Reconstructed image by computer tomography. The spine and ribs are bright white, the liver is the large organ on the left side of the image (patient's right) and the liquid and gas contents of the stomach are clearly resolved in the upper right (Courtesy of M. J. Bronskill)

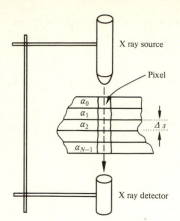

**Fig. 11.44.** Ray sum configuration

X ray source

Pixel

$\alpha_0$
$\alpha_1$
$\alpha_2$

$\Delta s$

$\alpha_{N-1}$

X ray detector

$\alpha_0, \alpha_1, \ldots, \alpha_{N-1}$ (Fig. 11.44). The power $I$ transmitted through the layers is

$$I = I_0 \exp\left[-(\alpha_0 + \alpha_1 + \alpha_2 + \ldots + \alpha_{N-1})\, \Delta s\right] \tag{11.49}$$

where $I_0$ is the incident x-ray power, and $\Delta s$ is the thickness of each layer. Equation (11.49) can be rewritten as

$$(\alpha_0 + \alpha_1 + \alpha_2 + \ldots + \alpha_{N-1})\, \Delta s = -\ln\frac{I}{I_0}. \tag{11.50}$$

The aim is to find the values of $\alpha_0, \alpha_1, \alpha_2, \ldots, \alpha_{N-1}$ by measuring $I/I_0$. In order to solve for the $N$ unknown values of $\alpha_0, \alpha_1, \ldots, \alpha_{N-1}$ uniquely, it is necessary to measure $I/I_0$ with $N$ different configurations. Expressed in integral form, (11.50) is equivalent to

$$\int f(x, y)\, ds = -\ln\frac{I}{I_0} \tag{11.51}$$

where $f(x, y)$ represents the two-dimensional (or cross-sectional) distribution of the attenuation associated with the layers under test.

Each small subdivision shown in Fig. 11.44 is called a picture element, or for short, pixel. Figure 11.45 shows one of the most common modes of scanning used in generating a computed tomogram. Each single scan is called a ray sum. A set of ray sums associated with one direction is called a projection. For example, in the diagram in Fig. 11.45, there are 4 projections, and 8 ray sums in each projection.

Computed tomography can be broadly classified into four categories according to the method of computation [11.20]:

a) Back projection method,
b) Algebraic reconstruction technique,

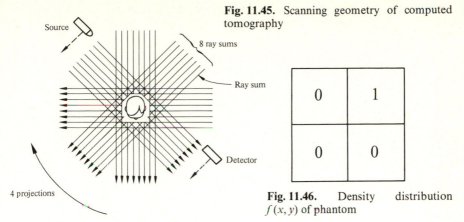

Fig. 11.45. Scanning geometry of computed tomography

Fig. 11.46.    Density    distribution $f(x, y)$ of phantom

c) Fourier transform method, and
d) Filtered back projection method.

Throughout the explanation of these four methods, the density distribution $f(x, y)$ shown in Fig. 11.46 will be used as an example. In the terminology of imaging, an object described by a density distribution, such as that shown in Fig. 11.46, is called *a phantom*.

**a) Back Projection Method.**    For simplicity of illustration, only the shadows projected in the two directions corresponding to the $x$ and $y$ axes are considered. The distributions of the projections are shown on the $x$ and $y$ axes in Fig. 11.47. Keep in mind that, in this example, a projected ray value (ray sum) of zero means the ray intensity suffered no attenuation in passing through the pixels in its path, while a projected ray value greater than zero indicates the degree of attenuation of the ray as it passed through the pixels. The purpose of this exercise is to extract the shape of the object from the projections. Using Fig. 11.47 as an example, the intensity distributions along the $x$ and $y$ axes are back projected along the $x$ and $y$ directions

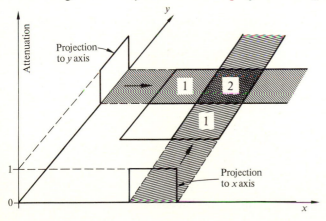

Fig. 11.47. Back projection method

as shown. The sections where the back projected shadows overlap denote the higher attenuation. The pattern thus obtained is shown by the shading in the squares in Fig. 11.47.

Computationally, one can achieve equivalent results to those displayed visually by the shaded areas in Fig. 11.47 as follows. Each pixel is assigned a value which is equal to the total of all the ray sums of the rays passing through it. For example, the upper right pixel in Fig. 11.47 has two rays passing through it, namely, the top 'row' ray and the right hand 'column' ray. Each of these rays has a ray sum of 1 so that the upper right pixel is assigned a value of 2. Comparing the results of the computation with the density distribution of Fig. 11.46, one finds that the computation correctly predicts that the darkest region is the upper right pixel. However, there is a fictitious attenuation present in two of the computed pixels. The back projection method is simple to implement but it has this disadvantage of creating unwanted streaks.

Fig. 11.48 a shows the geometry for deriving a more general expression for the projections. An expression for a straight line passing through a point $P$ at $(x, y)$ is sought. It is to be expressed in terms of $r$ and $\phi$ which are the perpendicular distance from the origin to the ray, and its angle to the $x$ axis, respectively. The expression is derived by an elementary vector analysis. Let the position vector of a point $P$ at $(x, y)$ be $P$, and let the unit vector of the perpendicular distance be $\hat{r}$. Then, $P = i x + j y$, and $\hat{r} = i \cos \phi + j \sin \phi$. Hence the projection of $P$ to $\hat{r}$ is given by the scalar product

$$r = P \cdot \hat{r} = x \cos \phi + y \sin \phi. \tag{11.52}$$

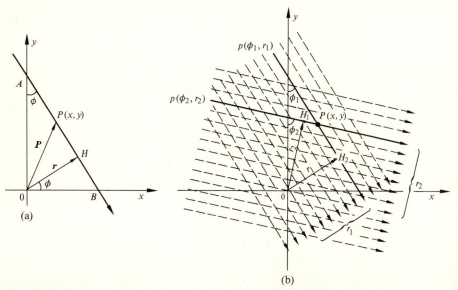

(a)

(b)

**Fig. 11.48.** Illustration of the back projection method of computed tomography

This means that for a given $(x, y)$ and $\phi$, the value of $r$ is determined by (11.52).

Let the intensity of the back projected image at a specific point $(x, y)$ be $\hat{f}(x, y)$. Here, the caret (ˆ) is used to signify the back projected image and, in general, this image will differ somewhat from the original object distribution $f(x, y)$. Notice that the rays in Fig. 11.48 b can be specified uniquely in terms of their $r$ and $\phi$ values. As the first step towards computing $\hat{f}(x, y)$, the computer stores all the ray sums in its memory banks, and at the same time, labels them according to the $r$ and $\phi$ values of the appropriate ray. For example, the ray sums passing through the point $P$ in Fig. 11.48 b would be stored as $p(r_1, \phi_1)$ and $p(r_2, \phi_2)$. For each point $(x, y)$ in $\hat{f}(x, y)$, the computer searches through the data in its memory to find the ray sums that have passed through this point. These ray sums are added together to form $\hat{f}(x, y)$ thus giving the intensity of the back projected image at the point $(x, y)$. Thus, the back projected image at $P$ in Fig. 11.48 b is $p(r_1, \phi_1) + p(r_2, \phi_2)$.

A general expression for the back projection method is

$$\hat{f}(x, y) = \sum_{i=0}^{N-1} p_{x,y}(r_i, \phi_i)$$

where $N$ is the number of projections passing through $(x, y)$. Noting that $r$ can be calculated from $x$, $y$, and $\phi$ using (11.52), the back projected image at $(x, y)$ can also be written as[2]

$$\hat{f}(x, y) = \int_{\phi_0}^{\phi_0 + \pi} p_{x,y}\{r(\phi), \phi\} \, d\phi. \tag{11.53}$$

These expressions will be used later.

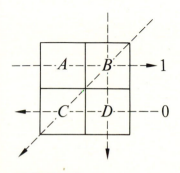

Fig. 11.49. Simultaneous equations to solve for pixel intensities

$$\begin{vmatrix} 1 & 1 & 0 & 0 \\ 0 & 1 & 0 & 1 \\ 0 & 0 & 1 & 1 \\ 0 & 1 & 1 & 0 \end{vmatrix} \begin{vmatrix} A \\ B \\ C \\ D \end{vmatrix} = \begin{vmatrix} 1 \\ 1 \\ 0 \\ 1 \end{vmatrix}$$

Solutions are $A = C = D = 0$, $B = 1$

---

2 All $H_i$ in Fig. 11.48 b are on the same circle and $\phi$ ranges from $\phi = \phi_0$ to $\phi_N$ covering $\pi$ radians.

The merit of this method is that the principle is straight forward but, unfortunately, it does not get rid of streaks in the reconstructed image.

**b) Algebraic Reconstruction Technique.**    If there are four unknown intensities of pixels, as shown in Fig. 11.49, four ray-sum equations are sufficient to find a unique solution.[3] While this is simple when only a small number of pixels are involved, the situation becomes more complicated as the number of pixels increases. For example, if there are $256 \times 256 = 65,536$ unknown pixels, this means that 65,536 simultaneous equations must be solved, clearly a formidable task! Instead of this brute force method, a more practical iterative method is considered. This iterative method is actually a combination of the back projection method and the simultaneous equation method.

The procedure for the zeroth iteration is exactly identical to the back projection method. The image obtained by the zeroth iteration is then used for the first iteration, as described in the following. New ray sums are calculated from the zeroth-order distribution. These calculated ray sums are then compared with the original experimentally obtained ray sums. The difference between two values is divided among all the pixels intercepted by the ray so that the modified ray sum exactly matches the experimentally obtained ray sum. Since a pixel is shared by more than one ray, the iteration in one ray undoes the iteration in the other ray that intercepts the same

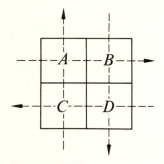

**Fig. 11.50.** Projections which do not give solutions

---

3 In formulating simultaneous equations, one should make sure that the equations are independent. For instance, simultaneous equations made by choosing such projections as shown in Fig. 11.50 do not have a unique solution because the equations are not all independent. This can be checked by the determinant, which must be non-zero for a unique solution to exist. Simultaneous equations made by the choice of projections shown in Fig. 11.50 are

$$\begin{vmatrix} 1 & 1 & 0 & 0 \\ 0 & 1 & 0 & 1 \\ 0 & 0 & 1 & 1 \\ 1 & 0 & 1 & 0 \end{vmatrix} \begin{bmatrix} A \\ B \\ C \\ D \end{bmatrix} = \begin{bmatrix} S_1 \\ S_2 \\ S_3 \\ S_4 \end{bmatrix}.$$

The determinant is zero. The determinant of the simultaneous equations in Fig. 11.49 is non zero.

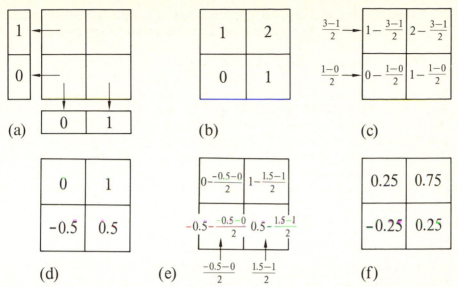

**Fig. 11.51 a – f.** Example for *ART* tomography. (**a**) Projection, (**b**) back projection. (**c**) First iteration to row pixels. (**d**) Result of the first iteration to row pixels. (**e**) First iteration to column pixels. (**f**) Result of the first iteration to column pixels

pixel. Iteration is repeated until a reasonably converged value is reached. Such an iterative method is sometimes called Algebraic Reconstruction Technique (ART).

An example of ART processing using the density distribution in Fig. 11.46 is shown in Fig. 11.51. The zeroth iteration is the same as that of the back projection indicated in chart (b). The first iteration is performed with the rows of the matrix. The sum of the first row of chart (b) is 3, and it is 2 in excess of the measured projected ray value shown in chart (a). This excess is evenly taken out from each pixel of the first row as shown in chart (c). The same procedure is repeated with the second row. The result with both rows iterated is shown in chart (d). Note that, as far as the sum of the pixels in each row of chart (d) is concerned, they match the measured projection shown in chart (a), but the sums in the columns do not match with the measured projection.

The same procedure is repeated with the columns of chart (d) as is illustrated by chart (e) with the result shown in chart (f).

Usually, the iteration in the columns undoes the result of the iteration in the rows, hence, only the result of either row or column, in general, matches with the measured projections. With this particular example, however, the values both in row and column match the measured projections and the differences to be used for the next iteration are all zero. The values have converged already after a small number of iterations.

As can be seen by comparing the final result in chart (f) of Fig. 11.51 with Fig. 11.46, the values obtained by iteration do not necessarily converge to the correct values even after the values both in row and column have matched the measured projections. This is one of the drawbacks of this method.

**c) Fourier Transform Method.** This analysis is made in the spatial frequency domain. First, $f(x, y)$ is transformed by using an $x-y$ rectangular coordinate system in which the orientation of the axis is taken in an arbitrary direction. The two-dimensional Fourier transform with respect to these coordinates is

$$F(f_x, f_y) = \int\int_{-\infty}^{\infty} f(x, y) \exp(-j2\pi f_x x - j2\pi f_y y) \, dx \, dy \tag{11.54}$$

where $f_x$, $f_y$ are the Fourier spectra in the $x$ and $y$ directions and where the $f_x$ and $f_y$ axes are parallel to the $x$, $y$ axes, respectively.

Special attention is focused on the value of $F(f_x, f_y)$ when $f_y = 0$. In that case (11.54) becomes

$$F(f_x, 0) = \int\int_{-\infty}^{\infty} f(x, y) \, e^{-j2\pi f_x x} \, dx \, dy$$

$$= \int_{-\infty}^{\infty} \left[ \int_{-\infty}^{\infty} f(x, y) \, dy \right] e^{-j2\pi f_x x} \, dx \, . \tag{11.55}$$

The meaning of (11.55) is that the integral in the square bracket is the projection of $f(x, y)$ along $x = x$ in the $y$ direction onto the $x$ axis. This means that *the value of the two-dimensional Fourier transform along the $f_x$ axis is obtained by Fourier transforming the projection $p$ with respect to $x$,* namely

$$F(f_x, 0) = \mathscr{F}\{p(x)\} \, . \tag{11.56}$$

In the above, the $x$ axis was taken in an arbitrary orientation with respect to the object, and the value of the Fourier transform along any other straight line can be found in the same manner. If the direction of the $x$ axis is rotated in steps through a total angle of $2\pi$ radians, the two-dimensional Fourier transform $F(f_x, f_y)$ is obtained for the entire $f_x - f_y$ plane. Once the two-dimensional Fourier transform in the entire plane is known, it is easy to reconstruct $f(x, y)$, which can be obtained by simply inverse Fourier transforming $F(f_x, f_y)$.

A slightly different explanation concerning the relationship between the Fourier transform and the projection can be given. A double integral is the limit of a sum of area elements, each element of which is multiplied by the value of the integrand at the location of the element. The integrand of

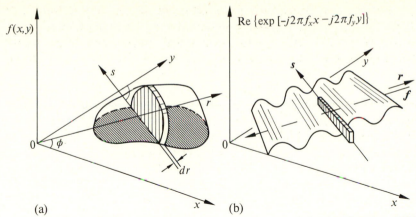

**Fig. 11.52 a, b.** Graphical illustration of the performance of the double integral $\mathscr{F}(f_x, f_y)$

$$= \iint_{-\infty}^{\infty} f(x, y) \exp(-j\,2\pi f_x\,x - j\,2\pi f_y\,y)\,dx\,dy.$$ **(a)** First factor $f(x, y)$. **(b)** Second factor $\exp[-j\,2\pi(f_x\,x + f_y\,y)]$

(11.54) can be separated into two factors; the first factor is $f(x, y)$ and the second factor $\exp(-j2\pi f_x\,x - j\,2\pi f_y\,y)$. These factors are indicated in Figs. 11.52a, b. Note that the second factor is precisely the expression of a plane wave propagating in the $-\boldsymbol{f}$ direction, or in the direction connecting the origin to the point at $(f_x, f_y)$ (Sect. 2.3). A new rotated coordinate system $(r, s)$ is introduced, and the $r$ axis is taken along the $\boldsymbol{f}$ direction,

$$
\begin{aligned}
\boldsymbol{f} &= -(f_x\,\boldsymbol{i} + f_y\,\boldsymbol{j}) = \hat{\boldsymbol{f}}f \\
f_x &= f\cos\phi, \quad f_y = f\sin\phi \\
|\boldsymbol{f}| &= \sqrt{f_x^2 + f_y^2} \\
\phi &= \tan^{-1}\left(\frac{f_y}{f_x}\right).
\end{aligned}
\right\} \tag{11.57}
$$

The expression for the second factor now can be written as $\exp(-j\,2\pi f\,r)$.

Since there is no restriction on how to slice the bounded area, one way of slicing is in the direction parallel to the equiphase line of the plane wave, as shown in Fig. 11.52a. The double integral can then be calculated by the following two steps. First, the calculation of the projection along the $s$ axis is performed,

$$p(r, \phi) = \int_s f(x, y)\,ds. \tag{11.58}$$

Next, all the projections are summed along the $r$ axis,

$$F(f_x, f_y) = P(f, \phi) = \int_{-\infty}^{\infty} p(r, \phi)\,e^{-j2\pi f r}\,dr. \tag{11.59}$$

The separation into two steps was possible because the value of $\exp(-j2\pi fr)$ was constant analog the $s$ axis. Thus, the single Fourier transform of the projection is the double Fourier transform along the projected line. By repeating the same procedures at each step of the rotation of $r$, the Fourier transform is obtained in the entire $f_x, f_y$ space.

The Fourier transform method is very elegant, but one difficulty with this method is that the points at which the two-dimensional Fourier transform is determined are in a polar pattern (Fig. 11.53 a). In order to reconstruct $f(x, y)$, $F(f_x, f_y)$ must be inverse Fourier transformed. This is often done by using the Discrete Fourier Transform (DFT), outlined in Sect. 7.1; however, the DFT applies to samples of $F(f_x, f_y)$ in a Cartesian pattern, as shown in Fig. 11.53 b. A possible solution is to interpolate or extrapolate in order to estimate the values in the rectangular pattern from those in the polar pattern.

The method of two point interpolation is illustrated in Fig. 11.54. The value $c$ at point $P$ is to be interpolated from the values $a$ and $b$ of the points located, respectively, $d_1$ and $d_2$ away from the point $P$. The idea is to weigh the values of $a$ and $b$ according to their proximity to $c$ as follows,

$$c = a\left(\frac{d_2}{d_1 + d_2}\right) + b\left(\frac{d_1}{d_1 + d_2}\right) = (b - a)\frac{d_1}{d_1 + d_2} + a = \frac{a/d_1 + b/d_2}{1/d_1 + 1/d_2}. \quad (11.60)$$

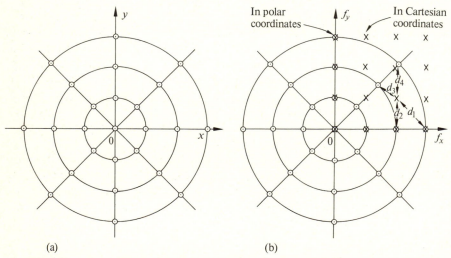

(a)                                                    (b)

**Fig. 11.53 a, b.** The available $DFT$ in polar coordinates is represented by ⊙ and those in cartesian coordinates by x. (a) Dots represent the location of the sampled points. (b) A rectangular grid in the spatial frequency domain (only the first quadrant for clarity) superimposed on the available DFT in Polar coordinates

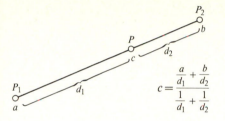

**Fig. 11.54.** Two-point interpolation for the value $c$ at $P$

$$c = \frac{\dfrac{a}{d_1} + \dfrac{b}{d_2}}{\dfrac{1}{d_1} + \dfrac{1}{d_2}}$$

In a similar manner, the interpolated value $F(P)$ at a point surrounded by the values $a$, $b$, $c$ and $d$ located $d_1$, $d_2$, $d_3$ and $d_4$ away, respectively, as shown in Fig. 11.53 b, is

$$F(P) = \frac{a/d_1 + b/d_2 + c/d_3 + d/d_4}{1/d_1 + 1/d_2 + 1/d_3 + 1/d_4}. \tag{11.61}$$

The interpolation needs sizable computation time and is therefore a drawback in this method.

An example of the Fourier transform method will be demonstrated in Fig. 11.55 by again using the same object distribution as in Fig. 11.46. From the projections to the $x$ and $y$ axis, the DFT's along the $f_x$ and $f_y$ axes are calculated[4] using (7.3) with $N = 2$. Simply using two projections 90 degrees apart is not enough to determine the value at $f_x = f_y = 1$ which is needed to find the inverse DFT later. The projection at 45° is used to determine the value near $f_x = f_y = 1$. In performing the Fourier transform in the diagonal direction, the spacing between the pixels is now $\sqrt{2}$ times as long as that of either in the $x$ or $y$ direction (Fig. 11.55 a). A property of Fourier transforms is that an elongation in one domain results in a contraction in the corresponding Fourier transform domain. Hence increasing the spacing by $\sqrt{2}$ in the spatial domain results in a contraction of $1/\sqrt{2}$ and expansion in the magnitude by $\sqrt{2}$ in the spatial frequency domain [Recall the similarity theorem: $\mathscr{F}\{f(s/\sqrt{2})\} = \sqrt{2}\,F(\sqrt{2}\,f)$]. The results of the Fourier transform are summarized in Fig. 11.55 b. The value of the DFT at $f_x = f_y = 1$ is approximated as $-0.8$ by using the interpolation formula (11.61).

It is interesting to compare this interpolated result with the exact value shown in Fig. 11.55 c obtained by Fourier-transforming $f(x, y)$ directly. It is evident that there is a sizable error in the interpolated value at (1, 1). Needless to say, it is not usually possible to make this comparison, simply because the object distribution is not known.

---

4 The easiest way to perform the two-dimensional Fourier transform is to first arrange the inputs in a matrix form and perform the one-dimensional transform along the row elements and put back the Fourier transformed results in the respective row of the matrix. Using these new matrix elements, the same process is then carried out for the column elements. The result is the two-dimensional Fourier transform.

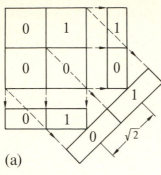

(a)

**Fig. 11.55 a – d.**   Charts of Fourier transform method of computer tomography. (**a**) Projections in $x, y$ and diagonal directions. (**b**) Fourier transform from the projections. (**c**) Exact Fourier transform of object distribution. (**d**) Inverse Fourier transform of (**b**)

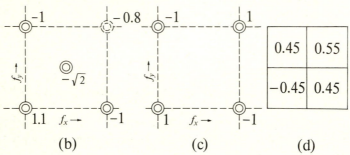

(b)              (c)              (d)

The expression $f(x, y)$ is finally obtained by inverse Fourier transforming the function $F(f_x, f_y)$ and the result is shown in Fig. 11.55 d. The sizable difference between the charts in Figs. 11.46 and 11.55 d is produced primarily by interpolation errors.

**d) Fourier Transform in Polar Coordinates and Filtered Back Projection.** These methods are based on Fourier analysis and back-projection. Writing the distribution function as a Fourier transform gives

$$f(x, y) = \int\limits_{-\infty}^{\infty}\int F(f_x, f_y)\, e^{j2\pi(f_z x + f_y y)}\, df_x\, df_y \,. \tag{11.62}$$

The Fourier transform operation will be manipulated in the polar coordinate system. Equation (11.62) is rewritten using (11.57) as

$$f(x, y) = \int\limits_{0}^{2\pi}\int\limits_{0}^{\infty} F(f \cos\phi, f \sin\phi)$$

$$\times \exp\left[j\, 2\pi f\, (x \cos\phi + y \sin\phi)\right] |f|\, df\, d\phi \,. \tag{11.63}$$

Recall that $F(f \cos\phi, f \sin\phi)$ was obtained from the single Fourier transform of the projection on the $r$ axis. The axis $r$ is oriented at an angle $\phi$ to the $x$ axis, as shown in Fig. 11.52. Now, (11.63) is further rewritten by using (11.52, 59) as

$$f(x, y) = \int\limits_{0}^{\pi} \left\{ \int\limits_{-\infty}^{\infty} \left[ \underbrace{\int\limits_{-\infty}^{\infty} p(r, \phi) \, e^{-j2\pi fr} \, dr}_{} \right] \underbrace{|f| \, e^{j2\pi fr} \, df}_{} \right\} d\phi. \qquad (11.64)$$

measured projection

"filter" in Fourier domain

One dimensional Fourier transform $P(f, \phi)$

Inverse Fourier transform

Addition from all directions passing through $(x, y)$

In order to express (11.64) exactly in the form of an inverse Fourier transform with respect to $f$, the lower limit of the integral with respect to $f$ was changed to $-\infty$ and the upper limit of the integral with respect to $\phi$ was changed to $\pi$, and $f$ in the integrand was changed to $|f|$. It is necessary to use $|f|$ rather than $f$ because the element of area $f \, df \, d\phi$ was defined for positive values of $f$.

There are two ways one can approach the calculation of (11.64). The more direct approach (Fourier Transform method in polar coordinates) is described first. This procedure is to obtain the one-dimensional Fourier transform $P(f, \phi)$ along the $r$ axis from the measured projections and then multiply the Fourier transformed value by $|f|$. The next step is to inverse Fourier transform the above quantity with respect to $f$. The last step is to sum up all the contribution from all angles $\phi$ to obtain the final result. In summary, the computation of (11.64) for one particular point $(x, y)$ can be separated into the following steps:

1) Obtain $P(f, \phi)$ by Fourier transforming the projection,
2) multiply $P(f, \phi)$ by $|f|$,
3) evaluate the inverse Fourier transform of $P(f, \phi) |f|$ at $r(\phi)$, which is the projection from the point $(x, y)$ to a particular direction $\phi$,
4) repeat step (3) for different $\phi$'s,
5) sum up all the values obtained in step (4) to find $f(x, y)$ at $(x, y)$ (in a way similar to the back projection), and
6) in order to fill up the values for the entire $x, y$ plane, repeat the same procedures for other points in the entire plane.

It should be recognized that, but for $|f|$, (11.64) is exactly the same as (11.53) of the back projection method. The factor $|f|$ is interpreted as a filter with a frequency characteristic of $|f|$. A filter with such a characteristic attenuates lower spatial frequency components and accen-

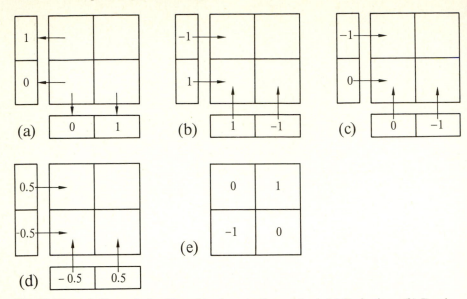

**Fig. 11.56 a – d.** Example for filtered back projection method. (a) Projections. (b) Fourier transform $P(f, \varphi)$ of the projections. (c) Product of $p(f, \varphi)|f|$. (d) Inverse Fourier transform of $P(f, \varphi)|f|$. (e) Back projection of (d)

tuates higher spatial frequency components. The intensity of a streak is constant across the image plane and it contains more of the lower spatial frequency components, whereas the edges of the contour of the images have abrupt changes and contain more of the higher frequency components. The present filter emphasizes the higher frequency components and is suitable for improving the quality of such an image.

The procedure will be illustrated using once again the distribution in Fig. 11.46. In step (1), $P(f, \phi)$ is obtained by Fourier transforming the projection in Fig. 11.56a. The result is shown in Fig. 11.56b. In the step (2), multiplication by $|f|$ is performed. The origin is taken at $(0, 0)$, and the result is shown in Fig. 11.56c. In step (3), $P(f, \phi)|f|$ is inverse Fourier transformed in the $\phi = 0°$ and $\phi = 90°$ directions and the result is given in Fig. 11.56d. In step (4), the final result (Fig. 11.56e) is obtained by back projecting the distribution (Fig. 11.56d). It is seen in this example that the contrast of the upper right pixel against the surrounding pixels is stronger than obtained by the method of back projection. Caution should be used when applying the DFT. In digital form, $|f|$ can be represented as in Fig. 11.57a. It extends from $l = -(N/2 - 1)$ to $l = N/2$. Now, suppose that $p_k$ are the points sampled from the projection and that the $P_l$ are the DFT of these points. If the procedure described in Chap. 7 is used to perform the DFT, the value of the $l$ index for $P_l$ will be either

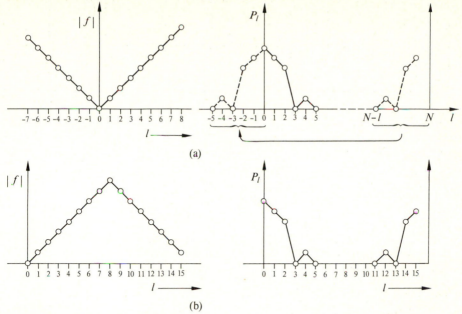

**Fig. 11.57 a, b.** Method of calculating product of $P_l$ and a filter $|f|$ (a) when $P_{-l}$ was generated from $P_{N-l}$ and (b) when $|f|$ was reformed

positive or zero. In other words, the $P_l$ values are normally plotted to the right of the origin (Fig. 11.57b). Since $|f|$ was plotted symmetrically with respect to the origin, this results in a possible source of confusion when one attempts to multiply $P_l$ by $|f|$. One approach for removing the confusion is to use the relationship given in Problem 7.2 in order to find the $P_l$ values for negative $l$. This relationship $P_{-l} = P_{N-l}$ allows the $P_l$ to be transferred to the left of the origin as shown in Fig. 11.57a. Alternatively, the left half of $|f|$ can be transferred to the right of the origin (Fig. 11.57b). In the case of the 2 point DFT, the transfer becomes ambiguous and $|f|$ was arbitrarily represented by the sequence of $\{0, 1\}$ in this example.

The second approach to calculating (11.64), called the filtered back projection, will now be described. The filtered back projection has the advantage of shortening the number of steps required through manipulation of (11.64). As far as the integral in $f$ is concerned, (11.64) is in the form of an inverse Fourier transform of a product, so that (11.64) can be rewritten as

$$f(x, y) = \int_0^\pi p'(r, \phi)\, d\phi \quad \text{where} \tag{11.65}$$

$$p'(r, \phi) = p(r, \phi) * h(r) \tag{11.66}$$

$$h(r) = \int_{-\infty}^{\infty} |f|\, e^{j2\pi rf}\, df. \tag{11.67}$$

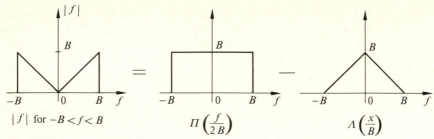

$|f|$ for $-B < f < B$     $\Pi\left(\dfrac{f}{2B}\right)$     $\Lambda\left(\dfrac{x}{B}\right)$

**Fig. 11.58.** Expression of $|f|$ when the range is restricted to $-B < f < B$

Because this expression is in the form of a convolution, the filtered back projection method is sometimes called the "convolution" method.

In a digital computation, the limit of integration $-\infty$ to $\infty$ in (11.67) is not practical. The assumption is then made that the image has a spatial frequency spectrum extending only up to $f = B$, so that the image is assumed to be band-limited. In that case, the limits of the integration can be safely replaced by $-B$ to $B$. With this assumption, (11.66) can be further calculated. The Fourier transform of (11.66) is

$$P'(f, \phi) = P(f, \phi) \cdot |f| .$$ (11.68)

Referring to Fig. 11.58, the function $|f|$ with the range limit of $B$ is represented by

$$|f| = B\left[\Pi\left(\frac{f}{2B}\right) - \Lambda\left(\frac{f}{B}\right)\right].$$ (11.69)

Inserting (11.69) into (11.68) gives

$$P'(f, \phi) = B P(f, \phi)\left[\Pi\left(\frac{f}{2B}\right) - \Lambda\left(\frac{f}{B}\right)\right].$$ (11.70)

Note that $P(f, \phi)$ is band-limited and is zero outside the band, hence the first term of (11.70) is simply $B P(f, \phi)$. Using (2.38), the inverse Fourier transform $p'(r, \phi)$ becomes

$$p'(r, \phi) = B p(r, \phi) - B^2 p(r, \phi) * \text{sinc}^2(B r) .$$ (11.71)

The computation of (11.71) is not as bad as it looks. If the spacing $\zeta$ of the sampled points is taken as $\zeta = 1/(2 B)$, which satisfies the sampling condition, then $r_n$, which is the $n$th sampled point, becomes $n/(2 B)$ and

$$\text{sinc}^2(B r_n) = \begin{cases} 1 & n = 0 \\ \dfrac{1}{(n \pi)^2}, & \text{odd } n \\ 0 , & \text{even } n \end{cases} .$$ (11.72)

Using (11.72), the calculation of (11.71) is straightforward.

The Filtered Back Projection method uses the same expression (11.64) as used for the Fourier transform method in polar coordinates. The difference is the means of computation. Equation (11.66) is the expression of the projection $p(r, \phi)$ filtered by $h(r)$ whose frequency spectrum is $|f|$. Equation (11.65) is the expression of the back projection.

The Filtered Back Projection method immediately filters the measured projected data and then back projects. More realistically, a band limited filter, such as expressed by (11.69), is used in (11.67). The computation procedure for the filtered back projection is

1) Calculate $h(r)$ using such an expression as (11.69).
2) Perform the convolution (11.66) between $p(r, \phi)$ and $h(r)$ which is a calculation in one dimension.
3) Use the result of the convolution for the operation (11.65) of the back projection.

As a matter of fact, the filtered back projection method is probably the most frequently used method by the manufacturers of computed tomography equipment.

# Problems

**11.1**   The operation of averaging is to be made by using the arrangement shown in Fig. 11.6. The spacing between the $A$'s is 0.5 mm. The wavelength of the light used is 0.63 µm and the focal length of the lens $L_1$ is 50 cm. What should the spacing $a$ between the pinholes in the mask be?

**11.2**   In Sect. 11.4, the number of delta functions used to represent the point response function $t(x, y)$ was 4. What would happen if 8 delta functions were used?

**11.3**   In processing the picture with quadruple exposure in Sect. 11.4, all $a_n$'s were assumed to be unity. What happens if the $a_n$'s are assigned the values: $a_0 = 1$, $a_1 = 2$, $a_2 = 3$, and $a_3 = 4$?

**11.4**   Explain why the tiny pinholes in the low pass filter of Sect. 11.5.3 should be randomly distributed. What happens if they are arranged periodically?

**11.5**   Two approaches were described as a means of removing the grid lines from a picture, such as shown in Fig. 11.35a. One method was to use a filter for periodic noise as described in Sect. 11.3.4, and the other employed a logarithmic filter. Compare the restriction on the value of the sampling period a.

**11.6** Figure 11.59 is a picture of a tiger in a cage. One wants to make a picture of the tiger without the cage from this picture. Design two coherent light systems and one incoherent light system for this purpose. The parameters are: $a = 0.4$ mm, $b = 0.5$ mm, $c = 20$ mm, and $d = 10$ mm. The highest spatial frequency in the picture is $f_B = 0.8$ lines/mm. The focal length of the lens to be used is 100 mm. The wavelength of the coherent light to be used is $\lambda = 0.6$ μm, and the incoherent light wavelength is centered at $\lambda = 0.6$ μm.

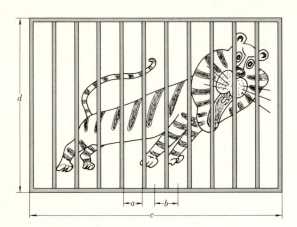

**Fig. 11.59.** A picture of a tiger in a cage

**11.7** Try all the computer tomographic methods using the density distribution shown in Fig. 11.60.

| 0 | 1 |
|---|---|
| 1 | 0 |

**Fig. 11.60.** Density distribution

**11.8** Derive (11.72) from (11.71).

# 12. Applications of Microwave Holography

The principle of holography is applicable not only to light waves but also to microwaves and acoustic waves. Light, microwave and acoustic waves are similar in that they are all wave phenomena but they are quite different in the way they interact with material media. For example, a medium which is opaque to light waves may be transparent to microwaves and acoustic waves. There are certain advantages in using microwaves in place of light waves in holography, one of which is that the wavelength of the microwave is about $10^5$ times larger than that of light, thus making it easier to manipulate the fringe pattern. Another advantage is that microwave sources radiating with long-term stability are readily available.

## 12.1 Recording Microwave Field Intensity Distributions

Microwave holography can be classified into three categories according to the methods of reconstructing the image. One method is to reconstruct the image by illuminating the photographically reduced microwave hologram with laser light. The second method is to illuminate the microwave hologram directly with microwaves without photographic reduction. The real image is then projected onto microwave sensitive sheets. The last method uses a computer to form the reconstructed image.

In the case of optical holography, recording the light intensity distribution is no problem; it is simply done by means of a photographic plate. However, in microwave holography, the recording of the microwave intensity distribution is a much more complex matter. Numerous procedures for recording the microwave intensity distribution have been proposed. In the following subsections, these procedures will be introduced.

### 12.1.1 Scanning Probe Method

Figure 12.1 illustrates a setup for scanning a probe antenna across a frame. The output from the probe antenna is detected and amplified. The amplified signal is used to modulate the intensity of a small lamp attached to the probe carriage. The probe carriage is scanned in the dark and a

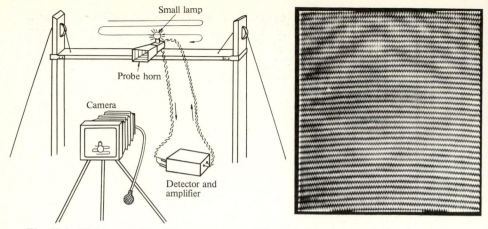

**Fig. 12.1.** Field mapping by a scanning probe

**Fig. 12.2.** Microwave field mapped by a scanning probe [12.2]

camera aperture is set open to take the entire intensity distribution of the microwave field. Figure 12.2 shows a microwave field mapped by this method. The sensitivity of the system can be varied by adjusting the gain of the amplifier. It is often $60-70$ dB higher than for most of the methods which will be described in this chapter. The limitation of the system is primarily due to mechanical vibration and a slow mapping speed [12.1-4].

In addition to planar scanning of the probe, there are also non-planar scanning methods where the probe movement is either spiral or cylindrical. Other variations include scanning the object while keeping the probe fixed, or scanning both.

The following methods are non-scanned, area mapping types. They all utilize the heat generated by the microwave dielectric loss.

### 12.1.2 Method Based on Changes in Color Induced by Microwave Heating

When a thin film is placed in an $E$ field, a current density $i = \sigma E$ is induced in the thin film. The film is heated in proportion to the square of this induced current. *The heat sensitive film will then change in color* [12.5-8].

The first method utilizes filter paper immersed in cobalt chloride $(CoCl_2-6H_2O)$. The color of this filter paper is light pink but the color changes to blue when the temperature is raised by microwave heating. Figure 12.3 shows an example of microwave intensity distributions recorded by this technique. The pattern is a mode pattern inside a horn antenna connected to a waveguide.

The second method utilizes a *temperature dependence of the developing time of a Polaroid film*. The Polaroid film starts the developing action as soon as the film is pulled through a narrow gap of rollers and the developing reagent fills a layer between the positive and negative. As soon

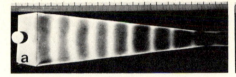

Fig. 12.3 a, b. Microwave field mapped by cobalt chloride sheet [12.5]. (a) Inside a microwave horn. (b) Inside a waveguide

Fig. 12.4. Microwave field mapped by Polaroid technique

as developing starts, the film is put into the microwave field in question. The developing reagent is a water based material with high dielectric loss, and a good microwave absorber. Those parts of the film irradiated by microwaves raise in temperature. The heated portions develop faster than the unheated portions. If positive and negative are separated at a premature stage of development, the fringes of the color density distribution representing the microwave field is obtained. Figure 12.4 shows a microwave field pattern at 35 GHz measured in this manner using type 58 Polaroid film. The sensitivity of the method is about $10 \sim 20$ mW/cm$^2$.

*The third method utilizes the dependence of color on temperature in liquid-crystal film.* The simplicity of the preparation is an advantage of this method. Preparation consists of merely painting the liquid crystal over a film of microwave absorber. The sensitivity of the method is about 10 mW cm$^{-2}$.

The fourth method utilizes the color fading rate of photochromic film. The photochromic film is transparent to visible light, but when illuminated with ultraviolet light it changes its color to dark blue. When the ultraviolet light is removed, the color starts fading and becomes transparent again. *The fading rate is accelerated by an increase in temperature.* By backing the photochromic film with a layer of microwave absorber, the variation of fading rate with temperature is used to map the microwave distribution. Since the photochromic is of non-granular texture, the advantage of this method is that the reconstruction of the microwave holographic image can be achieved by directly illuminating the photochromic sheet with laser light.

### 12.1.3 Method by Thermal Vision

A thermal vision camera can be used to map the heat distribution generated by the microwave illumination on a microwave absorber film [12.9]. The mechanism of the thermal vision is similar to a television camera except that the center of the sensitivity spectrum of the thermal vision camera is not in the visible region but in the infrared region. Figure 12.5 shows the appearance of the thermal vision camera system. It has the capability of detecting a temperature difference of 0.05 °C from a few meters away.

**Fig. 12.5.** Photo of the Thermal Vision of AGA Corporation

### 12.1.4 Method by Measuring Surface Expansion

When microwaves are directed onto a microwave absorber, a small heat expansion takes place according to the amount of microwave energy absorbed. This surface heating effect can be used to record the microwave intensity distribution, provided a highly sensitive measuring technique is available to detect the minute expansion [12.10, 11]. *Interferometric holography discussed in Sect. 8.8.6 is well suited for detecting such a minute*

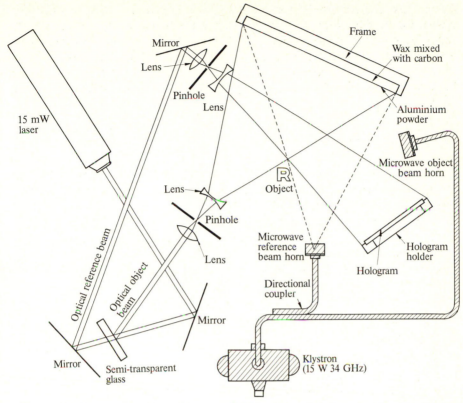

**Fig. 12.6.** Method of mapping microwave field by measuring the expansion distribution of the microwave illuminated absorber by means of an interferometric hologram

*change.* Best results are obtained when the microwave absorber is fabricated out of a material with a large heat expansion coefficient such as wax or silicone rubber mixed with carbon. Figure 12.6 is a block diagram describing the method. The shaded portions indicate the microwave holography and the remaining unshaded portions refer to the optical interferometric holography. The surface expansion can be recorded by making an interferometric hologram with and without the microwave source on. Figure 12.7a is the object "*R*" made out of aluminum sheet. Figure 12.7b is the pattern recorded by this interferometric method and Fig. 12.7c is the reconstructed image of the microwave hologram of the object "*R*".

Another highly sensitive technique for mapping the surface expansion is the *Moiré technique.* The Moiré pattern is best described by giving an example. Suppose one has two nylon stocking fabrics, and one nylon stocking fabric is placed over the other. If each one of the threads in one fabric is precisely laid over those of the other fabric, no pattern is seen, but as soon as one fabric moves even slightly, a big grid pattern which represents the

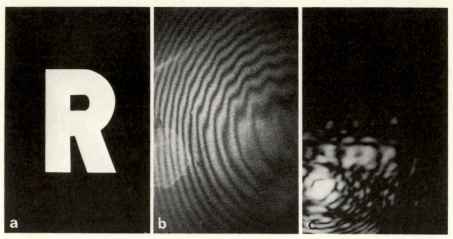

Fig. 12.7. (a) A target made of aluminium sheet. (b) A microwave hologram made from a surface expansion pattern measured by an optical interferometric hologram. (c) Reconstructed image obtained by illuminating a reduced microwave hologram with laser light

weaving pattern becomes visible. The pattern moves very rapidly with a small movement of the fabric and the amplified movement of the fabric is observed. The same amplification action as described in this simple fabric example is used to detect the minute surface expansion of a microwave absorber due to microwave heating. The microwave intensity distribution thus mapped is used for fabricating a microwave hologram. The optical image of a grid pattern is projected onto the surface of a microwave absorber.

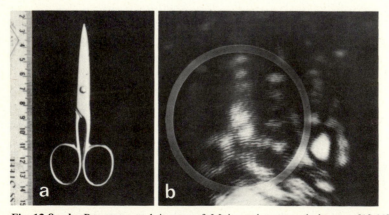

Fig. 12.8 a, b. Reconstructed image of Moire microwave hologram [12.11]. (a) Object. (b) Reconstructed image of Moire microwave hologram of the object

A double-exposed photograph of the grid pattern is made with the micro-wave source on and off. A Moiré pattern is observed in the portion where the expansion has taken place. The reconstructed image of a microwave hologram made by the Moiré technique is shown in Fig. 12.8. The object is a small pair of scissors.

## 12.2 Microwave Holography Applied to Diagnostics and Antenna Investigations

Microwave holography can be used to extract information about materials which is not obtainable by optical means alone. Microwave holography has also proven valuable in resolving dilemmas in antenna theory, as well as pro-viding new antenna designs.

### 12.2.1 "Seeing Through" by Means of Microwave Holography

Since microwaves can penetrate into several materials which are optically opaque, microwave holography has been used as a "seeing through" device [12.12–14]. An example of a large scale application is the investigation of the internal structure of the crust of the moon by holographically pro-cessing the signal transmitted and received by an orbiting space craft.

*The largest difference between microwave holography and x-rays as a "see through" device is the film position. The x-ray film must always be behind the object because the x-ray image is the shadow of the object, whereas the film position in a microwave hologram is almost arbitrary.* This flexibility in posi-tioning the film is clearly an advantage in such applications as nondestruc-tive testing and prospecting for the location of mining ore.

Figure 12.9 shows the detection of a coin and a triangular metal piece in a purse by means of microwave holography. In this example, the image is reconstructed by microwaves instead of by a laser beam, elaborated as fol-lows. The microwave field intensity distribution is first recorded as a liquid-crystal film color pattern (Sect. 12.1.2). By using properly set color filters, this color pattern is converted into a black and white pattern with an optical density proportional to the microwave field intensity. This black and white pattern is further changed into a screened photo consisting of tiny dots simi-lar to those used for half-tone photos in a newspaper. This screened photo is printed onto a printed circuit board. The printed circuit board is then used as a hologram when the image is reconstructed by an illuminating micro-wave. The transmittance of the microwave is low in the area of larger dots, and high in the area of smaller dots. When the printed circuit board is illu-minated by microwave, the real microwave image is reconstructed. This reconstructed microwave image can be observed as a color pattern by placing liquid crystal film at the location of the image formation.

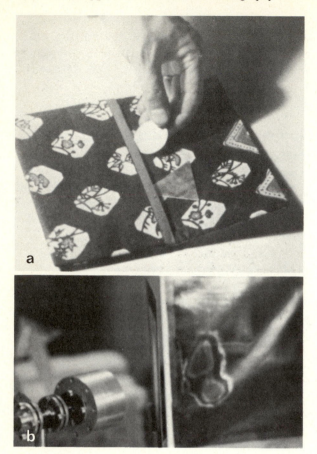

**Fig. 12.9 a, b.** Seeing through by microwave holography. (**a**) Inserting a Kennedy coin and a triangular metal sheet into a purse. (**b**) Result of seeing through the purse by microwave holography

## 12.2.2 Visualization of the Microwave Phenomena

Most of the time, the microwave hologram is photographically reduced by $10^{-4} \sim 10^{-5}$ and the image is reconstructed by laser light. This means that *phenomena occurring in the microwave wavelength region are translated to that of a visible wavelength region, as if our eyes had become sensitive to microwaves. Visualization of a radiating monopole is one example of this kind* [12.15].

This extended vision into the microwave region has helped resolve the two theories of interpretation of the mechanism of wave radiation from a linear antenna. One theory states that every portion of the wire is responsible for radiation. The other theory states that only the portions at the ends and the center driving-point, i.e. the areas of discontinuity, are responsible. In other words, the question is: does the reconstructed image of a monopole look like a corrugated sausage proportional to a sinusoidal curve, or does it consist of just two bright spots at both ends of the antenna? Figure 12.10 shows the result. The result confirms the latter theory. Once one hologram is fabricated, the radiation pattern at other arbitrary planes can be formed by the proper combination of the locations of a lens and a screen. This is another attractive feature of this method.

The radiation pattern of a half wave dipole is a figure eight and that of a longer antenna produces a pattern similar to a chrysanthemum flower. These radiation patterns can be nicely explained in terms of an interference pattern of three point sources, one at each of the two ends, and a point source in the center.

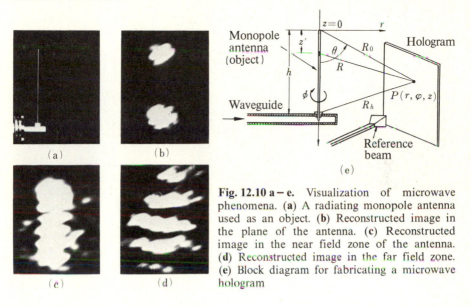

(a)      (b)

(e)

(c)      (d)

**Fig. 12.10 a−e.** Visualization of microwave phenomena. (**a**) A radiating monopole antenna used as an object. (**b**) Reconstructed image in the plane of the antenna. (**c**) Reconstructed image in the near field zone of the antenna. (**d**) Reconstructed image in the far field zone. (**e**) Block diagram for fabricating a microwave hologram

### 12.2.3 Subtractive Microwave Holography

Because the size of the microwave fringe pattern is about $10^5$ times larger than that of light, it is possible to manipulate the fringe patterns of the microwave hologram to play tricks. An example of one of these tricks is subtractive microwave holography. By making a contact print of the hologram, a hologram with reversed contrast is easily made. The shape of the reconstructed image from the hologram, after the reversal of contrast, is the same as before but the phase of the reconstructed image is reversed [12.16].

A subtractive microwave hologram is made by interlacing a reversed and a regular hologram to form a mosaic type hologram as shown in Fig. 12.11. This hologram is then photographically reduced in size and illuminated by laser light. The resulting reconstructed image represents the difference between the two original constituent images.

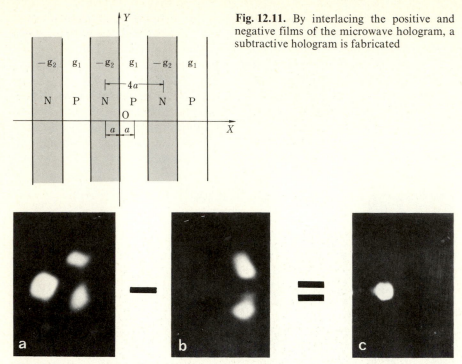

**Fig. 12.11.** By interlacing the positive and negative films of the microwave hologram, a subtractive hologram is fabricated

**Fig. 12.12 a – c.** Subtractive microwave holography. (a) Reconstructed image of three coins in a purse. (b) Reconstructed image of two coins in a purse. (c) The image of the difference of the above

In order to demonstrate this principle, two holograms were made. In the first, the object consisted of three coins, and in the second, only two coins were used. The phase of the latter hologram was reversed. The subtracted image is shown in Fig. 12.12.

By making a microwave hologram based upon this principle, the image of a plasma discharge, free from interference by the reflected wave from the container, is obtained, and the accuracy of the diagnosis is superior to other microwave diagnoses of the plasma discharge. The experimental arrangement is shown in Fig. 12.13 and the results are shown in Fig. 12.14. Figure 12.14a shows the reconstructed image of the glass tube only, Fig. 12.14b shows the complete cancellation of the image of the glass, and Fig. 12.14c shows the image of the plasma discharge only.

### 12.2.4 Holographic Antenna

A holographic antenna is an antenna designed to make use of the *duality principle of holography*. When a hologram is illuminated by the reference beam, the object beam is reconstructed and, when it is illuminated by the

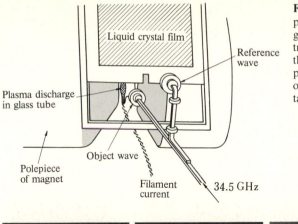

**Fig. 12.13.** Visualization of the plasma discharge enclosed in a glass tube. By means of subtractive microwave holography, the electron density of the plasma without the interference of the glass tube can be obtained

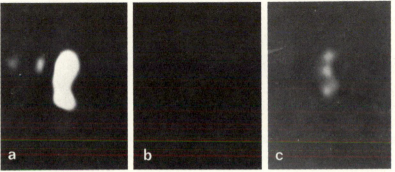

**Fig. 12.14 a–c.** Reconstructed image of the microwave hologram of a plasma tube. (a) Reconstructed image of only the glass tube. (b) Reconstruction of the subtracted image when the plasma discharge is not on. (c) Reconstruction of the subtracted image when the plasma discharge is on

object beam, the reference beam is reconstructed [12.17, 18]. This principle of holography is expressed by the formulae

reference beam + hologram → object beam
object beam + hologram → reference beam.

Figure 12.15 shows a procedure for designing a holographic antenna. As shown in Fig. 12.15a, a parallel beam from a paraboloid disc is used as a reference beam and a beam from a horn antenna is used as an object beam. Both beams illuminate the holographic plate. The field-intensity distribution across the holographic plate is then measured. Based upon this information, a microwave hologram is made. The microwave hologram in this case is simply a sheet of aluminum with many small windows. The size of the windows is modulated in such a way that its transmittance is proportional to the measured field intensity at that spot.

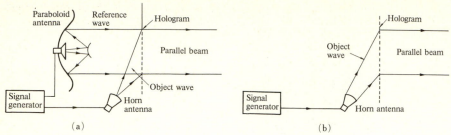

**Fig. 12.15 a, b.** Design principle of a holographic antenna. (**a**) Fabrication of hologram. (**b**) Formation of a parallel beam by the combination of the horn antenna and the hologram

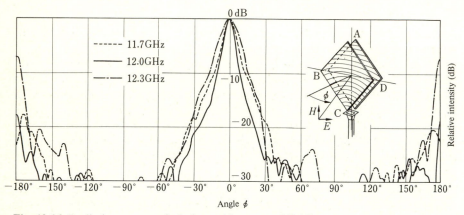

**Fig. 12.16.** Radiation pattern of a holographic antenna [12.18]

By illuminating this hologram with the diverging beam from the horn antenna, a parallel beam is obtained. Figure 12.16 shows the measured pattern of the holographic antenna fabricated in this manner.

### 12.2.5  A Method of Obtaining the Far-Field Pattern
####        from the Near Field Pattern

The radiation pattern of an antenna is normally measured in the far-field region of the antenna. The distance $R$, beyond which it is considered to be far field (Sect. 3.3) is

$$R = \frac{D^2}{\lambda},$$

where $D$ is the diameter of the antenna under test, and $\lambda$ is the wavelength of the radiating field. In the case of a very large aperture antenna, however, the value of $R$ which satisfies the far-field condition is so large that it

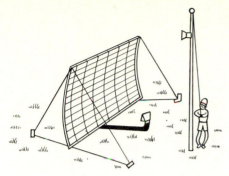

**Fig. 12.17.** Determination of the far-field pattern of a large aperture antenna

becomes too cumbersome to move the probe antenna around at such a large radius [12.19, 20].

By fabricating a hologram out of the microwave signal measured in the near field (Fig. 12.17), the far-field pattern can be optically observed. The measured amplitude of the near field and the theoretical amplitude of a reference beam are first added. After that, a fringe pattern is drawn with an intensity modulation given by the value of the sum, and this pattern is photographically reduced in size to make a hologram. The microwave image is optically reconstructed by illuminating the hologram by laser light. *The field intensity distribution across the antenna aperture is visualized when one looks into the hologram, and the far-field pattern is visualized when the transmitted light is projected onto a screen located some distance away.* The distance to the screen can be reduced by observing it at the back focal plane of a converging lens.

The optical system is not necessary for obtaining the final results. The results can be computed by using a simplified scattering theory. The far-field pattern of the antenna can be obtained by Fourier transforming the near-field distribution, and the antenna aperture field can be obtained by inverse Fresnel transforming the near-field distribution. The inverse Fresnel transform is explained as follows. From (3.34), the relationship between the field distribution $g(x, y)$ across the antenna aperture, and the field distribution $g_n(x, y)$ across a plane located a distance $d$ in front of the aperture, is given by

$$g_n(x, y) = g(x, y) * f_d(x, y) . \tag{12.1}$$

If the value of $g(x, y)$ is desired, it can be obtained by convolving (12.1) with $f_{-d}(x, y)$, namely

$$\begin{aligned} g_n(x, y) * f_{-d}(x, y) &= [g(x, y) * f_d(x, y)] * f_{-d}(x, y) \\ &= g(x, y) * [f_d(x, y) * f_{-d}(x, y)] \\ &= g(x, y) , \end{aligned} \tag{12.2}$$

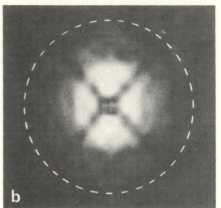

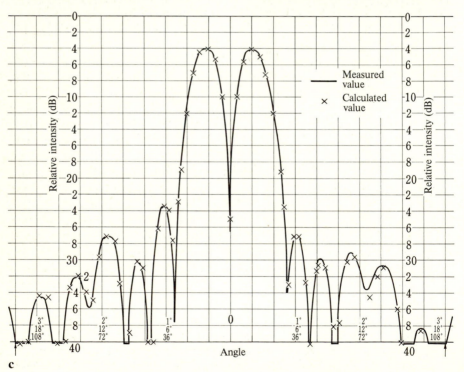

c

and therefore

$$g(x,y) = g_n(x,y) * f_{-d}(x,y) . \tag{12.3}$$

Figure 12.18 a shows the antenna used for this experiment, while Fig. 12.18 b shows the field distribution across the aperture obtained by the inverse Fresnel transform of the near field. Figure 12.18 c shows the comparison between the measured far-field pattern and the calculated far-field pattern obtained by Fourier transforming the near-field pattern by a computer. The agreement between the two values is quite good.

## 12.3 Side Looking Synthetic Aperture Radar

The side looking synthetic aperture radar is a radar with a power of resolution which is enhanced by processing the radar echo in a special manner [12.21−23]. The power of resolution of a radar is influenced by the beam width of the antenna. In order to obtain a narrow radiation beam and hence a high resolution, one generally needs an antenna with a large aperture.

A real aperture antenna funnels all the information into the system at once, but, for instance, if the target is a ground object that does not change with respect to time, it is not necessary to process the incoming information immediately. *In the side looking synthetic aperture radar, all the necessary information about the incoming signal is picked up by scanning a small probe antenna, where the distance scanned is called the synthetic aperture (Fig. 12.19). The resolution of this type of antenna is equivalent to that of an antenna with an aperture equal to the scanned distance.* For instance, if the total distance of the scanning is 1 km, and the wavelength is 10 cm, the beam width is found from (4.2) to be $1.14 \times 10^{-2}$ degrees, which is an extremely high power of resolution.

Unlike other radar antennas, the antenna for the side looking radar has a wide fan-shaped beam. When an aircraft uses this antenna to map the terrain below, the targets at larger distances are illuminated for a longer time than those at shorter distances. As a result, unlike the case of a pulse radar, *resolution does not deteriorate with the distance to the target.* This is a very desirable feature of the side looking synthetic aperture radar.

The recording of the radar signal is performed on a running film as a black and white fringe pattern. The image of the target is reconstructed

**Fig. 12.18 a − c.** A method of obtaining the far-field pattern out of the near-field pattern of a large aperture antenna [12.20]. (**a**) The 3 m paraboloid antenna under test. (**b**) From the measured distribution across a plane located some distance away from the paraboloid antenna under test, the distribution of the field on the paraboloid was calculated and mapped onto facsimile paper. (**c**) The far-field calculated from the near-field pattern is compared with the measured far-field [12.19]

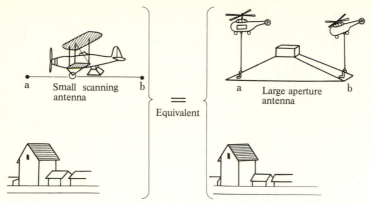

**Fig. 12.19.** The beam width of a synthetic aperture antenna is equivalent to that of a large real aperture antenna

from this film by means of an optical system. A disadvantage of the radar is that it takes some time before the image is displayed after the signal is received. Moreover, various mechanical devices have to be improvised, at high cost, to fly the probe antenna straight.

### 12.3.1 Mathematical Analysis of Side Looking Synthetic Aperture Radar

Suppose that an aircraft carrying a probe antenna flies over land to make a map. In order to make an accurate map, resolution is required in the direction of the flight path (azimuth) as well as in the direction perpendicular to it (ground range). The power of resolution of the side looking synthetic aperture radar is obtained by the pulse method in the ground range direction and by the synthetic aperture antenna in the azimuthal direction.

For simplicity, only one target, located at $(x_0, y_0)$ in the coordinate system in Fig. 12.20, is considered. Let the distance from the aircraft to the target be $r$, and let the wave transmitted from the aircraft be represented as

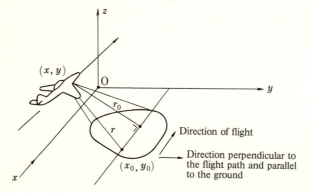

**Fig. 12.20.** Geometry of the synthetic aperture radar

$$S(t) = a\,e^{-j\omega t}. \tag{12.4}$$

The received signal $R(t)$ scattered from the target is

$$R(t) = a' \exp\left[-j\,\omega\left(t - \frac{2r}{c}\right)\right] \tag{12.5}$$

where $a'$ is a complex number which absorbs all physical constants including the electrical properties of the scatterer and the transmission loss.

The received signal $R(t)$ is first amplified and then mixed with the transmitter signal $S(t)$. The mixed signal is then square-law detected. The output $I(x)$ from the square-law detector is

$$I(x) = \tfrac{1}{2}[S(t) + R(t)][S(t) + R(t)]^*$$
$$= |a|^2\,(1 + \tfrac{1}{2}\,e^{j2kr} + \tfrac{1}{2}\,e^{-2jkr}) \tag{12.6}$$

where

$$k = \frac{\omega}{c} \tag{12.7}$$

and where, in order to simplify the expression, the amplifier is set such that $a'$ is equal to $a$.

Referring to Fig. 12.20, an approximate expression for $r$ is

$$r = \sqrt{r_0^2 + (x - x_0)^2} \doteq r_0 + \frac{1}{2}\frac{(x - x_0)^2}{r_0}.$$

Inserting (12.7) into (12.6) gives

$$I(x) = |a|^2\left[1 + \cos k\left(2r_0 + \frac{(x - x_0)^2}{r_0}\right)\right]. \tag{12.8}$$

Equation (12.8) is equivalent to the expression for a one-dimensional zone plate (Sect. 4.5 and Problem 4.5) with a focal length of $r_0/2$ centered at $x = x_0$. Consequently, a straight horizontal line drawn on a film with its opacity modulated in accordance with (12.8) has the properties of a one-dimensional zone plate. When the film is illuminated by a thin parallel beam, the transmitted light will be focused to one vertical line.

In reality, there will be other targets at distances other than $r_0$; and it is necessary to select only the signals associated with the distance $r_0$, otherwise the resultant signal will not have the form of (12.8).

Details of the selection process are illustrated in Fig. 12.21. The microwave pulse from the oscillator is radiated through the antenna. The returned echo from the targets is fed through a circulator to the detector amplifier. The output from the amplifier is used to modulate the raster of

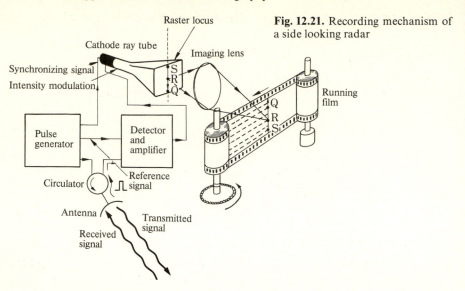

**Fig. 12.21.** Recording mechanism of a side looking radar

a scope. The raster is scanned only along one line. By using a converging lens, the image of the raster is projected onto film running perpendicular to the scanning direction of the raster.

As soon as a microwave pulse is emitted, the scanning of the raster on the scope starts from S to Q along a straight line. The intensity of the raster is stepped up as soon as the signal returns after being bounced from the target. The position of the brightened raster indicates the distance to the target. For instance, let us say that the raster was brightened at $R$, so that the length $\overline{SR}$ represents the time taken for the wave to make a round trip to the target. The further away the location of the target, the longer the length $\overline{SR}$; and the stronger the reflection from the target, the brighter the point $R$ becomes. Thus, the echoes from the targets at equidistances from the aircraft are recorded along a horizontal line on the film. The horizontal line from one target at the distance $r_0$, modulated according to (12.8), is represented by the line $R_0$ in Fig. 12.22. A cylindrical convex lens is placed in front of $R_0$ and illuminated from behind by a narrow parallel beam. The

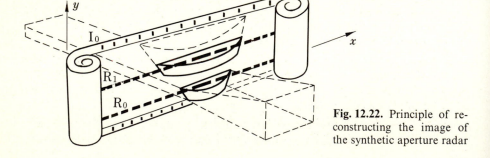

**Fig. 12.22.** Principle of reconstructing the image of the synthetic aperture radar

light passing through the cylindrical lens undergoes a phase shift of

$$\exp\left(-jk\frac{x^2}{2f}\right).$$

Light passing through both the lens and the film can be expressed as

$$I(x)\exp\left(-jk\frac{x^2}{2f}\right) = |a|^2\left(\exp\left(-jk\frac{x^2}{2f}\right)\right.$$

$$+\frac{1}{2}\exp\left\{jk\left[2r_0+\frac{x_0^2}{r_0}+\left(\frac{1}{r_0}-\frac{1}{2f}\right)x^2-\frac{2x_0}{r_0}x\right]\right\}$$

$$\left.+\frac{1}{2}\exp\left\{-jk\left[2r_0+\frac{x_0^2}{r_0}+\left(\frac{1}{r_0}+\frac{1}{2f}\right)x^2-\frac{2x_0}{r_0}x\right]\right\}\right) \tag{12.9}$$

where in order to simplify the analysis, the speed of the running film is taken to be the same as that of the airplane, and $I(x)$ itself is taken as the transmittance of the film disregarding the film characteristics. If the focal length $f$ of the lens is selected to be $r_0/2$, the second exponential term in (12.9) becomes

$$\frac{1}{2}\exp\left[jk\left(2r_0+\frac{x_0^2}{r_0}\right)\right]\exp\left(-jkx\sin\theta_0\right) \qquad \text{where} \tag{12.10}$$

$$\sin\theta_0 = \frac{2x_0}{r_0}.$$

This is an expression for a plane wave propagating at an angle $\theta_0$ with respect to $r_0$, which later becomes the optical axis $z$ in Fig. 12.23. All other terms in (12.9) represent non-planar waves. As far as the $y$ direction is concerned, there is no focusing action and the wave front is cylindrically divergent.

Next, the final stage of reconstructing the image from the film will be described using Fig. 12.23. The equi-phase surface of the light transmitted through the lens $L_1$ is at an angle $\theta_0$ to the $x$ direction and divergent in the $y$ direction. The shape of the equi-phase surface is similar to year markings on a tree-trunk cross-section in a wedge shape, with the center of the year markings as its apex. If the center of the year markings is placed at an angle $\theta_0$ to the $x$ axis, the year markings would then correspond to equiphase surfaces of the transmitted light.

If a cylindrical lens $L_2$ is placed with its front focal plane at $R_0$, the wave front of the wave from $L_2$ becomes parallel to both the $y$ direction and the direction angled at $\theta_0$ with respect to the $x$ axis, and can be focused to one point by the convex lens $L_3$.

The same can be said about the line $R_1$ in Fig. 12.22. If the focal length $f$ is selected to be $r_1/2$, the beam propagating in the $\theta_1$ direction is obtained.

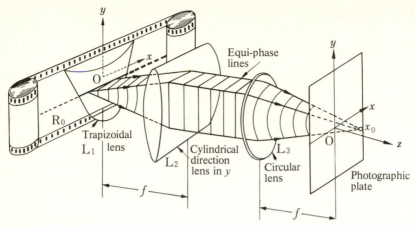

**Fig. 12.23.** Method of obtaining the image from the recording of a synthetic aperture radar by means of a lens system

**Fig. 12.24.** Reconstructed image of the glacier taken by a synthetic aperture radar (Courtesy of the Environmental Research Institute of Michigan)

As shown in Fig. 12.23, if a convex lens $L_1$ of a trapezoidal shape is placed such that the focal length decreases as the bottom end is approached, the transmitted wave across the film becomes a combination of plane waves emanating from different locations in the film and propagating in various directions. It can be used to focus multitargets to their respective points and the over-all image is formed. Figure 12.24 shows an image made by a side looking synthetic aperture radar.

## 12.4 HISS Radar

HISS radar (Holographic Ice Surveying Systems) is a short-range radar which measures distance by detecting the radius of curvature of the reflected wave [12.24]. Using Fig. 12.25 as an analogy, when a stone is thrown into a pond not too far from the bank, the radius of curvature of the wave ripple observed on the bank is small; on the other hand, when the stone is thrown far from the bank, the ripple wave front looks like a plane wave and the radius of the curvature approaches infinity. It is, therefore, possible to determine the distance by knowing the radius of curvature.

**Fig. 12.25.** Simplified illustration of the principle of HISS radar

HISS radar is based upon this principle. Figure 12.26 shows the block diagram of the radar. The principle of the HISS radar is quite different from that of a pulsed or an FM radar which measures distance by the lapse of time. The HISS radar measures distance by the shape of the wave front of the returned signal.

Referring to Fig. 12.27, a sharp beam of radiation is formed by the transmitting antenna array. This beam is reflected from all scattering centers $S_0(y)$ located underneath the top surface of the ice. The field distribution along the receiving antenna array from all these scatterers is given by

$$H(x) = K \int_{y_0}^{\infty} \frac{S_0(y)}{y^2} \exp{(j\, k\, \sqrt{y^2 + x^2})}dy \doteq K \int_{y_0}^{\infty} \frac{S_0'(y)}{y^2} \exp\left(j\, k\, \frac{x^2}{2\, y}\right) dy \qquad (12.10)$$

where $S_0(y)$ is the reflection coefficient of the target at $(0, y)$, and where the approximation, (12.7), was used. $S_0'(y)$ absorbed $\exp{(j\, k\, y)}$. The presence of the term $y^{-2}$ is due to the round-trip divergence of the beam. By changing variables

$$X = x^2 \qquad (12.11)$$

$$Y = (2\, y\, \lambda)^{-1} \qquad (12.12)$$

(12.10) becomes

$$H(X) = K' \int_{Y_0}^{\infty} S_1(Y)\, e^{j2\pi XY}\, dY . \qquad (12.13)$$

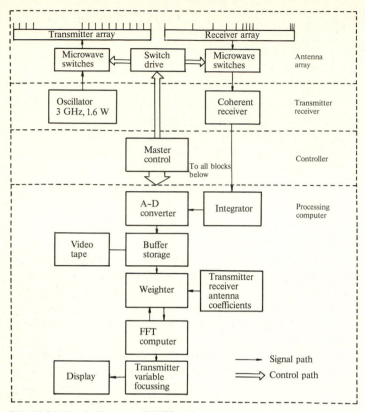

**Fig. 12.26.** Block diagram of HISS radar

Equation (12.13) is mathematically recognizable as a form of Fourier transform. $H(X)$ is the inverse Fourier transform of $S_1(Y)$. The distribution of the reflection coefficients $S_1(Y)$ of the targets can therefore be obtained by measuring $H(X)$ along the $X$ axis and calculating its Fourier transform.

The reason that the receiving antenna elements in Fig. 12.27 are not equally spaced is that, by arranging them quadratically, the transformation of (12.11) can be done automatically.

The other special feature of this radar is that the array elements need not be excited all at once and in fact, the performance can be improved if both transmitting and receiving elements are excited sequentially forming a so-called hologram matrix. Elements of the hologram matrix consist of the signals with every combination of the transmitter and receiver elements. The information about the distance is derived by double FFT of the elements of the hologram matrix. The next subsection explains how to manipulate the hologram matrix in a general manner to obtain the distance information.

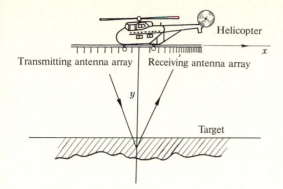

**Fig. 12.27.** Coordinate system of a HISS radar

### 12.4.1 Hologram Matrix

An ordinary hologram is made with a fixed position of illumination. The hologram matrix is generated by scanning the illuminating source as well as the receiver, thus acquiring information about the object with every geometric configuration of source, receiver and object. The concept of the hologram matrix is best realized in the case of microwave holography because the distribution of the phase of a microwave is much more easily measured than that of light.

Consider an antenna array equally spaced along the $X$ axis as shown in Fig. 12.28. Suppose only the $m$th antenna emits a spherical wave to illuminate the object and only the $n$th antenna records the signal of the complex scattered field which is denoted by $H_{nm}$. If the field $H_{nm}$ is recorded for each receiving antenna $n = 0, 1, 2, 3, \ldots, N-1$ with every transmitting antenna $m = 0, 1, 2, 3, \ldots, N-1$, one obtains an $N \times N$ matrix. This is called the holo-

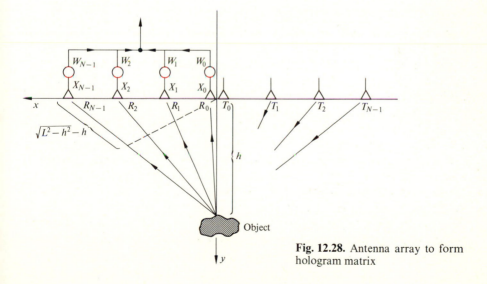

**Fig. 12.28.** Antenna array to form hologram matrix

gram matrix. Equation (12.14) is a $4 \times 4$ hologram matrix, and represents the case of 4 element transmitting and 4 element receiving antennas,

|       | $T_0$    | $T_1$    | $T_2$    | $T_3$    |
|-------|----------|----------|----------|----------|
| $R_0$ | $H_{00}$ | $H_{01}$ | $H_{02}$ | $H_{03}$ |
| $R_1$ | $H_{10}$ | $H_{11}$ | $H_{12}$ | $H_{13}$ |
| $R_2$ | $H_{20}$ | $H_{21}$ | $H_{22}$ | $H_{23}$ |
| $R_3$ | $H_{30}$ | $H_{31}$ | $H_{32}$ | $H_{33}$ |

$$(12.14)$$

The element $H_{13}$, for example, is the complex field when the field emitted by transmitter 3 is recorded by receiver 1.

The manner in which the HISS radar derives the depth information from the hologram matrix will be explained using Fig. 12.28. The signal received by $R_1$ is delayed from that by $R_0$ by

$$\Delta\phi = \frac{2\pi}{\lambda} \Delta l, \qquad (12.15)$$

where $\Delta l$ is the difference between the path from the target to $R_0$ and the path from the target to $R_1$. Those of $2, 3, \ldots, N-1$ are delayed by $2\Delta\phi, 3\Delta\phi, \ldots, (N-1)\,\Delta\phi$. However, the value of the phase shift $\Delta\phi$ decreases with the distance $h$ of the target and is given from Fig. 12.28 by

$$\Delta\phi = \frac{1}{N-1} \frac{2\pi}{\lambda} (\sqrt{h^2 + L^2} - h)$$

$$\doteqdot \frac{1}{N-1} \frac{\pi}{\lambda} \frac{L^2}{h}. \qquad (12.16)$$

The signals received by each antenna are

$$\left.\begin{aligned}
X_0 &= E_h \\
X_1 &= E_h\, e^{j\Delta\phi} \\
X_2 &= E_h\, e^{j2\Delta\phi} \\
&\;\vdots \\
X_n &= E_h\, e^{jn\Delta\phi} \\
X_{N-1} &= E_h\, e^{j(N-1)\Delta\phi}
\end{aligned}\right\} \qquad (12.17)$$

If the phase shifters $W_0, W_1, \ldots, W_{N-1}$ are installed, one to each port as indicated in Fig. 12.28, the output $y$ from the array is

$$y = \sum_{n=0}^{N-1} X_n\, W_n. \qquad (12.18)$$

The output $y$ will become a maximum if the values of the phase shifters are chosen in such way to exactly compensate the delays at each port, i.e.

$$
\left.\begin{aligned}
W_0 &= 1 \\
W_1 &= e^{-j\Delta\phi} \\
W_2 &= e^{-j2\Delta\phi} \\
&\vdots \\
W_n &= e^{-jn\Delta\phi} \\
W_{N-1} &= e^{-j(N-1)\Delta\phi}
\end{aligned}\right\} \tag{12.19}
$$

In other words, the antenna array is focused by the phase shifters to the distance $h$ specified by (12.16) and the focal distance can be controlled by the values of the phase shifters.

In fact, if a digital system is employed, the phase shifters, in a physical sense, are not needed. One can simply multiply numerically the input $X_n$ by the complex number $W_n$ specified by (12.19).

Notice that (12.17 and 19) are remarkably similar to certain expressions developed in connection with the DFT in Chap. 7. Equation (12.17) forms a geometric series which looks exactly the same as the inverse Discrete Fourier Transform and (12.19) forms a series which looks exactly the same as the weighting factors of the DFT. Let us take the DFT of (12.17),

$$
y_m = E_h \sum_{n=0}^{N-1} \exp\left[j\left(\Delta\phi - \frac{2\pi}{N}m\right)n\right]. \tag{12.20}
$$

At the particular distance $h$ [or $\Delta\phi$ which is related to $h$ by (12.16)] such that the innermost bracket in the exponent becomes zero, the value of $y_m$ peaks, and the antenna is focused to this distance. The term $(2\pi/N)\,m\,n$ in (12.20) can be considered as the $n^{\text{th}}$ phase shifter.

The $m$th spectrum of (12.20) represents the array output when the focused distance $h$ (or $\Delta\phi$) is

$$
\Delta\phi = \frac{2\pi}{N}m \qquad \text{and from (12.16)}
$$

$$
h_m = \frac{L^2 N}{2(N-1)\lambda}\frac{1}{m}. \tag{12.21}
$$

The spectrum $(y_0, y_1, y_2, y_m, y_{N-1})$ represents the outputs when the array is focused to the distances $(h_0, h_1, h_2, h_m, h_{N-1})$.

Now let us return our attention to the manipulation of the hologram matrix of (12.14). The DFT of the first column, denoted by $\bar{H}_{m0}$,

$$
\begin{vmatrix}
\mathscr{T}_0 \\
\bar{H}_{00} \\
\bar{H}_{10} \\
\bar{H}_{20} \\
\bar{H}_{30} \\
\bar{H}_{(N-1)0}
\end{vmatrix} \tag{12.22}
$$

represents the output from the array when the receiver array is focused to the distances $(h_0, h_1, h_2, h_m, \ldots, h_{N-1})$ and only one transmitter $T_0$ is used to irradiate the targets. Similarly, the DTF of the second column denoted by

$$
\begin{vmatrix}
\mathcal{T}_1 \\
\bar{H}_{01} \\
\bar{H}_{11} \\
\bar{H}_{21} \\
\bar{H}_{31} \\
\vdots \\
\bar{H}_{(N-1)1}
\end{vmatrix}
\tag{12.23}
$$

represents the output from the array when the receiver array is focused to similar distances but when the transmitter $T_1$ alone is used to irradiate the targets.

Exactly the same argument holds for the entire transmitter array except that the direction of propagation is reversed. The transmitting beam can be focused to various distances by installing the phase shifters. The focal distance $h$ and the amount of the phase shift is likewise determined by (12.16). Thus, the Fourier transform of the row elements of the matrix

$$
\mathcal{R}_0 \quad \bar{H}_{00}, \quad \bar{H}_{01}, \quad \bar{H}_{02}, \quad \bar{H}_{0m} \quad \bar{H}_{0N-1}
\tag{12.24}
$$

corresponds to the array output when the transmitter array is focused to the distances $(h_0, h_1, h_2, h_m, h_{N-1})$ and only one receiver $R_0$ is used to receive the reflected signal from the target.

Therefore, the double DFT of both column and row elements of the matrix represents the signal when both transmitting and receiving arrays are focused. The diagonal elements of

| | $\mathcal{T}_0$ | $\mathcal{T}_1$ | $\mathcal{T}_2$ | $\mathcal{T}_3$ |
|---|---|---|---|---|
| $\mathcal{R}_0$ | $\bar{H}_{00}$ | $\bar{H}_{01}$ | $\bar{H}_{02}$ | $\bar{H}_{03}$ |
| $\mathcal{R}_1$ | $\bar{H}_{10}$ | $\bar{H}_{11}$ | $\bar{H}_{12}$ | $\bar{H}_{13}$ |
| $\mathcal{R}_2$ | $\bar{H}_{20}$ | $\bar{H}_{21}$ | $\bar{H}_{22}$ | $\bar{H}_{23}$ |
| $\mathcal{R}_3$ | $\bar{H}_{30}$ | $\bar{H}_{31}$ | $\bar{H}_{32}$ | $\bar{H}_{33}$ |

$$\tag{12.25}$$

represent the output of the radar as the focii of both transmitter and receiver are moved together along the $y$ axis.

**Exercise 12.1**  Show the principle of operation of a hologram matrix radar taking the case of $N = 2$ as an example.

*Solution*  The geometry shown in Fig. 12.29 is considered. If the signal from $T_0$ to $R_0$ is taken as unity, the signal from $T_0$ to $R_1$ is $e^{j\Delta\phi} = A$, and that from $T_1$ to $R_1$ is $A^2$, where

**Fig. 12.29.** Two element hologram matrix radar

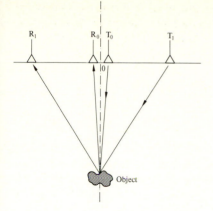

$$\Delta \phi = \frac{2\pi}{\lambda} (\sqrt{h^2 + L^2} - h) \doteq \frac{\pi}{\lambda} \frac{L^2}{h}. \tag{12.26}$$

The hologram matrix is then

|       | $T_0$ | $T_1$ |
|-------|-------|-------|
| $R_0$ | 1     | $A$   |
| $R_1$ | $A$   | $A^2$ |

$(12.27)$

The two point Fourier transform of the column is

|       | $\mathcal{T}_0$ | $\mathcal{T}_1$ |
|-------|-----------------|-----------------|
| $R_0$ | $1 + A$         | $A + A^2$       |
| $R_1$ | $1 - A$         | $A - A^2$       |

,

and the further transform in the row is

|                 | $\mathcal{T}_0$ | $\mathcal{T}_1$ |
|-----------------|-----------------|-----------------|
| $\mathcal{R}_0$ | $(1 + A)^2$     | $1 - A^2$       |
| $\mathcal{R}_1$ | $1 - A^2$       | $(1 - A)^2$     |

$(12.28)$

If the object is located at $h_0 = \infty$, then

$$A = 1$$

and the matrix elements become

|                 | $\mathcal{T}_0$ | $\mathcal{T}_1$ |
|-----------------|-----------------|-----------------|
| $\mathcal{R}_0$ | 4               | 0               |
| $\mathcal{R}_0$ | 0               | 0               |

$(12.29)$

The first element of the diagonal elements becomes large indicating the location of an object at $h = \infty$.

If, however, the object is located at $h = L^2/\lambda$ or $A = -1$ the matrix elements become

|           | $\mathcal{T}_0$ | $\mathcal{T}_1$ |
|-----------|-----------------|-----------------|
| $\mathcal{R}_0$ | 0 | 0 |
| $\mathcal{R}_0$ | 0 | 4 |

(12.30)

indicating that the location of the object corresponds to $m = 1$, and this location is given by (12.21) as $h_1 = L^2/\lambda$.

Figure 12.30 shows the appearance of the HISS radar and Fig. 12.31 shows the results obtained. Images of high quality can be obtained by this radar. An additional feature of this radar is that it needs neither pulse shaping nor frequency modulation as does a pulse or FM radar.

**Fig. 12.30.** Photograph of HISS radar

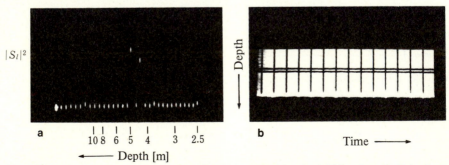

**Fig. 12.31 a, b.** Profile of the thickness of ice measured by the HISS radar. (a) Echo from top and bottom surfaces. (b) Intensity modulation of (a)

# 13. Fiber Optical Communication

Fiber optical communication is based on the principle of light transmission through a fine glass fiber by total internal reflection. A first demonstration of the guiding of light by total internal reflection was given by John Tyndall in 1870. In front of an audience of the Royal Academy of London, he

demonstrated that light illuminating the top surface of water in a pail can be guided along a semi-arc of water streaming out through a hole in the side of the pail. Tremendous progress has been made since then, and thin glass fiber is now a viable means of transmission of light for communications.

John Tindall's experiment in 1870

Figure 13.1 shows the basic components of a fiber optical communication system. At the transmitting end, an electrical signal is converted into a modulated light signal which is then transmitted through a fiber optical cable. At the receiving end, the signal is converted back into an electrical signal.

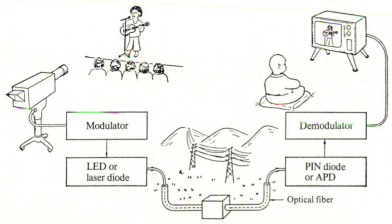

**Fig. 13.1.** Block diagram of fiber optical communication

# 13.1  Advantages of Optical Fiber Systems

The optical fiber is a very versatile transmission medium. The advantages itemized below explain why optical fibers have become such an attractive alternative to conventional transmission media in applications such as tele-communications.

### 13.1.1  Large Information Transmission Capability

Since the frequency of light is roughly $10^4$ times than that of microwave frequencies, the information transmitted by an optical fiber could theoretically be increased by four orders of magnitude. In other words, ten thousand microwave communication systems in parallel may be replaced by just one thin optical fiber. This large information capacity can best be utilized in the trunk lines of telecommunications.

### 13.1.2  Low Transmission Loss

In the case of ordinary window glass, it takes only a few centimeters of transmission for the light intensity to drop to half of the incident intensity. This attenuation is largely due to impurities in the glass. The manufacture of ultra-pure glass fibers has been refined to such an extent that a good optical fiber takes more than 10 kilometers of transmission before the intensity drops to a half. This low transmission loss means a longer span between repeater stations, thus reducing the maintenance cost in a long haul telecommunication network.

### 13.1.3  Non-Metallic Cable

Since the optical fiber transmission medium is non-metallic, it does not suffer from electromagnetic noise problems. With optical fibers, there is no stray electromagnetic pick-up, electromagnetic tapping, or ground potential problems. There is no risk of sparks or overheating which may occur with short-circuited metallic lines.

   These special features of a fiber optical link are utilized in many an adverse environment. The fiber optical link is used at electrical power stations where the high noise level associated with high-voltage corona discharge is a problem. The electrical power station also benefits from fiber-optic links in that these links help prevent faulty functioning of the control systems caused by a sudden ground potential rise associated with a surge current. The fiber optical link also provides an advantage in petrochemical factories where electrical sparks or over-heating of the communication cable could cause an explosion. Yet another industry where optical fibers would prove useful is the aircraft industry. Inside the fuselage of an aircraft a number of signal cables are fully packed. In order to avoid interference between the cables they are armored with heavy shielding metals. This weight can be substantially reduced by replacing them with fiber-optical links.

A disadvantage of fiber optical communication, however, is that even though the fiber is reasonably flexible, it is not as flexible as copper wire, so that splicing of the fiber cable is less easily achieved and needs more precision than connecting copper wires.

## 13.2  Optical Fiber

One of the parameters that specifies the optical fiber is its numerical aperture (NA) which is a measure of the maximum angle of acceptance for which light will be guided along the fiber when the fiber is illuminated at its end.

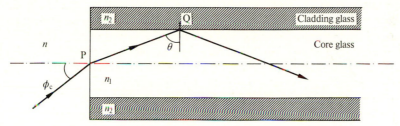

**Fig. 13.2.** Definition of numerical aperture: $NA = n \sin \varphi_c = \sqrt{n_1^2 - n_2^2}$

The step index fiber consists of core glass with a higher refractive index surrounded by cladding glass of a lower refractive index. Referring to Fig. 13.2, the incident light ray will be guided through the core if it undergoes total internal reflection at point Q on the core-cladding boundary. In order to determine the numerical aperture, the incident angle $\phi_c$ to the fiber end surface is calculated such that the angle $\theta$ at the point Q on the core-cladding boundary is the critical angle for reflection. When $\theta$ is the critical angle,

$$n_1 \sin \theta = n_2 \qquad (13.1)$$

where $n_1$ and $n_2$ are refractive indices of core and cladding. Snell's law is applied at P to obtain

$$n \sin \phi_c = n_1 \sin (90° - \theta) = n_1 \cos \theta = n_1 \sqrt{1 - \sin^2 \theta} \qquad (13.2)$$

where $n$ is the refractive index of the external medium in which the end of the optical fiber is immersed. Equations (13.1, 2) are combined to give

$$n \sin \phi_c = \sqrt{n_1^2 - n_2^2} = \text{NA}. \tag{13.3}$$

Equation (13.3) defines the numerical aperture (NA) and is an indicator of the fiber's light acceptance angle. The NA value is extremely important in optical fiber systems not only as an indication of the acceptance angle but also as a measure of the dispersion of the fiber, as described in the next section.

## 13.3 Dispersion of the Optical Fiber

When an input light pulse is launched into an optical fiber, the light energy is distributed among beams of various quantized angles (Chap. 5). Each beam has a different arrival time at the output end depending on these angles. This difference in arrival time is called dispersion and results in a spread in the received pulse width. Dispersion determines the highest rate of pulses (or the narrowest spacing between pulses) that can be transmitted per second through the fiber. In other words, if the rate is too high, two successive pulses can no longer be recognized as two separate pulses because of the spread. While there are no steadfast criteria, for the pulse spread of $\Delta T$, the minimum pulse width $T$ should satisfy

$$\Delta T < 0.7\, T. \tag{13.4}$$

The bit rate of transmission $B$ is defined as $B = 1/T$.

The amount of dispersion or pulse spreading is a function of the refractive index profile of the fiber core. It will be shown that the information transmission capability of the Selfoc fiber (Sect. 5.7) is much higher than that of a step-index multimode fiber. First, the dispersion of a step-index multimode fiber is considered. Referring to Fig. 13.3 it is easy to

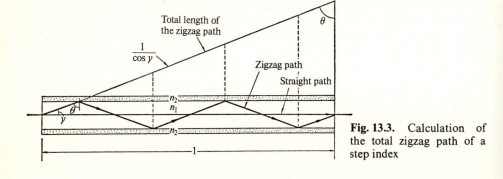

**Fig. 13.3.** Calculation of the total zigzag path of a step index

see that the total length of the zigzag path within a unit length of fiber is $1/\cos \gamma$, $\gamma$ being the angle of the ray to the fiber axis [13.1, 2]. The time that the light takes to travel a unit length of fiber is called the *group delay* $\tau$. The difference $\Delta\tau$ in the group delay of the zigzag path and that of the straight path is

$$\Delta\tau = \frac{n_1}{c}\left(\frac{1}{\cos \gamma} - 1\right). \tag{13.5}$$

The term $\Delta\tau$ is the modal pulse spread for a *unit length* of step-index multimode fiber. Since $\gamma$ is normally very small, $\cos \gamma \cong 1 - \frac{1}{2}\gamma^2$ and

$$\Delta\tau \cong \frac{n_1}{2c}\gamma^2. \tag{13.6}$$

When a ray following a zigzag path strikes the core-cladding boundary at the critical angle, (13.3) applies and

$$NA = n \sin \phi_c = n_1 \sin \gamma \qquad \text{hence}$$

$$\gamma \cong \frac{NA}{n_1}. \tag{13.7}$$

Inserting (13.7) into (13.6), $\Delta\tau$ finally becomes

$$\Delta\tau \cong \frac{1}{2 c n_1}(NA)^2. \tag{13.8}$$

Thus, the modal pulse spread $\Delta\tau$ of a step-index multimode fiber is proportional to $(NA)^2$. This means that the *capacity for transmission of information through the fiber is inversely proportional to the square of the numerical aperture.* The value of NA, however, cannot be made indefinitely small because, as seen from (13.3), the acceptance angle of the incident beam decreases with the value of NA and the difficulty associated with launching the incident light into the fiber increases. Thus, the number of bits that can be sent through a step index fiber can be obtained from (13.4, 8).

Next, the pulse spreading of a graded-index fiber, specifically of a Selfoc fiber, will be calculated in order to compare this result with that of the step-index multimode fiber. The pulse spreading of the Selfoc fiber can easily be found using the earlier result for the group delay derived in Chap. 5. From (5.155), the group delay difference is

$$\Delta\tau = \frac{n_c}{c}\left(\frac{1 + \cos^2 \gamma}{2 \cos \gamma} - 1\right)$$

$$= \frac{n_c}{c}\frac{(1 - \cos \gamma)^2}{2 \cos \gamma}. \tag{13.9}$$

Using (13.7), and the small-angle approximation for $\cos \gamma$ gives

$$\Delta\tau \approx \frac{1}{8c} \frac{(NA)^4}{n_c^3}.$$ (13.10)

Thus, the *pulse spreading of the Selfoc fiber is proportional to the fourth power of NA* and the dispersion of the Selfoc fiber is extremely small since its NA is typically $0.1 \sim 0.3$.

From (13.4), the information carrying capacity of the fiber is inversely proportional to the pulse spreading. By taking the ratio of (13.8) and (13.10), it is seen that the information carrying capacity of the Selfoc fiber is $4(n_c^3/n_1) NA^{-2}$ times ($10 \sim 100$ times) larger than that of a step-index multi-mode fiber for the same values of NA. *The Selfoc fiber is thus well suited to long-distance, high density information applications.*

The pulse spreading mentioned above is called *mode dispersion* because it is caused by differential delays between the straight path of the lowest-order mode and the zigzag paths of the higher-order modes. It should be noted that mode dispersion is not the sole source of pulse width spreading. There are two other mechanisms for dispersion: *material dispersion* and *waveguide dispersion*. Such dispersion takes place only when the light is composed of more than one wavelength. In contrast, mode dispersion will occur even if the light is purely monochromatic. Material dispersion is caused by the change in the refractive index of glass as a function of the wavelength of the light. Waveguide dispersion is caused by the fact that the quantized angles of the zigzag path change with respect to the wavelength of the light. In fact, the two dispersions cannot be clearly separated because a change in the refractive index also changes the quantized zigzag angles.

The pulse spreading due to the spread in frequency, $k = \omega/c$, is

$$\Delta\tau = \frac{d\tau}{dk} \Delta k.$$ (13.11)

From Fig. 13.3 the group delay of a step index multimode fiber is

$$\tau = \frac{n_1}{c} \frac{1}{\cos \gamma}.$$ (13.12)

Inserting (13.12) into (13.11) gives

$$\Delta\tau = \underbrace{\frac{1}{c} \frac{dn_1}{dk} \frac{1}{\cos \gamma} \Delta k}_{\text{material dispersion}} + \underbrace{\frac{n_1}{c} \frac{d}{dk}\left(\frac{1}{\cos \gamma}\right) \Delta k}_{\text{waveguide dispersion}}.$$ (13.13)

One approach to reducing dispersion is to use *a single-mode fiber*. As mentioned in Chap. 5, the core size of a single-mode fiber is reduced so that only the lowest-order mode (straight path) can be excited, thus eliminating mode dispersion completely. The other alternative is to use a light source with a narrow spectral width, such as a laser diode, so as to minimize the material and waveguide dispersions. Another approach is to operate within a wavelength region where the first and second terms of (13.13) cancel each other. For instance, with a single-mode fiber made of fused silica, the first term changes sign near 1.27 μm and the dispersion effect is significantly reduced.

## 13.4 Fiber Transmission Loss Characteristics

The causes of attenuation in an optical fiber can be broadly grouped into material-related and structure-related areas [12.3, 4]. First, the material-related causes are described. Typical curves of the dependence of fiber loss on wavelength are shown in Fig. 13.4. The curves are approximately V-shaped, and can be divided more or less into four regions according to the attenuation mechanisms which are

 i) Rayleigh scattering loss,
 ii) OH⁻ ion absorption loss,
iii) inherent absorption loss,
iv) impurity absorption loss.

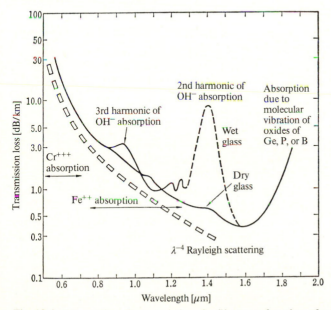

**Fig. 13.4.** Attenuation characteristics of a fiber as a function of wavelength

i) The *Rayleigh scattering loss* is caused by irregularity in the refractive index of the fiber with dimensions of the order of 0.001 to 0.01 μm, which are produced in the fiber drawing process. An approximate expression for the scattered intensity from objects whose size is much smaller than the wavelength was derived by Rayleigh who showed that this intensity is proportional to $\lambda^{-4}$. This type of scattering, also known as Rayleigh scattering, is predominant in the shorter wavelength region and follows very closely the $\lambda^{-4}$ line.

ii) The *OH⁻ ion absorption loss* arises from the presence of water molecules in the glass. The $OH^-$ ion has its fundamental absorption line at $\lambda = 2.8$ μm with second and third harmonics at 1.4 and 0.95 μm, respectively. The second and the third harmonics of this absorption line are within the range shown in Fig. 13.4. These peaks disappear as the content of $OH^-$ ions becomes less than 1 ppb (parts per billion). The glass with lower peaks is sometimes called "dry glass" and the glass with higher peaks, is then refered to as "wet glass".

iii) The third region, i.e. the *inherent absorption region,* is on the right shoulder of the V-shape. This loss is caused by the absorption bands of molecules that are an essential part of the fiber's composition (as opposed to unwanted impurities) and hence the term inherent loss. The refractive index profiles of the core and cladding are controlled by adding dopants to the glass. Additives such as germanium and phosphorus raise the refractive index, while additives such as boron lower the refractive index. These dopants have molecular absorption bands in the inherent loss region.

iv) The *impurity absorption loss* is due to impurities introduced unintentionally during the process of fabrication. They are primarily transition metal ions. The absorption band of $Fe^{++}$ ions ranges from 0.7 to 1.3 μm and $C_r^{+++}$ ions ranges from 0.6 to 0.7 μm.

We consider next the structure-related losses, which are summarized in Fig. 13.5. When the cable is bent excessively, some of the light rays will strike the core-cladding boundary at angles larger than the critical angle and these rays will be lost in the cladding, as indicated in Fig. 13.5a. Micro-bending is a special case of bending. When a protective plastic coating is applied to the bare fiber, the loss increases. This is due to small bends located at intervals of about one millimeter due to irregular stresses built in during the curing process. This loss is called microbending loss. Even though the angles involved are small, the number of bends are large enough so that the total loss cannot be ignored.

Irregularities in the boundary surface between the core and the cladding glass also acts as a loss mechanism. Any ray scattered at an angle larger than the critical angle will be lost into the cladding glass, as shown in Fig. 13.5b. This loss is smaller in graded-index fibers than in step-index fibers. This happens because not all the rays in the graded-index fiber reflect at the outer region of the fiber, so that irregularities in the outer region do not contribute as much to the loss.

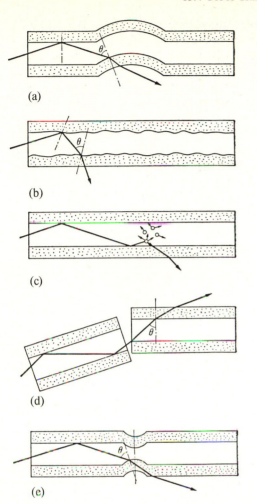

**Fig. 13.5 a – e.** Structure related loss of the optical fiber due to (**a**) bending, (**b**) boundary irregularities, (**c**) irregularities in the core, (**d**) misalignment, (**e**) imperfect splicing

Any irregularities in the core glass made during the fabrication of the fiber obviously scatter the light, as shown in Fig. 13.5c, and contribute to the loss.

In many practical situations it is necessary to join two fibers together. If the cores of the two fibers are misaligned, this will give rise to a loss at the connection (Fig. 13.5d). Misalignment either axially or angularly causes the loss.

A splice refers to a special type of connection in which two fiber ends are fused or bonded together. This joining process may result in deformations at the location of the splice. Imperfect splicing, as shown in Fig. 13.5e, creates scattering and leads to loss. Likewise, core diameter variations along the fiber will contribute to fiber loss in a manner similar to that illustrated in Fig. 13.5e.

## 13.5 Types of Fiber Used for Fiber Optical Communication

Step-index fiber, graded-index fiber and single-mode fiber are most widely used for fiber-optical communication systems. Figure 13.6 shows the light path and the radial distribution of the refractive index of these three optical fibers.

The *step-index multimode fiber* has a large core diameter (50 ~ 75 µm) which has the advantage that coupling light into the fiber and splicing fibers together is much easier. However, the information transmission capacity is low, because hundreds of modes are excited in the fiber and the pulse spreading is as large as 50 ns/km, which means that the signal bit rate is of the order of Mb/km (13.4).

The *graded-index* fiber has much less dispersion than a step-index multimode fiber. According to the conclusion of Sect. 13.3, the information carrying capacity of a graded index fiber is $4(n_c^3/n_1)$ NA$^{-2}$ times larger than that of a step-index multimode fiber. With a fiber of NA = 0.2, the capacity of the graded-index fiber is approximately 200 times higher. The group delay

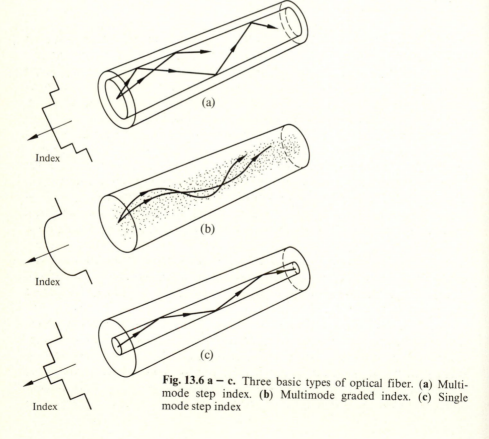

**Fig. 13.6 a – c.** Three basic types of optical fiber. (**a**) Multimode step index. (**b**) Multimode graded index. (**c**) Single mode step index

Index

Index

Index

spread $\Delta\tau$ of the graded-index fiber is of the order of 1 ns/km and the bandwidth is approaching 1 GHz/km.

The *single-mode fiber* permits only one mode to propagate, and the mode dispersion is completely eliminated, thus achieving a bandwidth of several GHz/km. The disadvantage of single-mode fibers is the difficulty of coupling and splicing as a consequence of its small core diameter.

## 13.6 Receivers for Fiber Optical Communications

The detectors used for fiber optical communication are, in a sense, photon-electron converters [13.5−12]. Solid-state devices are by far the most common detectors because they have such good features as small size, long life, durability, and low electrical power consumption [13.6]. The PIN photodiode and the avalanche photodiode (APD) are the major solid state detectors used.

### 13.6.1 PIN Photodiode

A heuristic explanation of the principle of operation of the PIN photodiode will be made using the microscopic views in Fig. 13.7. A PIN photodiode consists of three regions: a $p$-type region, an intrinsic region and an $n$-type region [13.7]. Figures 13.7a and b show the microscopic views of these regions before and after they are put together to complete the fabrication. The crystal structure of each region will be explained separately.

First, the intrinsic region in the middle section is addressed. It is an uncontaminated region of silicon (Si) atoms. Each silicon atom has four outer shell electrons of its own and the atoms are held together to form a crystal by sharing the outer shell electrons of four neighboring silicon atoms. The silicon atoms strive to fill their outer shells and form a complete set of eight electrons by this valence electron sharing. In the intrinsic region, the valence electrons in the outer shell of the atoms are trapped in the crystal lattice and cannot freely move around, so that the conductivity of the intrinsic region is very low. These valence electrons, however, can be broken loose to produce free electrons when the crystal is exposed to either high thermal agitation or irradiation of light. These freed electrons play an important role in photo detection.

Next, in the $n$-type region, group V atoms like arsenic atoms are doped into the intrinsic crystal. The arsenic (As) atoms have five valence electrons. When this atom forms a lattice with neighboring silicon atoms having four valence electrons, one electron is left unpaired since only eight electrons are needed to complete the outer shell. This excess electron is only loosely attached to the As atom. Even at room temperature, thermal energy will

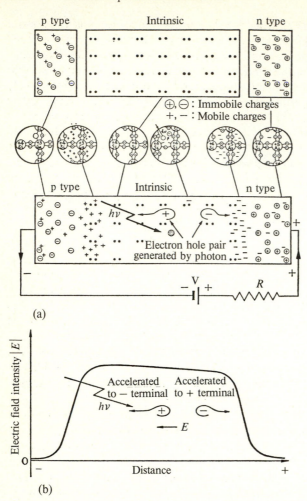

loosen some of these weakly bound electrons so that they become free electrons. Although the electron becomes mobile, the As atom remains immobile and a plus charge $\oplus$ is created at the site of the As atom after its electron has left. In order to differentiate the mobile (free) charges from the immobile charges, the latter charges are circled in the figures. The fact that one type of charge is mobile and the other is not is essential to the functioning of semiconductor devices.

In the $p$-type region, group III atoms such as gallium (Ga) are doped into the intrinsic crystal. The gallium atoms have only three valence electrons. As a result, when such an atom forms a lattice with neighboring silicon atoms having four valence electrons, the outer shell is incomplete since one electron is lacking. Even at room temperature, thermal energy may loosen a valence electron from a neighboring silicon atom in order to fill in the vacancy in the silicon-gallium pair leaving again a missing valence electron behind. As far as the gallium atom is concerned, when the missing valence electron is filled an extra immobile negative charge ⊖ is created because Ga was electrically neutral originally. As far as the neighboring Si atom that has lost the valence electron is concerned, a + charge was effectively created. The missing electron can be readily filled by electrons from further neighboring Si atoms, and this mobile effective + charge is called a hole.

Next, these three regions are put together to form a PIN photodiode. As soon as they are joined together, the highly polulated free electrons in the $n$-type region start to diffuse and spread into the intrinsic region, as do the holes in the $p$-type region. This diffusion process, however, does not go on indefinitely. For every electron that departs from the $n$-type region, an immobile ⊕ charge is left behind in the $n$-type region. The amount of ⊕ charge increases as the number of departing free electrons increases.

In a similar manner, as the holes depart from the $p$-type region, immobile ⊖ charges build up in the $p$-type region. As a result, a potential difference exists between the two regions. This potential is observed externally as a contact potential. An electric field $E$, which is the slope of the potential versus distance curve, is established as shown in Fig. 13.7b. The force exerted on the electrons by this electric field $E$ is directed toward the $n$-type region (to the right) while the force exerted on the holes by $E$ is directed toward the $p$-type region (to the left), exactly in directions opposite to the directions that the electrons and holes want to diffuse. The diffusion stops when the diffusion tendency and the force due to this electric field $E$ are balanced.

The density of mobile charges drops rapidly as one moves away from the $p-i$ and $n-i$ boundaries toward the middle of the intrinsic region. That portion of the intrinsic region which has a scarcity of mobile charges is called the depletion region.

The PIN photodiode is usually operated with a bias voltage. A positive potential is applied to the $n$-type region and a negative one to the $p$-type region (reverse bias). With this bias, an additional electric field is created which reinforces the internal electric field and forces the mobile electrons and holes back towards their native $n$ and $p$ regions. The depletion region expands as a result of the reverse bias.

When a photon $h\nu$ is injected into the intrinsic region, there is a high probability that the photon will be absorbed by a silicon atom. As a result of the photon absorption, the silicon atom gains energy. If the photon energy exceeds a certain critical value, then the energy gained by the silicon atom will liberate one of the valence electrons creating a mobile electron. A mobile hole is created simultaneously as a result of the vacancy in the silicon valence shell. This process is called *electron-hole pair production*. While the conductivity of the intrinsic region is low, that of either the $p$- or $n$-type regions is high because of an abundance of mobile charges. When an external potential is applied, this conductivity distribution causes the electric field to build up primarily in the intrinsic region and not in the *p*- or *n*-type regions. This applied $E$ field swiftly sweeps the newly produced electrons to the *n*-type region and the holes to the *p*-type region creating a signal current in the external circuit.

The intrinsic region is made long because the light absorption coefficient of silicon is rather small so that long distances are needed for the majority of incident photons to participate in electron-hole production.

In conventional PIN photodiodes, the photons enter the intrinsic region through the *p*-type region. The *p*-type region is made thin because any electron-hole pairs generated in this region diffuse in random directions as a consequence of the absence of an electric field due to high conductivity. The electrons generated in this region take a long time to reach the positive terminal. This not only deteriorates the response time but also becomes a source of noise.

### 13.6.2 Avalanche Photodiode

The avalanche photodiode (APD) utilizes the avalanche multiplication process to achieve higher electric output. The structure of the APD is shown in Fig. 13.8 a. The APD resembles a PIN photodiode except for the in-

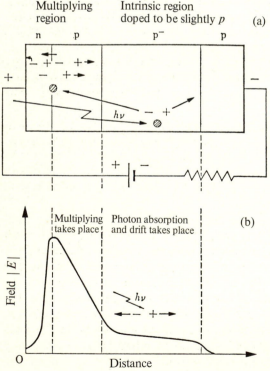

**Fig. 13.8 a, b.** Principle of the avalanche photodiode (APD). (**a**) Mechanism of the avalanche effect. (**b**) Field distribution

sertion of an additional *p*-type multiplying region. Most of the incident photons are absorbed in the intrinsic region and electron-hole pairs are first generated in this region as in the PIN photodiode. As before, the holes drift toward the *p*-type region and the electrons toward the *n*-type region. As shown in Fig. 13.8, the electrons, before reaching the *n*-type region, pass through a *p*-type multiplying region. If the electrons have undergone sufficient acceleration on reaching the multiplying region, new electron-hole pairs are generated by the collision-ionization process. The newly generated electron-hole pairs themselves generate more electron-hole pairs, thus initiating the process known as *avalanche multiplication*. It may produce 10 to 100 or even 1000 times more electron-hole pairs than is produced by photon absorption alone.

The intrinsic region is slightly doped to reduce its resistivity to around $300\ \Omega\mathrm{cm}$, so that the region of the highest resistivity is concentrated near the junction between the *n*-type region and the multiplying *p*-type region when reverse bias is applied. The highest field concentrates in the region of the highest resistivity when an external potential is applied and fields higher than $10^5$ V/cm can be established in the vicinity of the junction where most of the avalanche multiplication takes place. The *n*-type region is made thinner but is more highly doped than the multiplying region. The field distribution in the APD is shown in Fig. 13.8 b.

Next, an expression for the output signal current from the photo-detector is calculated. The number of photons in the light having total energy $E$ is $E/h v$ where $h v$ is the energy of a single photon, $h = 6.63 \times 10^{-34}$ J s is Planck's constant, and $v$ is the frequency of the light. The total charge generated in the detector is $\eta\,(E/h\,v)\,e\,M$, where $\eta$ is the quantum efficiency, $e$ is the elementary charge of $1.6 \times 10^{-19}$ C, and $M$ is the multiplication factor which is unity for a PIN photodiode, and $10-1000$ for an APD. The quantum efficiency is the probability that an incident photon will generate an electron-hole pair in the intrinsic region and is a number less than or equal to unity. The output current $i$ is thus obtained by dividing the charge by time, namely

$$i = \eta\, p\, \frac{e}{h\,v}\, M \quad [\mathrm{A}] \tag{13.14}$$

where $p$ is the optical power. The value of $h v$ is $2 \times 10^{-19}$ J for light with $\lambda = 1\ \mu\mathrm{m}$. The value of $M$ depends critically on the applied voltage and the temperature. As a rule of thumb, the output current from a PIN photodiode is $i = 0.5$ A/W and an average APD has an output current of $i = 50$ A/W.

Two major sources of noise in the PIN photodiode and the APD are thermally generated noise and shot noise. *Thermal noise is independent of signal current whereas the shot noise is dependent on the current.* Thus, both the current and the signal bandwidth must be specified in order to specify the photodiode's signal to noise ratio (S/N ratio).

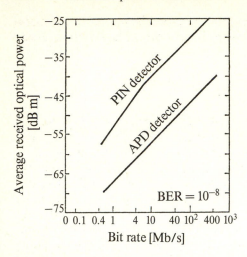

**Fig. 13.9.** Average optical power required to maintain BER $10^{-8}$ [13.17]

The S/N ratio is normally used for analog signals. In the case of digital signals, the *bit error rate* (BER) is used to specify the quality of the system instead of the S/N ratio. The bit error rate is the ratio of the number of error pulses in the receiver to the number of transmitted pulses. In general, the BER is a function of the rate at which pulses are transmitted, as well as being a function of optical power. Figure 13.9 shows the optical power needed as a function of pulse rate in order to maintain a BER of $10^{-8}$.

### 13.6.3 Comparison Between PIN Photodiode and APD

The PIN photodiode is more stable than the APD in the presence of fluctuations of temperature and applied voltage. The fact that the APD is very sensitive to changes in applied voltages can be used advantageously to control the output of the APD by feeding an automatic gain control (AGC) signal directly to the applied voltage. The gain of the APD is $10-1000$ times higher than that of the PIN photodiode because of the internal avalanche effect. Another advantage of the PIN photodiode is that it requires less applied voltage, typically $15-40$ V. The applied voltage of an APD is $100-150$ V.

## 13.7 Transmitters for Fiber Optical Communications

By virtue of their small size, high reliability and stability, modest power supply requirements, long lifetime, ease of modulation and low cost, solid-state light sources are the preferred light sources for fiber optical communication systems [13.13]. These same considerations are likewise respon-

sible for the common use of solid-state receivers. The solid-state trans-
mitters which are discussed in this section are the light emitting diode
(LED) and the laser diode (LD).

### 13.7.1 Light Emitting Diode (LED)

The LED and PIN photodiode are photon-electron converters having
exactly opposite functions. With the PIN photodiode, the input is photons
and the output is electrical current. The input photons liberate valence electrons from the outer shell of the atoms of the crystal to create free electrons.
The free electrons are the output signal. By contrast, in an LED, electrons
and holes are injected into the junction region of the $p-n$ junction diode by
forward biasing the LED (recall that the PIN photodiode and the APD are
reverse biased), as shown in Fig. 13.10. When the injected electrons recombine with the holes, i.e., when the electrons settle down into the positions of
missing valence electrons in the crystal, the electron's energy is released.
The released energy $E_g$ is used for either emitting photons (radiative
recombination) or for heating the crystal lattice (non-radiative recombination). The wavelength of the emitted light is $\lambda = h\,c/E_g$.

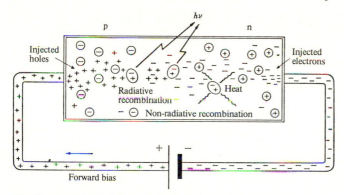

**Fig. 13.10.** Light emitting diode (LED)

The conversion rate of the LED is about $10 - 20$ mW/A and the output is
reasonably linear with respect to the injected current. The maximum light
output is a few milliwatts. The highest frequency of modulation is determined by the recombination time, which is the time the injected electrons
take to recombine. The highest practical frequency of modulation is several
hundred Megahertz.

Not all the recombination takes place with the release of the same
amount of energy because of the non-uniformity of the energy of the
electrons. *This non-uniformity results in a spread in the frequency spectrum.*
The half power width of the LED is as wide as a few hundred Angstroms
($1\,\text{Å} = 10^{-4}\,\mu\text{m}$), as shown in Fig. 13.11, and as a result, the LED is useful

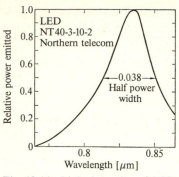

**Fig. 13.11.** Light intensity of LED as a function of wavelength

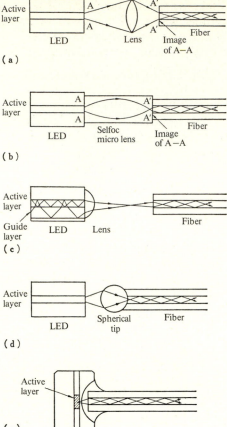

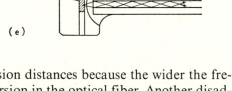

**Fig. 13.12 a − e.** Various methods of coupling the emission of an LED to the optical fiber. (**a**) Formation of image of source by a lens. (**b**) Formation of image of source by a Selfoc micro lens. (**c**) Auxiliary optical guide along active layer. (**d**) Spherical tip. (**e**) Coupled from close to active region (Burrus type)

for only short to moderate transmission distances because the wider the frequency spread, the greater the dispersion in the optical fiber. Another disadvantage of the LED is that the light radiation is emitted over a large range of angles. This makes efficient coupling of the light into an optical fiber difficult. Lenses of various types, such as shown in Fig. 13.12, are used to couple the output of the LED to the fiber.

In Sect. 13.6, silicon was chosen as the intrinsic semiconductor, and gallium and arsenic were chosen as dopants but a number of other elements are equally well suited to constructing semiconductor devices. Germanium can be used instead of silicon as an intrinsic semiconductor. Boron B, aluminum Al, and indium In can replace gallium as *p*-type dopants. Nitrogen N, phosphorus P, and antimony Sb can replace arsenic as *n*-type dopants. These are shown in Table 13.1.

In addition to using elements as semiconductor building blocks, certain compounds can also be used. For example, the compound gallium arsenide

GaAs which is made up of the group III element Ga and the group V element As can be used as a semiconductor. By changing the proportions of the elements in the compound, the wavelength of the emission can be varied. The combinations shown by the lines in Table 13.1 are used for fabricating semiconductors.

**Table 13.1.** Summary of Compounds and elements used for semiconductors [13.3]

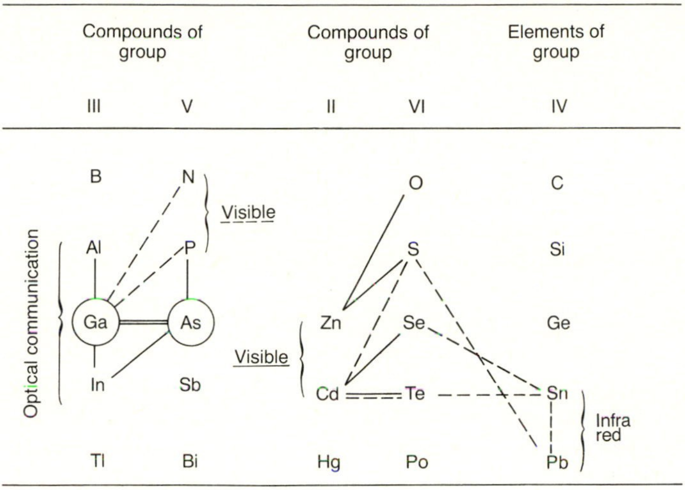

| Compounds of group | | Compounds of group | | Elements of group |
|---|---|---|---|---|
| III | V | II | VI | IV |
| B | N | | O | C |
| Al | P | | S | Si |
| Ga | As | Zn | Se | Ge |
| In | Sb | Cd | Te | Sn |
| Tl | Bi | Hg | Po | Pb |

(Left axis: Optical communication. Annotations: Visible, Visible, Infra red)

### 13.7.2  Laser Diode (LD)

Emission of light with frequency $v_{21} = (E_2 - E_1)/h$, occurs when a transition from an upper energy state $E_2$ to a lower one $E_1$ takes place. On the other hand, when the transition of the reversed direction, namely from $E_1$ to $E_2$ takes place, energy is absorbed. There are two types of emission of light; one is stimulated emission and the other, spontaneous emission. *Stimulated emission* takes place when the light whose wavelength exactly corresponds to $v_{21}$ (seed photons) impinges onto the atoms. The *spontaneous emission* is the emission which takes place regardless of the external illumination.

The degree of the stimulation is proportional to the intensity $P_i$ of the illuminating light. The expression for the total emission is

$$P_0 = A N_2 + P_i (B_{21} N_2 - B_{12} N_1)$$

where $A$ is Einstein's constant, $N_1$, $N_2$ are the electron populations in the lower and upper energy levels, $B_{21}$ and $B_{12}$ are the transition probabilities from $E_2$ to $E_1$ and that from $E_1$ to $E_2$, respectively. If equal degeneracies in these two levels is assumed, the transition probabilities from $E_2$ to $E_1$

and that from $E_1$ to $E_2$ are equal[1] and $B = B_{21} = B_{12}$. The expression of the total emission becomes

$$P_0 = A N_2 + P_i (N_2 - N_1)$$

Thus, the populations of the electrons determine whether emission or absorption takes place. In the case when the population of the upper state is larger and $N_2 > N_1$, emission occurs, and when the population of the lower state is larger and $N_2 < N_1$, we have absorption. The populations are related by Maxwell-Boltzmann statistics at thermal equilibrium as

$$N_2 = N_1 \exp\left[- (E_2 - E_1)/k T\right].$$

Thus at thermal equilibrium, power is always absorbed.

In the laser diode, $N_2$ is larger than $N_1$ due to the injection of mobile electrons (upper state electrons), thus maintaining emission. In photo-detection, $N_2$ is kept smaller than $N_1$ by removing the mobile electrons to the external circuit by reverse bias so that the incident photons are readily absorbed.

Radiation from the LED is predominantly due to spontaneous emission, whereas that from the LD is due to stimulated emission, amplified by positive feedback in a laser cavity. Both ends of the LD are highly polished so that light is reflected back and forth through the active region of the LD. Each time light passes through the diode, it stimulates emission. As the emission increases, the number of photons available to stimulate more emission also increases (positive feedback), so that the light radiation builds up very quickly.

Another important difference between spontaneous and stimulated emission is that the photons emitted spontaneously have no preferred phase whereas photons emitted by stimulated emission have not only the same frequency but also follow the same phase as the seed photon.

The wavelength spectrum of the LD is very narrow and suitable for high information density, long-transmission distances because a narrower

---

1   In a degenerate system, the transition probabilities are related as [13.14, 15]

$$g_1 B_{12} = g_2 B_{21}$$

where $g_1$ and $g_2$ are the degeneracies of the levels 1 and 2. A degeneracy $g_i$ means that there are $g_i$ independent states (ways) in which the level $i$ of the system can have the same energy. A good example is the spin of an electron. There are two ways for an electron to spin, i.e., clockwise or counterclockwise about its axis (2 states). Both states can have the same energy, and in this case, the degeneracy of the level associated with these spinning electrons is two. In fact, this degeneracy can be removed by applying an external magnetic field. A spinning electron can be considered as a small magnet whose polarity is determined by the direction of the electron spin. The magnetic energy of a small magnet in an external magnetic field depends upon the polarity of the magnet. Thus, in the presence of a magnetic field, the two electron spin states no longer have the same energy, and the degeneracy is removed.

wavelength spread means less dispersion in the optical fiber. In the following subsections, the LD is described in more detail.

### 13.7.3 Laser Cavity and Laser Action

Figures 13.13a,b show the geometry of a laser diode. The length of the laser cavity $L$ is a few hundred micrometers and is formed by cleaving the crystal along two parallel crystalline planes. The region in which lasing takes place is called the active region. For instance, when the refractive index of the active layer of the LD is 3.5, and using Fresnel's formula for the reflection coefficient at normal incidence, $r = (n-1)/(n+1)$, the cleaved ends each have a reflection coefficient $r = 0.56$.

The refractive index $n_1$ of the active layer is chosen higher than that of the cladding layer $n_2$ so as to form a light guide. The thickness $2d$ of the active layer is made thin $(0.3-0.4\,\mu m)$ so that only one mode is excited in the transverse (thickness) direction. The transverse direction is taken as the $x$-direction in Fig. 13.13. The electrode is stripe shaped so that the laser action is confined to a narrow region $W$ of the order of $10\,\mu m$ (in the $y$-direction).

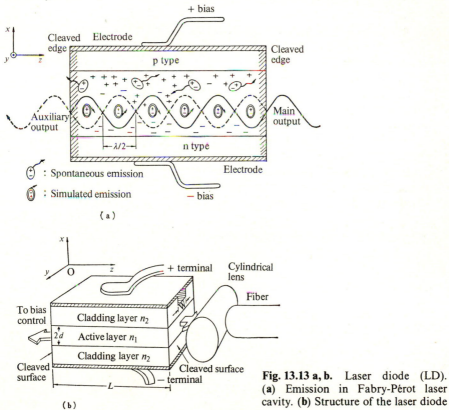

**Fig. 13.13 a, b.** Laser diode (LD). (a) Emission in Fabry-Pérot laser cavity. (b) Structure of the laser diode

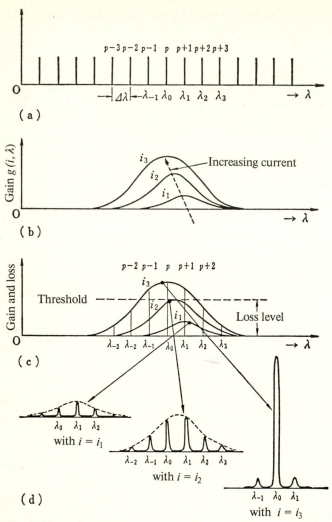

**Fig. 13.14 a – d.** Principle of laser action. (a) Laser cavity resonance. (b) Spontaneous emission with current as a parameter. (c) Threshold of lasing. (d) Spectrum of laser output

The resonance condition of the cavity in the longitudinal $z$-direction is

$$\beta_N L = p \pi \tag{13.15}$$

where $p$ is an integer, and $\beta_N$ is the propagation constant of the $N$th mode in the longitudinal direction. If the thickness $2d$ is made small enough that only the lowest mode is excited, then

$$\beta_N \cong \frac{2\pi}{\lambda} n_1 . \tag{13.16}$$

From (13.15, 16), the lasing wavelength must satisfy the condition

$$\frac{\lambda}{2 n_1} p = L .$$

(13.17)

As indicated in Fig. 13.14a, there are a number of wavelengths that satisfy (13.17) and these wavelengths are known as resonant wavelengths. The spacing $\Delta\lambda$ between adjacent resonant wavelengths is

$$\Delta\lambda = \frac{\lambda^2}{2 n_1 L} .$$

(13.18)

The shorter the cavity length is, the wider the spacing $\Delta\lambda$ becomes.

Next, the condition for lasing will be found. Lasing occurs when the gain in light intensity per round trip in the cavity exceeds the loss per round trip. For each round trip in the cavity, the amplitude decays by $r^2 \exp(-2\alpha L)$, $\alpha$ being the amplitude attenuation constant. The gain in each round trip is $\exp[2g(i, \lambda) L]$, $g(i, \lambda)$ being the amplitude gain per unit distance; it is a function of the current $i$ and wavelength $\lambda$. The threshold condition of oscillation is

$$r^2 e^{-2\alpha L + 2g(i,\lambda) L} = 1 .$$

(13.19)

The threshold gain is

$$g(i, \lambda) = \alpha + \frac{1}{L} \ln \frac{1}{r} .$$

(13.20)

The gain $g(i, \lambda)$ increases with current. Taking $i$ as a parameter, the spectral distribution of the spontaneous emission of the LD is shown in Fig. 13.14b. In Fig. 13.14c the loss level, gain curve, and resonant wavelengths are all combined in one diagram. At the current $i_1$ which is below the threshold, no laser action takes place, the light intensity is low and the spectrum is wide, as shown on the left side in Fig. 13.14d. With a further increase in current to just below the threshold $i_2$, the output increases and the spectrum spreads, as shown in the center in Fig. 13.14d. There is no laser action yet. As soon as the current exceeds the threshold value, laser action starts taking place. Stimulated emission, which is initiated by the presence of spontaneously emitted photons in the cavity, begins to predominate. It is the nature of stimulated emission that the particular resonant wavelength that first starts to predominate becomes the dominant wavelength of emission. More emission means more stimulation. Emission of one wavelength predominates, as shown on the right side in Fig. 13.14d. Although the mechanisms are completely different, analogies can be drawn between stimulated emission of light and the rupturing of a strained river dam. An initial small starting spot rapidly leads to a rupture of that portion. In reality, emission from the adja-

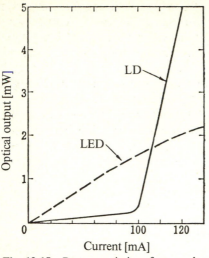

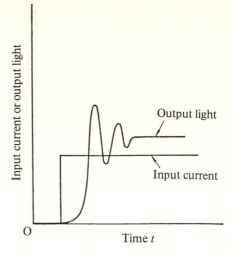

**Fig. 13.15.** Power emission from a laser diode compared with a light emitting diode

**Fig. 13.16.** Delay in the start of lasing

cent resonant wavelengths is also present. The adjacent spectra can however be suppressed by making $\Delta\lambda$ large by choosing a shorter cavity length $L$ in (13.18). The price that must be paid is that the threshold gain in (13.20) increases as $L$ is shortened.

Typical output power characteristics as a function of current are shown for the LD and LED in Fig. 13.15. The output power of the LD increases very slowly until the threshold current of 100 mA is reached, and then the optical power rises steeply to 5 mW with an additional 20 mA of current. In comparison, the output power of the LED increases steadily with increasing current. The LED curve was added in the figure for comparison.

The threshold current of the LD is strongly influenced by temperature. A special control circuit is necessary to ensure that the LD is operated in the linear region of the output power versus current curve. The control signal is usually derived from the laser beam emitted from the unused end of the device.

Another characteristic of laser oscillation is the delay between the applied current and the start of the emission, as shown in Fig. 13.16. It takes time to reach the threshold density of injected electrons needed to start lasing. Besides the delay, the light intensity goes through relaxation oscillations at the starting time of lasing. These set the upper limit on the modulation frequency to the order of a Gigahertz.

The mechanism for the relaxation oscillation is explained as follows. Soon after the injection of electrons into the laser cavity, the electron density surpasses the threshold value and lasing begins. As soon as lasing starts, the electron density begins to decrease. The reduced electron density means reduced emission, and this again allows the build up of the electron density. After a few cycles of this competition between electron density and light emission, the intensity of the emission reaches a steady state value. One way of increasing the upper limit of the modulation frequency is to apply a dc bias current just below the threshold value. In the case of an on-off modulation, however, the dc bias deteriorates the on-off ratio of the light emission because of the non-zero light at "off" time.

The beam radiation pattern of the LD shown in Fig. 13.13 is wide in the vertical $x$-direction and narrow in the horizontal $y$-direction because the beam width is roughly inversely proportional to the dimensions of the source (Chap. 4). A rather interesting practical idea for injecting the LD beam into a fiber guide is shown in Fig. 13.13b. Here, a short piece of optical fiber is used as a cylindrical lens for focusing the LD beam.

### 13.7.4 Temperature Dependence of the Laser Diode (LD)

The resonance condition of the laser cavity is significantly influenced by temperature [13.16]. Both the refractive index and the dimensions of the cavity change as the temperature varies. As far as the changes in the refractive index are concerned, the temperature influences the refractive index in two ways. First of all, the temperature directly changes the refractive index. Secondly, the temperature changes the wavelength of operation and this change in lasing wavelength in turn changes the refractive index. This cascade can be expressed by the formula

$$dn = \frac{\partial n}{\partial T} dT + \frac{\partial n}{\partial \lambda} \frac{d\lambda}{dT} d\lambda . \tag{13.21}$$

The resultant change in the wavelength of operation due to the changes in both dimension and the refractive index is obtained by differentiating both sides of (13.17) with respect to $\lambda$ and using (13.21) to obtain

$$\frac{d\lambda}{dT} = \lambda \frac{\frac{1}{n}\frac{\partial n}{\partial T} + \frac{1}{L}\frac{dL}{dT}}{1 - \frac{\lambda}{n}\frac{\partial n}{\partial \lambda}} . \tag{13.22}$$

It should be remembered that the $g(i, \lambda)$ curves in Fig. 13.14 are also very critically influenced by temperature. The relative positions of the curves in Fig. 13.14 change as the temperature changes. It is difficult to achieve a stability better than $1\,\text{Å/C}°$.

### 13.7.5 Comparison Between LED and LD

This subsection summarizes the comparisons between the LED and the LD mentioned above. One of the most striking differences is their wavelength spectra. While the wavelength spectrum of the radiation from the LED is a few hundred Angstroms wide, that of the center spectrum of the LD is only a few Angstroms wide. The peak power of the LED is about one milliwatt but that of the LD is several milliwatts. The radiation beam pattern of the LD is much narrower than the radiation beam pattern of the LED. The output of the LED is linear with bias current, and the ratio of output optical power to input current is approximately 0.01 W/A. The output from the LD is nonlinear near the threshold current. Moreover the output exhibits a time lag. The LED is simpler in structure, simpler to modulate, and has a longer life than the LD.

## 13.8  Connectors, Splices, and Couplers

In order to build a complete fiber optical communication system, the transmitter and receiver must be interfaced with the optical fiber. In this section, the various components and techniques for linking transmitter, fiber and receiver are described [13.5, 11, 12].

### 13.8.1  Optical Fiber Connector

Connectors are installed at the end of the optical fiber for the purpose of connecting two fibers together or for the purpose of connecting the fiber to some other component. For connections between two fibers, the smaller the fiber core diameter is, the tighter the alignment tolerance between the two centers has to be. The alignment tolerance for 50 μm core diameter fibers is of the order of a few micrometers. Both the positioning of the center and the parallelism of the axis are important.

One approach to the problem of aligning the core is to align the cladding glass under the assumption that the core and the cladding are concentric. The other approach is the alignment of the core itself. In the former approach, the glass fiber is inserted into a watchmaker's jewel secured in a metal ferrule. Figure 13.17a shows an assembly of this type. Even though the connector can be assembled in the field, it is difficult to reduce the insertion loss to less than 1 dB since the connection relies on the intrinsic tolerance of the fiber. This type of connector is sometimes called the FA-type connector (field assembly). The other approach is a C-type (C stands for concentric) connector. The fiber is first put into a sheath to increase the diameter for better handling. Then the sheathed fiber is posi-

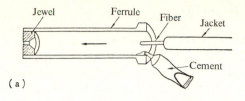

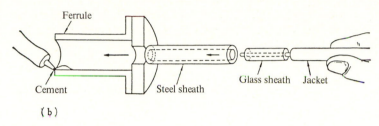

(a)

Jewel  Ferrule  Fiber  Jacket  Cement

(b)

Ferrule  Cement  Steel sheath  Glass sheath  Jacket

tioned by a microscope so that the fiber core is accurately centered in a plug. While positioned in the center, the sheathed fiber is cemented to the plug to complete the connector.

It is important that the connection loss be the same every time the connection is made. It is much preferable to make a connection by splicing if a permanent connection is intended because *splicing is more reliable in the long run than a connector.*

## 13.8.2 Splicing

A permanent joint between two fiber ends is often needed. The joints are first aligned and then are bonded either by a cement or by melting the glass. The alignment is usually made by relying upon the quality of the fiber. Figure 13.18 illustrates the common methods of splicing with bonding agents. The ends of the fiber are first cleanly cleaved and then brought

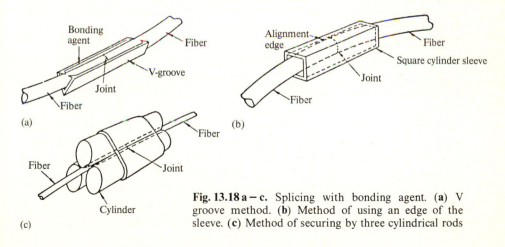

**Fig. 13.18 a – c.** Splicing with bonding agent. (**a**) V groove method. (**b**) Method of using an edge of the sleeve. (**c**) Method of securing by three cylindrical rods

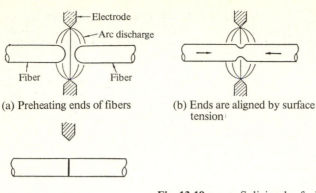

(a) Preheating ends of fibers

(b) Ends are aligned by surface tension

(c) Splicing is completed

**Fig. 13.19 a – c.** Splicing by fusion. (**a**) Preheating ends of fibers. (**b**) Ends are aligned by surface tension. (**c**) Splicing is completed

together. It is important to prevent the fibers from touching with any force. Even with a very light force, the pressure per unit area is very large and the ends are easily abraded. A spacing as large as 100 μm practically makes no difference in the loss. The alignment is achieved by either a V groove, the corner of a square tube or round cylinders, as shown in Fig. 13.18. They are permanently secured by a bonding agent.

Figure 13.19 shows the procedure for splicing by the fusion method. The ends of the fiber are heated by a controlled electric arc. The ends are preheated to the extent that the shape of the end becomes a hemisphere, as shown in Fig. 13.19a. Then the ends are brought together until they touch. At the moment that the ends touch each other, the connection evolves from the center of the core glass, as shown in Fig. 13.9b, thus eliminating the chance that air bubbles will become trapped inside the fused region. The surface tension of the glass also helps to align the fiber ends. The fiber ends are brought together at exactly the precalculated distance to give the minimum deformation of the fiber's cylindrical shape at the location of the splice. Since the diffusion constant of the glass is very low, the original distribution of the refractive index is preserved during the fusion, and the splicing loss is typically less than 0.2 dB.

### 13.8.3 Fiber Optic Couplers

A coupler is a device which couples light from the optical fiber to another fiber or to other optical components, such as monitoring photodiodes. There are three major types of couplers: (i) monitoring couplers, (ii) branching couplers, and (iii) directional couplers.

The monitoring couplers are shown in Fig. 13.20a – c. Since the purpose of these couplers is to monitor the optical power in the transmission line, it is desirable that the coupled power be independent of the type of modes

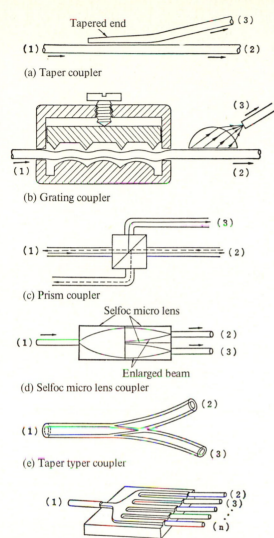

Tapered end (3)

(1) (2)

(a) Taper coupler

(b) Grating coupler

(c) Prism coupler

Selfoc micro lens

(d) Selfoc micro lens coupler

Enlarged beam

(e) Taper typer coupler

(f) Planar multiple coupler

**Fig. 13.20 a – f.** Optical power divider coupler. All couplers have isolation between (2) and (3) and transmission between (1) and (2). (a) Taper coupler. (b) Grating coupler. (c) Prism coupler. (d) Selfoc micro lens coupler. (e) Tapered butt-joint type. (f) Planar multiple coupler

excited in the main line. The ability to adjust the amount of coupling is also desirable. The grating coupler has the advantage of a controllable coupling coefficient, but some modes can be coupled out more than the others and the monitored power represents the power of only a portion of the modes in the main line. The prism coupler can pick up the power of all modes but the amount of coupling cannot be varied.

Figures 13.20 d–f are diagrams of the branching couplers which are used for distributing the same information to all branches of the subsystems. Equal light power is distributed among the branches.

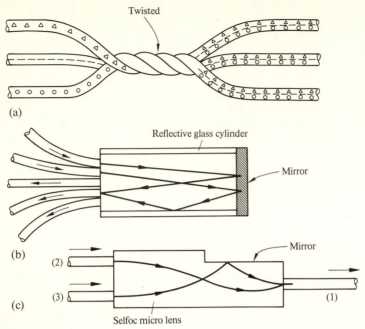

**Fig. 13.21 a – c.** Light signal scrambler. (**a**) Star type. (**b**) Glass cylinder type. (**c**) Reflection type

Figures 13.21 a−c are diagrams of light scramblers. Each input port carries a different signal. The scrambler adds the input port signals together so that each output carries a fraction of the addition of all the input port signals.

Figures 13.22 a, b show optical directional couplers. When the input light is transmitted from port (1) to port (2), some of the input signal from port (1) is coupled into port (3). When the direction is reversed and the light is transmitted from port (2) to port (1), port (3) is isolated and no signal comes out of port (3). The prism coupler shown earlier in Fig. 13.20 c also has a directional property and can be used as a directional coupler. Both the beam splitter and Selfoc half-mirror type directional couplers in Fig. 13.22 can couple all modes without any prejudice, but the star directional coupler in Fig. 13.22 b does not couple all the modes evenly.

## 13.9 Wavelength Division Multiplexing (WDM)

The wavelength division multiplexing technique is used to transmit more than one wavelength of light in one optical fiber [13.4]. The signal transmission capacity of the fiber increases $m$-fold if an $m$-fold wavelength

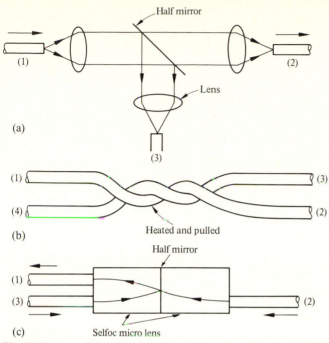

(a)

(b)

(c)

**Fig. 13.22 a − c.** Directional coupler. When transmitting from (*1*) to (*2*), (*3*) is coupled but when transmitting from (*2*) to (*1*), (*3*) is isolated. (**a**) Beam splitter type. (**b**) Star directional coupler. (**c**) Selfoc with half mirror coupler

multiplexing is employed. For instance, both analog and digital signals can be transmitted in one fiber. Bi-directional transmission also becomes possible.

Figure 13.23 shows the most commonly used components for wavelength division multiplexing. The prism type multiplexing shown in Fig. 13.23 a is simple in structure, and the number of available channels can be large. The interference filter type multiplexing shown in Fig. 13.23 b is limited in the number of available channels. The grating type shown in Fig. 13.23 c requires high precision but can handle many channels. Needless to say, the same element can be used for both multiplexing and demultiplexing the signal.

Figure 13.24 shows examples of bi-directional transmission in (a) and wavelength division multiplexing in (b). The bidirectional system in Fig. 13.24 a uses the filter $F_1$ with pass band at $\lambda_1$ to separate the transmitting channel from the receiving channel. The system in Fig. 13.24 b uses filter $F_1$ with pass band at $\lambda_1$, and filter $F_3$ with pass band at $\lambda_3$. Neither $F_1$ nor $F_3$ can pass $\lambda_2$. Using these two filters, the lights of three different wavelengths are combined and sent through one optical fiber. At the receiving end, they are separated using an arrangement identical to the transmitting end.

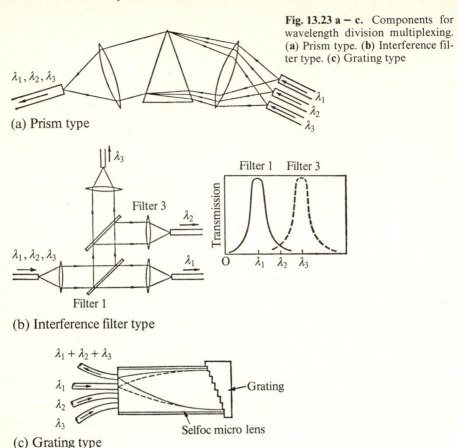

$\lambda_1, \lambda_2, \lambda_3$

$\lambda_1$
$\lambda_2$
$\lambda_3$

(a) Prism type

$\lambda_3$

Filter 1    Filter 3

Transmission

Filter 3

$\lambda_2$

$\lambda_1, \lambda_2, \lambda_3$

$\lambda_1$

O    $\lambda_1$  $\lambda_2$  $\lambda_3$

Filter 1

(b) Interference filter type

$\lambda_1 + \lambda_2 + \lambda_3$

$\lambda_1$

$\lambda_2$

$\lambda_3$

Grating

Selfoc micro lens

(c) Grating type

## 13.10 Optical Attenuators

There are discretely variable and continuously variable attenuators. The attenuation of the discretely variable attenuator is changed by changing the combination of the individual attentuation cells, as shown in Fig. 13.25 a. The attenuator film in each of the attenuation cells in Fig. 13.25 a is tilted so as to minimize the light reflected back to the incident port. Selfoc micro lenses are used to widen the beam so that any spatial non-uniformity in the attenuator film is averaged out.

Figure 13.25 b shows a continuously variable attenuator which is useful as a measuring device. The required value of attenuation is obtained by the combination of the step attenuator films and the continuously varying attenuator film.

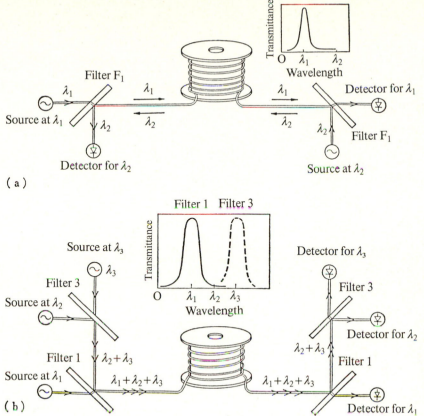

**Fig. 13.24 a, b.** Wavelength division multiplexing systems. (**a**) Bi-directional transmission of light in one optical fiber. (**b**) Three wavelenghts of light are combined and transmitted in one fiber and separated at the receiver end

## 13.11 Design Procedure for Fiber Optical Communication Systems

In this section, a step by step procedure for designing a fiber optical communication system is presented along with an example problem [13.17, 18]. The procedure is divided into six steps. Steps 1 to 4 consider the distribution of the light power levels at various components, and Steps 5 and 6 consider the signal frequency bandwidth of the system. The normal procedure is to start at the receiver end and work back to the transmitter.

Step 1:  Determine the required light power to the receiver using Fig. 13.9. The required light power depends on the choice of detectors (PIN photodiode or avalanche photodiode (APD)), the bit rate of the signal, and the allowable bit error rate (BER).

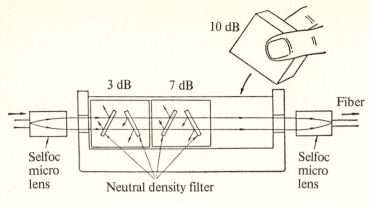

(a)

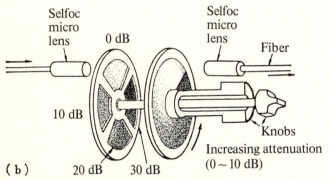

(b)

**Fig. 13.25 a, b.** Optical attenuators. (a) Discretely variable attenuator. (b) Continuously variable attenuator

Step 2: Calculate the total transmission losses which include optical fiber loss, connector loss and splicing loss.

Step 3: Make allowances for degradation of the optical power of the source or the sensitiviy of the receiver with either temperature or time.

Step 4: Select the source.

Step 5: Calculate the required rise time of the system from the required bit rate.

Step 6: Determine the rise time of the entire system using the empirical formula for finding the rise time of tandem circuits:

$$t_r = 1.1 \sqrt{t_1^2 + t_2^2 + t_3^2 + \ldots}$$  (13.23)

where $t_1 \ldots t_n$ are rise times of components such as the source, optical fiber and detector [13.19].

The details of the procedure will be explained using an exercise.

**Exercise 13.1**   Design a fiber optical communication system to connect two computers located in buildings 1.5 km apart. The bit rate is 10 Mb/s and the type of coding used is return to zero code (RZ code). The BER should be less than $10^{-10}$.

*Solution*

Step 1: If one uses an APD as a detector, the required average optical power is $-39$ dB m. The values in Fig. 13.9 are for a BER of $10^{-8}$, but a general rule of thumb is that for each decimal of BER an additional dB is required. Hence the required average light power is $-37$ dB m.

Step 2: If 10 dB/km loss fiber is used, the total loss for 1.5 km is 15 dB. Let us say that 2 splices are required to connect the ends of the spools of the optical fiber and allow one additional splice for a possible breakage during installation. The loss of each splice is 0.2 dB so that the total splice loss is 0.6 dB. Connectors are used at both terminals. The connector loss for each is 1.0 dB and the total connector loss is 2 dB.

Step 3: The degradation allowances for temperature and time are both 3 dB each and the total allowance is 6 dB.

Step 4: The total loss is

| | |
|---|---|
| fiber loss | 15  dB |
| splice loss | 0.6 dB |
| connector loss | 2  dB |
| degradation allowances | 6  dB |
| | 23.6 dB |

This means that one has to select a source whose average output power is larger than $-37 + 23.6 = -13.4$ dB m. This is within the power range of an LED. Sometimes, the LED is rated in terms of peak power. An example of an RZ signal is shown in Fig. 13.26. "1" is represented by a pulse that comes down to zero at the center of the clock pulse. Thus, if there are equal numbers of "1" s and "0" s, the peak power is 6 dB above the average power and an LED whose peak power is higher than $-7.4$ dB m has to be selected.

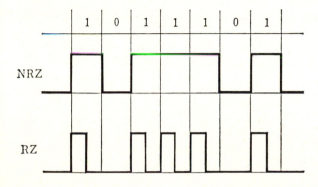

**Fig. 13.26.** Shapes of pulse of non-return to zero (NRZ) and return to zero (RZ) coding

Step 5: The most commonly used criterion for estimating the required rise time of a pulse transmission system is that the system rise time has to be faster than 70% of the input pulse width. The pulse width of a 10 Mb/s signal is 50 ns. Thus the rise time of the system has to be less than 35 ns. If, however, the coding were NRZ, the required rise time would have been 70 ns.

Step 6: The rise time of each component is normally found in the manufacturer's specification sheets. A typical rise time of an LED is $10-20$ ns and that of an LD is $1-2$ ns. The rise time of a PIN photodiode is 1 ns and that of an APD is 2 ns. The rise time of the optical fiber has a wide range of values depending on the type of the fiber. Assume that a graded index fiber with 3 ns/km was selected.

| Components | Rise time | Rise time square |
|---|---|---|
| LED | 15 ns | 225 |
| Fiber | 3 ns | 9 |
| PIN photodiode | 1 ns | 1 |
| | | 235 |

The rise time of the system $= 1.1 \times \sqrt{235} = 16.86$ ns which is well within the rise time required.

## Problems

**13.1**   a) What is the modal pulse spread for 1 km of a step index multimode fiber with NA $= 0.2$ and refractive index of the core glass $n_1 = 1.48$?

b) What is the maximum pulse rate that can be sent if the dispersions other than mode dispersion can be ignored? Non-return to zero (NRZ) pulses are used.

**13.2**   a) What is the theoretical limit ($\eta = 1.0$) for the output current from a PIN photodiode when the power of the incident light is 1 mW and the wavelength is $\lambda = 0.84$ μm?

b) What value was assigned to the quantum efficiency to obtain the rule of thumb $i = 0.5$ A/W, which expresses the output current from a PIN photodiode?

**13.3**   From Fig. 13.9, prove that number of photons that have to be contained during one pulse to maintain a given BER value is practically independent of pulse rate. How many photons have to be contained in one pulse for an APD to maintain a BER of $10^{-8}$? (The wavelength of the light is $\lambda = 0.84$ μm.)

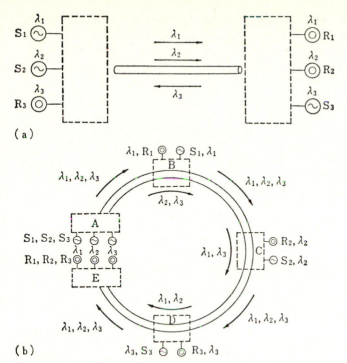

**Fig. 13.27 a, b.** Wavelength division multiplexing systems. **(a)** Wavelength division multiplexing in both directions. **(b)** Wavelength division multiplexing loop

**13.4**   Figures 13.27 a, b are partially drawn diagrams of a wavelength multiplexing system. Complete the diagram by filling in necessary optical components in the areas *A*, *B*, *C*, *D*, and *E* enclosed by the dotted lines. Complete the diagram using only two kinds of filters: filter $F_1$ with pass band at $\lambda_1$ and filter $F_3$ with pass band at $\lambda_3$. The characteristics of these filters are shown in Fig. 13.24.

**13.5**   A fiber optical communication system which is capable of transmitting pulses at a rate of 10 Mbits per second is to be designed. The mode of the pulse modulation is return to zero mode (RZ mode). The bit error rate (BER) has to be smaller than $10^{-8}$. Find the spacing between the repeaters if only the system margin of power is considered (dispersion is not considered). A PIN photodiode is used as the detector and an LED whose peak power is $-14$ dB m is used as the transmitter. The loss of the optical fiber is 5 dB/km. The connector loss is 1 dB each. Splice loss is 0.3 dB for one. The length of one spool of fiber cable is 5 km. Allow for one additional splice for an accidental break of the cable. The allowance for temperature and time degradations combined is 6 dB.

# 14. Electro and Accousto Optics

This chapter reviews the propagation of light waves in anisotropic media because most of the media used for controlling light propagation are anisotropic. This review will be extended to include devices that use the bulk effect of these phenomena. The bulk phenomena often become the basis of thin-film integrated-optics devices, which are described in Chap. 15.

## 14.1 Propagation of Light in a Uniaxial Crystal

When an electric field is applied to an electrooptic material, electrons in the crystal are perturbed from their equilibrium distributions, and the polarizability of the crystal changes, so that the refractive index changes. In addition to the change in refractive index in the direction of the applied field, it is possible for the refractive index to change also in other directions. A simple analog is the soft pencil eraser which deforms in every direction when it is pressed by a thumb in one direction. A property common to almost all of these crystals is that their refractive indices are anisotropic, even in the absence of external fields. (Gallium arsenide, however, becomes anisotropic only when an external electric field is applied to the crystal.) They are sometimes called doubly refractive or birefringent, which implies that the refractive index for one direction of light polarization is different from that for another direction of polarization. To understand the basics of wave propagation in a birefringent crystal, it is instructive to present here a brief review on birefringence.

As is well known, the shape of a wave front diverging from a point source in an isotropic medium is a sphere. The wave front, however, is no longer a sphere if the source is in an anisotropic medium because the phase velocity depends upon the direction of polarization of the wave. Figure 14.1a shows wave fronts of two waves propagating in the same direction in a plane containing the crystal axis but polarized in two perpendicular directions. One of the waves is polarized perpendicular to the page (represented by dots), and the other, in the plane of the page (represented by bars). The rate of advance of the wave front of the former is the same in all directions in the page and the leading wave front lies in a circle, whereas that of the latter depends upon its direction of propagation and the wave front lies in an

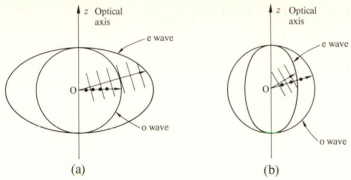

**Fig. 14.1 a, b.** Wave front of a point source in a birefringent crystal. (**a**) Negative birefringence. (**b**) Positive birefringence

ellipse. The former wave is called the ordinary wave or o wave, whereas the latter, is called the extraordinary wave or e wave. There is one direction for which both o and e waves have the same phase velocity, and the circle and ellipse intercept[1]. This direction is called the optical axis of the crystal.

The refractive index of the crystal for the o wave is called the ordinary index $n_o$, and that for the e wave whose direction of polarization is parallel to the optical axis is called the extraordinary index $n_e$. There are two types of birefringence. The birefringence is called positive birefringence when $n_e - n_o$ is positive, and it is called negative birefringence when $n_e - n_o$ is negative. Figure 14.1a illustrates the case of negative birefringence. (Note that what is shown is the developed wave front some time after the emission, and that the distribution of the refractive index is inversely proportional to the radius.) Figure 14.1b is the corresponding situation for positive birefringence. Crystals such as quartz ($SiO_2$), and lithium tantalate $LiTaO_3$ show positive birefringence while lithium niobate $LiNbO_3$, calcite ($CaCO_3$), potassium dihydrogen phosphate ($KH_2PO_4$) KDP, and ammonium dihydrogen phosphate ($NH_4H_2PO_4$) ADP exhibit negative birefringence.

In nature there are two kinds of crystals: uniaxial crystals which have only one optical axis and biaxial crystals which have two optical axes. The analysis of the phase front in a biaxial crystal is much more complicated than that of a uniaxial crystal. Most of the birefringent crystals found in nature are biaxial but fortunately, almost all the crystals used for electro-optic devices are uniaxial. Therefore, subsequent descriptions will concentrate solely on the uniaxial crystal.

In order to deal with the general case, the concept of the optical indicatrix, or index ellipsoid, has been found useful. For the uniaxial crystal,

---

1 A crystal in which the circle and ellipse do not quite intercept is called an optically active crystal and is a very important class of crystal that will be discussed in Sect. 14.4.1.

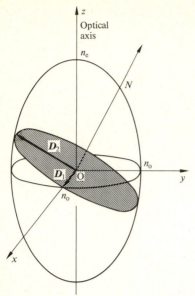

**Fig. 14.2.** Indicatrix of a uniaxial crystal

the optical indicatrix is an ellipsoid of revolution (Fig. 14.2). Such an ellipsoid is expressed by the equation

$$\frac{x^2}{n_o^2} + \frac{y^2}{n_o^2} + \frac{z^2}{n_e^2} = 1 \ . \tag{14.1}$$

The manner in which the indicatrix is used is explained in the following. Consider light propagating with its wave normal in the $\overline{ON}$ direction. The intercept of the plane perpendicular to $\overline{ON}$ (hatched section) with the ellipsoid generates an ellipse with minor axis $D_1$ and major axis $D_2$. When the light propagates in the $\overline{ON}$ direction, the wave *must be* polarized in either $D_1$ or $D_2$ with the index of refraction represented by the length of $D_1$ or $D_2$. The former is the o wave, and the latter is the e wave. No matter what the direction of $\overline{ON}$ is, the vector $D_1$ is always perpendicular to the optical axis, and its magnitude is always $n_o$. The vector $D_2$ however is different. Its direction changes and approaches the optical axis as $\overline{ON}$ moves away from the optical axis, and its magnitude increases as $D_2$ approaches the optical axis. The magnitude is maximum (positive birefringence is assumed) when $D_2$ coincides with the optical axis.

It is important to note that, when the wave front is propagating along any given direction $\overline{ON}$, *there are only two allowed directions of polarization, these being the directions of the minor and major axes of the ellipse, $D_1$ and $D_2$. No other direction of polarization is permitted,* except when $\overline{ON}$ is along the optical axis, in which case the intercept is a circle, and any direction is allowed.

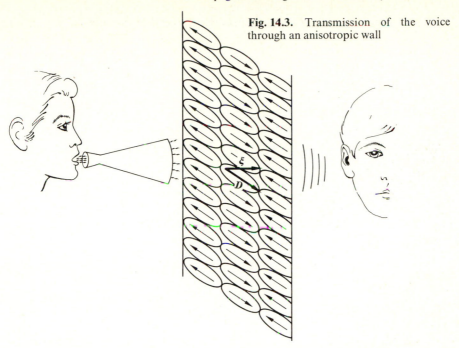

**Fig. 14.3.** Transmission of the voice through an anisotropic wall

A heuristic explanation of this reasoning is given in Fig. 14.3. Imagine a voice is transmitted through a fictitious wall made up of rectangular pebbles arranged neatly like a crystal lattice with their axes slanted. These pebbles are constrained so that they can move only along the length axis. Now, an incident sound wave tries to vibrate the wall in the $\xi$ direction but, due to the anisotropy of the pebbles, a displacement $D$ is allowed only in the direction of the length axis of the pebble. Regardless of the direction of $\xi$, the allowed direction of $D$ is always the same. It may be added that the sound energy mainly transmits in the direction straight across the wall and is not greatly influenced by the direction of $D$. This corresponds to the fact that the direction of the energy transmission is not necessarily the same as that of the wave normal in an anisotropic crystal.

Even when the light incident upon the anisotropic crystal is polarized in the direction of neither $D_1$ nor $D_2$, the light still transmits through the crystal. In this case, the electric field of the light is decomposed into components parallel to $D_1$ and to $D_2$, and each component is treated separately. The final result is obtained by combining the two results.

When the incident wave is linearly polarized in a direction bisecting $D_1$ and $D_2$, the two decomposed components each parallel to $D_1$ and $D_2$ have equal magnitudes but they propagate at different phase velocities so that the resultant wave becomes elliptically polarized. At a certain distance, the phase difference becomes $\pi/2$ radians, and the wave becomes circularly polarized. A plate cut and polished at this particular length is called a quarter wave plate. It is used for generating a circularly polarized wave from a linearly polarized wave.

Figure 14.4 shows two examples of wave front propagation in planes perpendicular and parallel to the optical axis. They were obtained using the indicatrix in Fig. 14.2. When the direction $\overline{ON}$ of propagation is in the plane perpendicular to the optical axis, the wave front behaves, as shown in Fig. 14.4a. The polarization of the e wave is parallel and the o wave is perpendicular to the optical axis, but the orientation of both polarizations with respect to the optical axis remains the same, regardless of the direction of $\overline{ON}$, as long as it stays in this plane. Figure 14.4b shows the wavefront when the direction of $\overline{ON}$ is in a plane containing the optical axis ($y-z$ plane). Referring to the indicatrix in Fig. 14.2, it is seen that as $\overline{ON}$ approaches the $y$ axis, $D_2$ increases while $D_1$ is invariant. Thus with a scan of $\overline{ON}$, the locus of the phase front of the e wave is an ellipse, while that of the o wave is a circle.

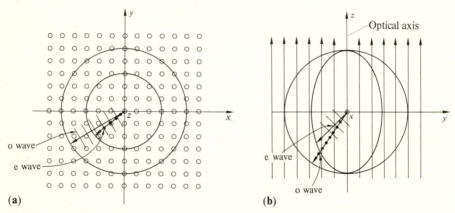

**Fig. 14.4 a, b.** Wave front of light in a uniaxial crystal. (**a**) Wave front of light whose direction of propagation is in a plane perpendicular to the optical axis. (**b**) Wave front of light whose direction of propagation is in the plane of the optical axis (in $y-z$ plane)

## 14.2 Field in an Electrooptic Medium

The change in the refractive index $n$ due to an applied electric field $\mathscr{E}$ is expressed as

$$\Delta\left(\frac{1}{n^2}\right) = r\mathscr{E} + g\,\mathscr{E}^2 . \tag{14.2}$$

The first term is linearly proportional to the applied electric field and is known as the *Pockels effect*. The second term, which has a quadratic dependence on the applied electric field, is known as the *Kerr electrooptic effect.* While the Pockels effect depends upon the polarity of the applied electric field, the Kerr electrooptic effect does not.

The most popular crystals which display the electrooptic effect and are used for electrooptic devices are: lithium niobate ($LiNbO_3$), lithium tantalate ($LiTaO_3$), potassium dihydrogen phosphate better known as KDP ($KH_2PO_4$), ammonium dihydrogen phosphate better known as ADP ($NH_4H_2PO_4$), and gallium arsenide (GaAs).

The electrooptic effect can be best represented as a deformation of the indicatrix. If the major or minor axes are lined up with the $x$, $y$ and $z$ axes, the ellipsoid is expressed in the compact form

$$\frac{x^2}{n_x^2} + \frac{y^2}{n_y^2} + \frac{z^2}{n_z^2} = 1 .\tag{14.3}$$

However, once the ellipsoid is deformed and these axes move away from the $x-y-z$ axes, the cross terms $yz$, $zx$ or $xy$ become necessary to represent the new ellipsoid. The general expression for such an ellipsoid is

$$a_{11}\,x^2 + a_{22}\,y^2 + a_{33}\,z^2 + 2\,a_{23}\,yz + 2\,a_{31}\,zx + 2\,a_{12}\,xy = 1 .\tag{14.4}$$

The values of the coefficients $a_{11}$, $a_{22}$, $a_{33}$, $a_{23}$, ... change according to the applied electric fields. They can be represented by

$$\begin{bmatrix} a_{11} - 1/n_x^2 \\ a_{22} - 1/n_y^2 \\ a_{33} - 1/n_z^2 \\ a_{23} \\ a_{31} \\ a_{12} \end{bmatrix} = \begin{bmatrix} r_{11} & r_{12} & r_{13} \\ r_{21} & r_{22} & r_{23} \\ r_{31} & r_{32} & r_{33} \\ r_{41} & r_{42} & r_{43} \\ r_{51} & r_{52} & r_{53} \\ r_{61} & r_{62} & r_{63} \end{bmatrix} \begin{bmatrix} \mathcal{E}_x \\ \mathcal{E}_y \\ \mathcal{E}_z \end{bmatrix}\tag{14.5}$$

where $r_{11}$, $r_{12}$, $r_{13}$, ... are called electrooptic (or Pockels) constants; $\mathcal{E}_x$, $\mathcal{E}_y$, $\mathcal{E}_z$ are the components of the applied electric field in the $x$, $y$, $z$ directions; and $n_x$, $n_y$, $n_z$ are the refractive indices in the $x$, $y$, $z$ directions when the applied field is absent. Note that when the applied field is zero, (14.4) reduces to (14.3).

Equation (14.5) looks rather cumbersome, but many of the 18 elements of the matrix are zero and some of the non-zero elements have the same value. Their values depend upon the symmetry of the crystal. Lithium niobate, for instance, has only 4 different values for its 8 non-zero elements. Gallium arsenide has only 1 value for its 3 non-zero elements. Electrooptic constants of representative crystals and their matrices are tabulated in Table 14.1 a, b.

**Table 14.1 a.** Electrooptic constants [14.1 – 3]

| Material | Crystal symmetry | $r\,[10^{-12}\,\mathrm{m/V}]$ | $\lambda\,[\mu m]$ | Refractive index |
|---|---|---|---|---|
| LiNbO$_3$ | 3 m | $r_{13} = 8.6$ <br> $r_{33} = 30.8$ <br> $r_{22} = 3.4$ <br> $r_{51} = 28$ | 0.63 | $n_x = n_y = 2.286$ <br> $n_z \quad\;\; = 2.200$ |
| LiTaO$_3$ | 3 m | $r_{13} = 7.9$ <br> $r_{33} = 35.8$ <br> $r_{22} = 1$ <br> $r_{51} = 20$ | 0.63 | $n_x = n_y = 2.176$ <br> $n_z \quad\;\; = 2.180$ |
| GaAs | $\bar{4}\,3\,m$ | $r_{41} = 1.2$ | 0.9 | $n_x = n_y = n_z = 3.42$ |
| HgS | 32 | $r_{11} = 3.1$ <br> $r_{41} = 1.4$ | 0.63 | $n_x = n_y = 2.885$ <br> $n_z \quad\;\; = 3.232$ |
| CdS | 6 mm | $r_{13} = 1.1$ <br> $r_{33} = 2.4$ <br> $r_{51} = 3.7$ | 0.63 <br> 0.63 <br> 0.59 | $n_x = n_y = 2.46$ <br> $n_z \quad\;\; = 2.48$ |
| KDP | $\bar{4}\,2\,m$ | $r_{41} = 8.6$ <br> $r_{63} = 10.6$ | 0.55 | $n_x = n_y = 1.51$ <br> $n_z \quad\;\; = 1.47$ |
| ADP | $\bar{4}\,2\,m$ | $r_{41} = 28$ <br> $r_{63} = 8.5$ | 0.55 | $n_x = n_y = 1.52$ <br> $n_z \quad\;\; = 1.48$ |
| SiO$_2$ (Quartz) | 32 | $r_{11} = 0.29$ <br> $r_{41} = 0.2$ | 0.63 | $n_x = n_y = 1.546$ <br> $n_z \quad\;\; = 1.555$ |

**Table 14.1 b.** Electrooptic matrices

$$3\,m \qquad\qquad \begin{pmatrix} 0 & -r_{22} & r_{13} \\ 0 & r_{22} & r_{13} \\ 0 & 0 & r_{33} \\ 0 & r_{51} & 0 \\ r_{51} & 0 & 0 \\ -r_{22} & 0 & 0 \end{pmatrix}$$

$$\bar{4}\,3\,m \qquad \begin{pmatrix} 0 & 0 & 0 \\ 0 & 0 & 0 \\ 0 & 0 & 0 \\ r_{41} & 0 & 0 \\ 0 & r_{41} & 0 \\ 0 & 0 & r_{41} \end{pmatrix}$$

$$\bar{4}\,2\,m \qquad \begin{pmatrix} 0 & 0 & 0 \\ 0 & 0 & 0 \\ 0 & 0 & 0 \\ r_{41} & 0 & 0 \\ 0 & r_{41} & 0 \\ 0 & 0 & r_{63} \end{pmatrix}$$

$$6\,mm \qquad\qquad \begin{pmatrix} 0 & 0 & r_{13} \\ 0 & 0 & r_{13} \\ 0 & 0 & r_{33} \\ 0 & r_{51} & 0 \\ r_{51} & 0 & 0 \\ 0 & 0 & 0 \end{pmatrix}$$

$$32 \qquad \begin{pmatrix} r_{11} & 0 & 0 \\ -r_{11} & 0 & 0 \\ 0 & 0 & 0 \\ r_{41} & 0 & 0 \\ 0 & -r_{41} & 0 \\ 0 & -r_{11} & 0 \end{pmatrix}$$

### 14.2.1 Examples for Calculating the Field in an Electrooptic Medium

Examples which show how to analyze the electrooptic effect are presented. The general steps to solve the problems are:

1) find the indicatrix under the applied electric field,
2) find the allowed directions of polarization for a given direction of propagation,
3) find the refractive indices in the allowed directions, and
4) decompose the incident light wave into components along the allowable directions of polarization.

The sum of the components of the emergent wave is the desired solution.

**Exercise 14.1** An external electric field $\mathcal{E}_z$ is applied along the optical axis $z$ of the uniaxial crystal LiNbO$_3$, and the light is incident along the $y$ axis.

a) Find the allowable directions of polarization and their refractive indices.

b) Find the emergent light when the incident light is along the $y$ axis with its polarization at $45°$ from the $x$ axis in the $xz$ plane. What is the condition for the emergent light to be circularly polarized? The length of the crystal is $L$, the applied voltage is $V$, and the spacing between the electrodes is $d$.

*Solution* The matrix for the electrooptic constants of lithium niobate is

$$
\begin{pmatrix}
0 & -r_{22} & r_{13} \\
0 & r_{22} & r_{13} \\
0 & 0 & r_{33} \\
0 & r_{51} & 0 \\
r_{51} & 0 & 0 \\
-r_{22} & 0 & 0
\end{pmatrix} . \tag{14.6}
$$

With $n_x = n_y = n_o$, $n_z = n_e$ and $\mathcal{E}_x = \mathcal{E}_y = 0$, (14.4–6) lead to

$$(n_o^{-2} + r_{13}\mathcal{E}_z)\, x^2 + (n_o^{-2} + r_{13}\mathcal{E}_z)\, y^2 + (n_e^{-2} + r_{33}\mathcal{E}_z)\, z^2 = 1 . \tag{14.7}$$

From Table 14.1a, the electrooptic coefficients $r$ are expressed in units of $10^{-12}$ m/V. For applied electric fields of practical interest, the product $r\mathcal{E}$ is a number much less than unity. Since in each bracket the second term is much smaller than $n_o$ or $n_e$, (14.7) can be approximated by using the binomial expansion[2] as

---

2 Note

$$(n_o^{-2} + r_{13}\mathcal{E}_z) = 1/(n_o + \Delta)^2 \doteq n_o^{-2}[1 - 2(\Delta/n_o)]$$

and solve for $\Delta$.

$$\frac{x^2}{\left(n_o - \dfrac{n_o^3}{2} r_{13} \mathscr{E}_z\right)^2} + \frac{y^2}{\left(n_o - \dfrac{n_o^3}{2} r_{13} \mathscr{E}_z\right)^2} + \frac{z^2}{\left(n_e - \dfrac{n_e^3}{2} r_{33} \mathscr{E}_z\right)^2} = 1. \tag{14.8}$$

Equation (14.8) is the expression for the indicatrix.

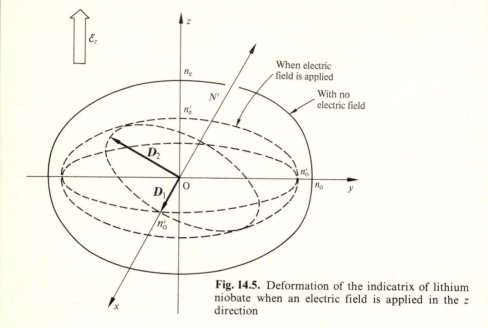

**Fig. 14.5.** Deformation of the indicatrix of lithium niobate when an electric field is applied in the $z$ direction

As shown in Fig. 14.5, the size of the indicatrix shrinks with an increase in the external field. Since the electrooptic constants of lithium niobate are $r_{33} = 30.8 \times 10^{-12}$ m/V and $r_{13} = 8.6 \times 10^{-12}$ m/V, the shrinkage in the $z$ direction is more than 3 times that in the $x$ or $y$ directions. It also should be recognized that the amount of change in the refractive index is, in general, very small compared to $n_o$ or $n_e$.

Next, the allowable directions of polarization for light propagating in the $y$ direction are considered. The cross-section of the indicatrix cut by a plane normal to $\overline{ON}$ is an ellipse. The ellipse is in the $xz$ plane since $\overline{ON}$ is along the $y$ axis. The allowable directions of polarization are the major and the minor axes $D_1$ and $D_2$ of this ellipse. The corresponding refractive indices are

$$\left. \begin{aligned} n_x &= n_o - \tfrac{1}{2} r_{13} n_o^3 \mathscr{E}_z \\ n_z &= n_e - \tfrac{1}{2} r_{33} n_e^3 \mathscr{E}_z \end{aligned} \right\}. \tag{14.9}$$

b) Finally, the emergent field is considered. The incident field has to be decomposed into components parallel to the allowed directions of polarization. The incident wave is decomposed into $E_x$ and $E_z$ components. The phases of these components at the exit are

$$\phi_x = \frac{2\pi}{\lambda} n_o L - \frac{\pi}{\lambda} n_o^3 r_{13} \frac{V}{d} L$$

$$\phi_z = \frac{2\pi}{\lambda} n_e L - \frac{\pi}{\lambda} n_e^3 r_{33} \frac{V}{d} L \quad .$$

(14.10)

Because of the phase difference $\Delta\phi = \phi_z - \phi_x$, the emergent wave is in general elliptically polarized. In particular, it is circularly polarized when $\Delta\phi = \pi/2$, and it is linearly polarized when $\Delta\phi = \pi$.

**Exercise 14.2**  Consider a situation identical to Exercise 14.1, but with GaAs in place of LiNbO$_3$.

a)  Find the direction of the optical axes when the field is applied.
b)  Find the allowable directions of polarization and their refractive indices.

*Solution*  The matrix for the electrooptic constants of GaAs from Table 14.1 is

$$\begin{pmatrix} 0 & 0 & 0 \\ 0 & 0 & 0 \\ 0 & 0 & 0 \\ r_{41} & 0 & 0 \\ 0 & r_{41} & 0 \\ 0 & 0 & r_{41} \end{pmatrix} .$$

With $n_x = n_y = n_z = n_o$, and using (14.4, 5), the expression for indicatrix is

$$n_o^{-2} x^2 + n_o^{-2} y^2 + n_o^{-2} z^2 + 2 r_{41} \mathscr{E}_z x y = 1 .$$

(14.11)

Notice that, in the absence of an external field $\mathscr{E}_z$, the indicatrix is a sphere, which simply means that the GaAs crystal is isotropic.

In order to see the shape of the indicatrix more closely, the coordinate system is rotated by 45 degrees around the $z$ axis into a new coordinate system $(x', y', z)$ using the rotation matrix,

$$\begin{pmatrix} x \\ y \end{pmatrix} = \frac{1}{\sqrt{2}} \begin{pmatrix} 1 & -1 \\ 1 & 1 \end{pmatrix} \begin{pmatrix} x' \\ y' \end{pmatrix} .$$

It is interesting to see how the $x y$ term in (14.11) is uncoupled by the magic power of a change of coordinates, i.e., $x y = (x' - y')(x' + y')/2 = (x'^2 - y'^2)/2$. The result of the coordinate rotation is

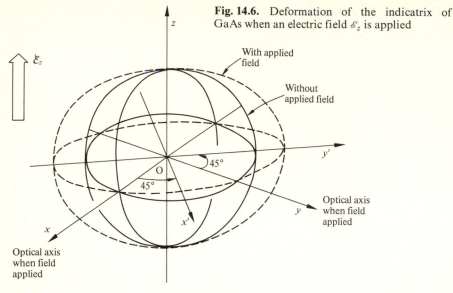

**Fig. 14.6.** Deformation of the indicatrix of GaAs when an electric field $\mathscr{E}_z$ is applied

$$\frac{x'^2}{\left(n_o - \dfrac{n_o^3}{2} r_{41} \mathscr{E}_z\right)^2} + \frac{y'^2}{\left(n_o + \dfrac{n_o^3}{2} r_{41} \mathscr{E}_z\right)^2} + \frac{z^2}{n_o^2} = 1. \tag{14.12}$$

As indicated by the dotted line in Fig. 14.6, the indicatrix changes from a sphere to an ellipsoid when an external electric field $\mathscr{E}_z$ is applied. The refractive indices in these directions are

$$\left.\begin{aligned}
n_{x'} &= n_o - \frac{n_o^3}{2} r_{41} \mathscr{E}_z \\[2mm]
n_{y'} &= n_o + \frac{n_o^3}{2} r_{41} \mathscr{E}_z \\[2mm]
n_z &= n_o
\end{aligned}\right\}. \tag{14.13}$$

From Fig. 14.6, the cross-section with the $y = 0$ plane is a circle, and the $y$ axis is the optical axis. [It can also be seen directly by setting $y = 0$ in (14.11)]. Similarly, the $x$ axis is also the optical axis. The refractive indices in these directions are $n_o$.

It may be added that, if the direction of propagation were along the $y'$ axis, the allowed directions of polarization would be the $x'$ and $z$ axes with refractive indices given by (14.13).

**Exercise 14.3**  Consider the case of KDP with an external electric field $\mathscr{E}_z$ in the $z$ direction. The light is incident at an angle $\theta$ from the $x$ axis in the $xy$ plane. Find the allowed directions of polarization and their refractive indices.

*Solution*   The matrix for KDP is

$$
\begin{pmatrix}
0 & 0 & 0 \\
0 & 0 & 0 \\
0 & 0 & 0 \\
r_{41} & 0 & 0 \\
0 & r_{41} & 0 \\
0 & 0 & r_{63}
\end{pmatrix}
$$

with $n_x = n_y = n_o$ and $n_z = n_e$.
Inserting the matrix into (14.4, 5) leads to

$$
n_o^{-2} x^2 + n_o^{-2} y^2 + n_e^{-2} z^2 + 2\, r_{63}\, \mathscr{E}_z\, x\, y = 1. \tag{14.14}
$$

The cross-section of the indicatrix cut by the plane perpendicular to the direction of light is needed. It can be found by rotating the $x$ axis to the $x'$ axis, which coincides with the direction of light, and then setting $x' = 0$. The rotation of the axis by $\theta$ degrees around the $z$ axis can be executed by a rotation matrix

$$
\begin{pmatrix} x \\ y \end{pmatrix} = \begin{pmatrix} \cos\theta & -\sin\theta \\ \sin\theta & \cos\theta \end{pmatrix} \begin{pmatrix} x' \\ y' \end{pmatrix}.
$$

The application of the matrix to (14.14) with $x' = 0$ leads to

$$
\frac{y'^2}{\left( n_o + \dfrac{n_o^3}{2} r_{63}\, \mathscr{E}_z \sin 2\theta \right)^2} + \frac{z^2}{n_e^2} = 1. \tag{14.15}
$$

Thus, the allowed directions of polarization are $y'$ and $z$ with refractive indices

$$
n_{y'} = n_o + \frac{n_o^3}{2} r_{63}\, \mathscr{E}_z \sin 2\theta, \qquad n_z = n_e. \tag{14.16}
$$

Note, if the direction of propagation is either along $x$ or $y$ axis there is no influence of $\mathscr{E}_z$.

**Exercise 14.4**   Find the optical axes of a crystal whose indicatrix axes are $(n_x, n_y, n_z)$.

*Solution*   The problem reduces to cutting the indicatrix in such way as to create a cross-section of circular shape. First, pick the shortest axis. Let us say $n_x$ is the shortest axis of the indicatrix drawn in Fig. 14.7. The cross-section of the indicatrix cut by the plane normal to the direction of light $\overline{ON}$ along the $z$ axis is an ellipse. The shape of the ellipse becomes more circular as $\overline{ON}$ starts to tilt in the $x\, z$ plane away from the $z$ axis. The minor

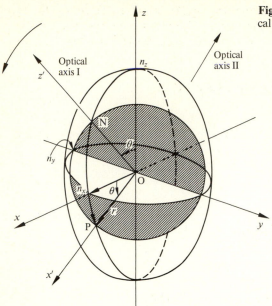

**Fig. 14.7.** Method of finding the optical axis of an indicatrix $(n_x, n_y, n_z)$

axis $r$ starts to increase from $n_x$, whereas the major axis $n_y$ is fixed. There must be a particular angle $\theta$ of inclination such that $r$ equals $n_y$. That is the direction of the optical axis.

A point P in the $y = 0$ plane is represented by

$$\left.\begin{array}{l} x = r\cos\theta \\ z = r\sin\theta \\ y = 0 \end{array}\right\} .$$

(14.17)

If P were to stay on the ellipse in $y = 0$ plane, (14.17) should satisfy

$$\frac{r^2\cos^2\theta}{n_x^2} + \frac{r^2\sin^2\theta}{n_z^2} = 1, \qquad \text{hence}$$

$$\frac{1}{r^2} = \frac{1}{n_x^2} + \left(\frac{1}{n_z^2} - \frac{1}{n_x^2}\right)\sin^2\theta$$

and the angle $\theta$ that $r = n_y$ is

$$\sin\theta = \pm\frac{n_z}{n_y}\sqrt{\frac{n_y^2 - n_x^2}{n_z^2 - n_x^2}} .$$

(14.18)

Notice that there are two angles which satisfy (14.18), and there are two optical axes which are symmetric with respect to the $z$ axis. Such a crystal is called a biaxial crystal.

An alternate method of solving the problem is rotation of the co-ordinates $x$, $z$ into $x'$, $z'$ around the $y$ axis. The cross-section cut by $z' = 0$ is the ellipse concerned. The angle of rotation $\theta$ is adjusted so that the major axis equals the minor axis.

## 14.2.2 Applications of the Electrooptic Bulk Effect

A prism deflector, such as shown in Fig. 14.8 a, uses the bulk electrooptic effect. The direction of the deflected beam is controlled by an external electric voltage. The angle of deflection can be increased by staggering a number of prisms, as illustrated at the bottom of the figure.

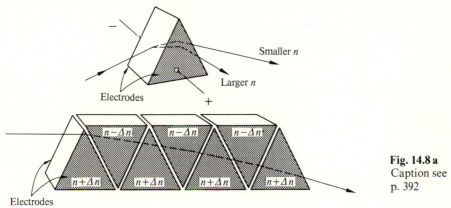

Fig. 14.8 a
Caption see
p. 392

Figure 14.8 b shows a light deflector which scans in both the horizontal and vertical directions. The angle of deflection is proportional to grad $n$ (Sect. 5.4), and $\Delta n$ is proportional to the $E$ field which is $-$ grad $V$. This deflector with its quadrupole shape deflects the light in linear proportion to the applied voltage [14.4].

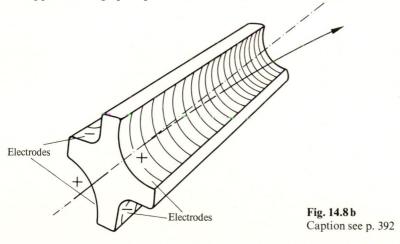

Fig. 14.8 b
Caption see p. 392

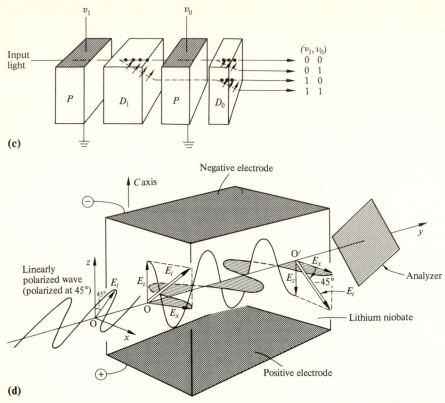

**Fig. 14.8 a – d.** Electrooptic devices. (**a**) Prism type electrooptic deflector. (**b**) Quadrupole type deflector. (**c**) Electrooptic digital light deflector. *P*: Electrooptic polarizer (LiNbO$_3$) to rotate the direction of polarization by 90° when the electric field is applied. *D*: Birefringent crystal like calcite to defelct light. (**d**) Electrooptic effect used for a light switch or a light modulator

Figure 14.8c shows a digital electrooptic deflector. A birefringent crystal like calcite is combined with an electrooptic polarizer like lithium niobate. There are two distinct directions of refraction depending upon the direction of the polarization of the incident light. The direction of the polarization of the incident light is switched between two directions 90 degrees apart by switching on and off the voltage to the electronic polarizer. With the arrangement shown in the figure, the light can be shifted to 4 different locations by various combinations of the control voltages $v_0$ and $v_1$.

A deflector with $N$ number of stages can direct the light to $2^N$ number of locations. A binary coded voltage can be directly used to control the light. This kind of deflector is called a digital light deflector because the light is directed to only discrete locations, whereas deflectors such as the previously described prism and quadrupole gradient deflectors are called analog light deflectors because the amount of deflection is continuously changed by the control voltage.

The geometry of an electrooptic modulator using lithium niobate is shown in Fig. 14.8 d. The crystal axis (most often called $C$ axis) is placed along the $z$ axis of the coordinate system. The direction $\overline{ON}$ of propagation is taken along the $y$ axis. The external electric field $\mathscr{E}_z$ is in the $z$ direction. This is a geometry similar to Exercise 14.1. This time, however, the explanation will make use of a diagram rather than the formulae.

The allowed directions of polarization are the $x$ and $z$ axes. The input polarization has to be decomposed into $x$ and $z$ components. Consider an incident wave which is linearly polarized at an angle of 45° with respect to the $z$ axis. At the entrance, $E_x$ and $E_z$ reach their maximum values at the same instant in time or in other words, both oscillate in phase. But once inside the crystal, the velocities for $E_x$ and $E_z$ are different due to anisotropy, and one starts to lag behind the other. There is a location O′ where $E_z$ becomes a negative maximum at the instant when $E_x$ becomes a positive maximum. Namely, the phase of $E_z$ lags 180 degrees behind that of $E_x$. At this location, the resultant $E_t$ of $E_x$ and $E_z$ points at $-45°$ and the resultant field vibrates along a line which is rotated 90 degrees from the direction of the incident polarization. If one arranges a polarized sheet at the location O′ with its direction of polarization perpendicular to $E_t$, then the light can be completely switched off. The device can be used as an optical switch. The location O′ can be calculated using (14.10), which stipulates that the required length $L$ has to satisfy

$$\left[ \frac{\pi}{\lambda} (n_e^3 r_{33} - n_o^3 r_{13}) \frac{V}{d} \right] L = \pi , \tag{14.19}$$

where a constant value $(2\pi/\lambda) (n_e - n_o)$ was suppressed. The same device can be used as a light modulator if $V$ is modulated.

Electrooptic materials are generally quite temperature sensitive, and special precaution has to be taken to keep the temperature constant. A fluctuation of 1 °C results in a change of 0.26 radians in a 1 mm long lithium niobate crystal at 1 μm wavelength of light [14.5].

The upper and lower electrodes of the modulator in Fig. 14.8 d form a capacitor and this usually sets the upper limit on modulation. When the wavelength of the modulating signal is short compared to the length of the electrode, the applied field becomes non-uniform due to the presence of a standing wave. The polarities of two points spaced half a wavelength apart oppose one another and efficiency is reduced. One way to avoid this and raise the frequency limit is to treat the parallel electrodes as a parallel plate transmission line to match the velocity of the modulating signal along the transmission line with that of the light in the crystal. It is also important to terminate it with a load impedance which is the same as the characteristic impedance of the line.

## 14.3 Elastooptic Effect

Materials such as water, lithium niobate, rutile ($TiO_2$), and Lucite change their refractive index when they are under stress. With many crystals, the change in refractive index in one direction is different from that in other directions. In other words, the relationship between the amount of change in refractive index and the strain is expressed by strain-optic constants in a tensor form. Since an acoustic wave in a medium is a particular type of strain wave, the terms "acoustooptic effect" and "elastooptic effect" are often used interchangeably.

### 14.3.1 Elastooptic Effect in an Isotropic Medium

With such a medium as water whose elastooptic effect is isotropic, the change $\Delta n$ in refractive index due to the strain is expressed by the scalar equation

$$\Delta n = -\frac{n_o^3}{2} p S \qquad (14.20)$$

where $n_o$ is the refractive index under no stress, $p$ is the elastooptic constant, and $S$ is strain defined by the deformation divided by the total length.

A light deflector, such as shown in Fig. 14.9, is one of the applications of the acoustooptic effect of fluid. An acoustic wave is launched from the bottom of the deflector cell containing an elastooptic medium. A periodic strain distribution is established in the cell. Since the velocity of the acoustic wave is much slower than that of the light waves, the cell can be treated as if a stationary diffraction grating were occupying the cell.

When the cell is thin, and the intensity of the acoustic wave is low, the change of direction of the ray during transmission through the cell is small. The emergent wave is still considered to be a plane wave, but with its phase spatially modulated. The diffraction pattern under such a condition is called Raman-Nath diffraction, and was discussed in Problem 4.2. The angle $\theta$ of the direction of the maxima of the emergent wave was given by

$$\sin \theta - \sin \theta_0 = m \frac{\lambda}{\lambda_a}, \qquad (14.21)$$

where $\theta_0$ is angle of incidence, $\lambda_a$ is wavelength of the acoustic wave, and $m$ is an integer. Angles are taken as positive when they are counter-clockwise from the normal to the cell surface. As shown in Fig. 14.9, on both sides of a line extending from the direction of the incident ray, a number of maxima associated with $m = 0, \pm 1, \pm 2, \pm 3, \dots$ appear. The angles of diffraction can be steered by changing the frequency of the acoustic wave. The

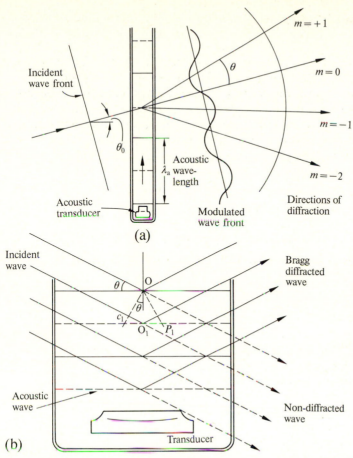

Fig. 14.9 a, b. Acoustooptic deflector. (a) Raman-Nath diffraction. (b) Bragg diffraction

advantage of such a deflector over a deflector using a mechanically moving mirror is that the former can steer the beam with greater speed.

When the thickness of the cell and the intensity of the acoustic wave is increased, as shown in Fig. 14.9b, the mechanism of diffraction of the beam becomes slightly different and is called Bragg diffraction. The cell is interpreted as a stack of reflecting layers spaced by the wavelength of the acoustic wave. A similar analysis is used for treating x-ray diffraction from the layers of atoms. When the path of the ray becomes such that the difference in the optical path between the adjacent layers in the figure equals an integral multiple of the light wavelength, the light rays reflected from each layer will be in phase and the reflected light intensity displays a peak. The Bragg diffraction condition is

$$\sin \theta = m \frac{\lambda}{2 \lambda_a}. \tag{14.22}$$

The Bragg condition is a special case of the Raman-Nath condition. Since with Bragg diffraction, each layer is considered as a mirror, the additional condition of

$$-\theta_0 = \theta \tag{14.23}$$

is imposed over the Raman-Nath condition. Insertion of (14.23) into (14.21) leads to the Bragg condition (14.22). It should be noticed that, in the case of Raman-Nath diffraction, several discrete peaks can be observed in the emergent wave corresponding to the various diffraction orders $m = 0, \pm 1, \pm 2$, and so forth. However, when light is incident at the Bragg angle, a single peak in the emergent wave dominates all others, and this peak is the Bragg diffracted wave. In fact, when certain parameters such as the acoustic intensity, frequency and cell length are properly chosen, a very high percentage of the incident wave can be converted into the Bragg diffracted wave. The efficiency of converting the incident to the Bragg diffracted wave decreases dramatically if the direction of incidence strays from the Bragg angle defined by (14.22).

It may be added that, in the case of x-ray diffraction, if the setting of the *incident* beam to the crystal is arbitrary, absolutely no diffraction pattern is observed. The pattern appears only when the *incident* beam satisfies (14.22, 23). In x-ray crystallography [14.6−8], the Laue pattern meets this condition by continuously varying the wavelength, whereas the rotating-crystal method and powder method (that uses a powdered sample rather than a lump sample) meet the condition by varying the incident angle.

By turning on and off the acoustic wave, the acoustooptic deflector can be used to switch directions between the beam which passes straight through the cell, and the beam or beams which are diffracted by the cell. The deflector shown in Fig. 14.9 can also be used as a modulator. The amount of light diffracted by the cell depends on the amplitude of the acoustic wave. Therefore, an amplitude modulation of the acoustic wave will produce an amplitude modulation on each of the emergent light beams. Another characteristic of the acoustooptic diffraction is that the diffracted beams are inherently shifted in frequency by an integral multiple of the acoustic frequency. One interpretation of this phenomenon is that the light is being reflected by a moving reflecting plane, and a shift of frequency accompanies. Frequency modulation of the acoustic wave results in a frequency modulation of the deflected waves, so that the device can also be used as a frequency modulator.

Compared with the bulk electrooptic modulator, which needs several kilovolts to operate, a bulk acoustooptic modulator needs only several volts. However, the response time is much slower than the electrooptic device because it needs time for the acoustic wave to propagate across the cell. Quite often, the frequency band characteristics are limited by the frequency response of the transducer. If one wants to keep the incident Bragg diffraction angle constant, the range of the modulation frequency has to be restricted.

### 14.3.2 Elastooptic Effect in an Anisotropic Medium

Such elastooptic materials as lithium niobate or tellurium dioxide ($TeO_2$) are crystals and the elastooptic effect is anisotropic. The relationship between the applied strain and the change in the refractive index is found by calculating the deformation of the indicatrix [14.9]. Almost the same expression as the electrooptic effect (Sect. 14.2) is used for the elastooptic effect. The only difference is that the size of the matrix is $6 \times 6$ rather than $6 \times 3$, namely

$$
\begin{bmatrix}
B_{11} - 1/n_1^2 \\
B_{22} - 1/n_2^2 \\
B_{33} - 1/n_3^2 \\
B_{23} \\
B_{31} \\
B_{12}
\end{bmatrix}
=
\begin{bmatrix}
p_{11} & p_{12} & p_{13} & p_{14} & p_{15} & p_{16} \\
p_{21} & p_{22} & p_{23} & p_{24} & p_{25} & p_{26} \\
p_{31} & p_{32} & p_{33} & p_{34} & p_{35} & p_{36} \\
p_{41} & p_{42} & p_{43} & p_{44} & p_{45} & p_{46} \\
p_{51} & p_{52} & p_{53} & p_{54} & p_{55} & p_{56} \\
p_{61} & p_{62} & p_{63} & p_{64} & p_{65} & p_{66}
\end{bmatrix}
\begin{bmatrix}
s_1 \\
s_2 \\
s_3 \\
s_4 \\
s_5 \\
s_6
\end{bmatrix}
\tag{14.24}
$$

where $B_{11} x^2 + B_{22} y^2 + B_{33} z^2 + 2 B_{23} y z + 2 B_{31} x z + 2 B_{12} x y = 1$.

There are two types of strain. One type is the deformation which takes place uniformly across the plane, as shown in Fig. 14.10a. Such a strain is

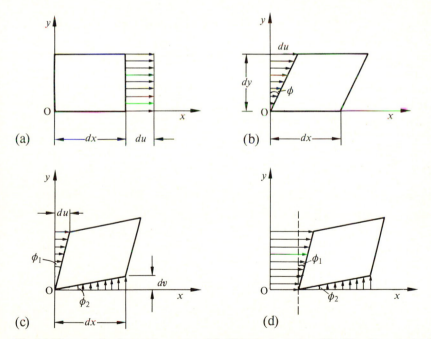

**Fig. 14.10 a – d.** Strain components. (**a**) Principal strain. (**b**) Shearing strain in $y$. (**c**) Shearing strain in $x$ and $y$. (**d**) Combination of principal strain and shearing strain

called the principal strain and is defined as

$$S_{xx} = \frac{du}{dx},$$ (14.25)

where $du$ is the deformation in the $x$ direction in a line element of length $dx$. The other type is the deformation which is not uniform across the plane, such as shown in Fig. 14.10b. Here, the deformation $du$ varies with $y$. Such a deformation is called the shearing strain. The shearing strain $S_{xy}$ in Fig. 14.10b is defined as $\partial u/\partial y$. Figure 14.10c shows the shearing strain in the general case. The shearing deformation takes place in both the $x$ and $y$ directions. The shearing strain in this case is defined as

$$S_{xy} = \frac{1}{2}\left(\frac{\partial u}{\partial y} + \frac{\partial v}{\partial x}\right)$$ (14.26)

where $dv$ is the deformation in the $y$ direction.

Note that $\partial u/\partial y$ is approximately $\phi_1$ which is the angle associated with the deformation in the $x$ direction. Similarly $\partial v/\partial x$ is approximately $\phi_2$ which is associated with the deformation in the $y$ direction. Thus, $S_{xy}$ is one half of the angle associated with the total shearing strain taking place in the $x y$ plane.

The general case of deformation in the $x y$ plane (Fig. 14.10d) can be represented by a combination of these two types of strain. The components of strain are summarized as

$$\left.\begin{array}{ll} S_{xx} = \dfrac{\partial u}{\partial x}, & S_{yz} = \dfrac{1}{2}\left(\dfrac{\partial v}{\partial z} + \dfrac{\partial w}{\partial y}\right), \\[3mm] S_{yy} = \dfrac{\partial v}{\partial y}, & S_{xz} = \dfrac{1}{2}\left(\dfrac{\partial u}{\partial z} + \dfrac{\partial w}{\partial x}\right), \\[3mm] S_{zz} = \dfrac{\partial w}{\partial z}, & S_{xy} = \dfrac{1}{2}\left(\dfrac{\partial u}{\partial y} + \dfrac{\partial v}{\partial x}\right). \end{array}\right\}$$ (14.27)

These can be expressed by one formula

$$S_{ij} = \frac{1}{2}\left(\frac{\partial u_i}{\partial x_j} + \frac{\partial u_j}{\partial x_i}\right),$$ (14.28)

where $S_{ij}$ is sometimes called Cauchy's infinitesimal strain tensor. A common convention is to express the subscripts by numerals as

$$\begin{array}{cccccc} S_{xx} & S_{yy} & S_{zz} & S_{yz} & S_{xz} & S_{xy} \\ \downarrow & \downarrow & \downarrow & \downarrow & \downarrow & \downarrow \\ S_1 & S_2 & S_3 & S_4 & S_5 & S_6. \end{array}$$

**Exercise 14.5**    An acoustic shear wave whose direction of propagation is in the $z$ direction and whose direction of oscillation is in the $y$ direction is launched into a lithium niobate crystal. Find the expression of the indicatrix under the influence of the acoustic wave. The matrix of the elastooptic constants of lithium niobate is

$$\begin{bmatrix} p_{11} & p_{12} & p_{13} & p_{14} & 0 & 0 \\ p_{12} & p_{11} & p_{13} & -p_{14} & 0 & 0 \\ p_{31} & p_{31} & p_{33} & 0 & 0 & 0 \\ p_{41} & -p_{41} & 0 & p_{44} & 0 & 0 \\ 0 & 0 & 0 & 0 & p_{44} & -p_{41} \\ 0 & 0 & 0 & 0 & p_{14} & p_{66} \end{bmatrix}$$

with $n_x = n_y = n_o = 2.286$, $n_z = n_e = 2.2$, $p_{11} = -0.02$, $p_{12} = 0.08$, $p_{13} = 0.13$, $p_{14} = -0.08$, $p_{31} = 0.17$, $p_{33} = 0.07$, $p_{41} = -0.15$, $p_{44} = 0.12$, $p_{66} = -0.05$.

*Solution*    Figure 14.11 shows an exaggerated view of the deformation of the elastooptic medium due to the acoustic wave. The strain created by the acoustic wave is in the $z$ direction and is a function of $y$, as shown in Fig. 14.11. Such a strain is the shearing strain taking place in the $yz$ plane. Thus, only $S_4$ is nonzero, and the expression for the indicatrix becomes

$$(n_o^{-2} + p_{14} S_4) x^2 + (n_o^{-2} - p_{14} S_4) y^2 + n_e^{-2} z^2 + 2 p_{44} S_4 \, y \, z = 1. \qquad (14.29)$$

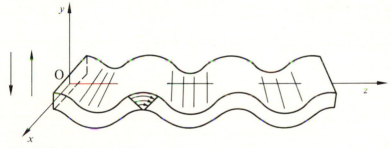

**Fig. 14.11.** Excitation of shear wave

One method of eliminating the $yz$ cross term is to rotate the coordinates around the $x$ axis by such an angle that this term disappears, but here another method is employed.

First consider a general expression of the form

$$A y^2 + 2B \, y z + C z^2 = 1. \qquad (14.30)$$

It is more complicated to find the major axes when $A \neq C$ as in (14.30). The axis will be found by an eigenvalue method. The normal to this ellipse is found from the gradient, $\nabla g(y, z)$, where $g(y, z)$ is the left-hand side of (14.30). A vector $N$ in the direction of the normal is

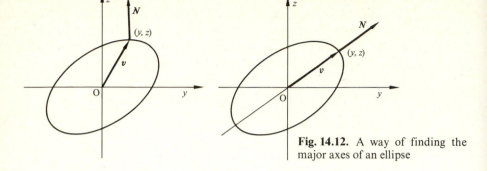

**Fig. 14.12.** A way of finding the major axes of an ellipse

$$N = \begin{pmatrix} A\,y + B\,z \\ B\,y + C\,z \end{pmatrix} = \begin{pmatrix} A & B \\ B & C \end{pmatrix} \begin{pmatrix} y \\ z \end{pmatrix} \tag{14.31}$$

where the factor of 2 contained in $\nabla g\,(x,\,y)$ is immaterial for expressing the direction and is suppressed. As shown in Fig. 14.12, a vector $v = \begin{pmatrix} y \\ z \end{pmatrix}$ connecting the origin to the point $(y,\,z)$ on the ellipse lines up with the normal $N$ from that point if the point is on either the major or minor axis. When both vectors are lined up,

$$\begin{pmatrix} A & B \\ B & C \end{pmatrix} \begin{pmatrix} y \\ z \end{pmatrix} = \lambda \begin{pmatrix} y \\ z \end{pmatrix} \tag{14.32}$$

should be satisfied. In order that a non-trivial solution exist, the determinant of

$$\begin{vmatrix} A - \lambda & B \\ B & C - \lambda \end{vmatrix}$$

should vanish so that

$$(A - \lambda)(C - \lambda) - B^2 = 0$$

and hence,

$$\lambda_{1,2} = \tfrac{1}{2}\,[(A + C) \pm \sqrt{(A + C)^2 - 4\,(AC - B^2)}\,]. \tag{14.33}$$

The major and minor axes[3] of the ellipse are $\lambda_1^{-1/2}$ and $\lambda_2^{-1/2}$, and the unit vectors expressing the axis are obtained from (14.32, 33) as

$$u_1 = \frac{1}{\sqrt{B^2 + (\lambda_1 - A)^2}} \begin{pmatrix} B \\ \lambda_1 - A \end{pmatrix},$$

$$u_2 = \frac{1}{\sqrt{B^2 + (\lambda_2 - A)^2}} \begin{pmatrix} B \\ \lambda_2 - A \end{pmatrix}. \tag{14.34}$$

This procedure is the same as finding eigenvalues and eigenvectors of a matrix. Finally, the expression of the indicatrix in (14.29) becomes

$$\lambda_0\, x^2 + \lambda_1\, Y^2 + \lambda_2\, Z^2 = 1, \quad \text{where}$$
$$\lambda_0 = n_0^{-2} + p_{14}\, S_4,$$

and $\lambda_1$ and $\lambda_2$ are calculated using (14.33) with

$$A = n_0^{-2} - p_{14}\, S_4, \quad B = p_{44}\, S_4, \quad C = n_e^{-2}.$$

## 14.4 Miscellaneous Effects

Besides the above-mentioned Pockels, Kerr, and elastooptic effects, there are a few other effects that are important.

### 14.4.1 Optical Activity

With materials such as a quartz crystal or a sugar solution, the phase front surfaces for the o and e waves shown in Fig. 14.1 do not touch each other even when propagation takes place along the optical axis. When a linearly

---

3 Note that (14.30) can be rewritten as an inner product of the normal vector $N$ and the position vector $v$,

$$(y, z)\, M \begin{pmatrix} y \\ z \end{pmatrix} = 1, \quad \text{where} \quad M = \begin{pmatrix} A & B \\ B & C \end{pmatrix}.$$

If the position vector $v$ is expressed by new coordinates $(Y, Z)$ along $u_1$ and $u_2$ such that

$$v = \begin{pmatrix} y \\ z \end{pmatrix} = Y\, u_1 + Z\, u_2, \quad \text{then}$$

$$\begin{aligned} v^t M v &= (Y\, u_1^t + Z\, u_2^t)\, M(Y\, u_1 + Z\, u_2) \\ &= (Y\, u_1^t + Z\, u_2^t)(\lambda_1\, Y\, u_1 + \lambda_2\, Z\, u_2) \\ &= \lambda_1\, u_1^t u_1\, Y^2 + \lambda_2\, u_2^t u_2\, Z^2, \end{aligned}$$

where the superscript $t$ means transpose. Therefore

$$\lambda_1\, Y^2 + \lambda_2\, Z^2 = 1,$$

where the eigenvector relationship (14.32) and the orthogonality condition $u_1^t\, u_2 = u_2^t\, u_1$ $= 0$ were used. The orthogonality can be easily checked by calculating this product using (14.33, 34).

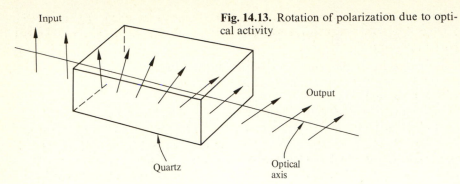

**Fig. 14.13.** Rotation of polarization due to optical activity

polarized wave is incident along the optical axis of such a material, the direction of polarization rotates as it propagates, as shown in Fig. 14.13. This rotation of polarization takes place even without an applied field. Such a phenomenon is called optical activity.

Fresnel's explanation of the effect is that it is due to the difference between the refractive indices $n_l$ and $n_r$ for left and right circularly polarized waves. A linearly polarized wave can be decomposed into two circularly polarized waves rotating in opposite senses, as shown in Fig. 14.14. In other words, the resultant of the two circularly polarized waves rotating in opposite directions always oscillates along a fixed line, provided that the speed of rotations of both circularly polarized waves are the same.

When one of the circularly polarized waves rotates slightly faster than the other, the polarization of the resultant wave rotates as it oscillates. The amount of rotation of the resultant wave per unit length of material is known as the rotary power. The fact that the sense of rotation and the

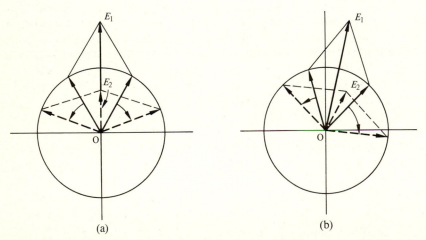

**Fig. 14.14 a, b.** Rotation of a linearly polarized wave. (**a**) When $n_l = n_r$. (**b**) When $n_l = n_r$

rotary power are peculiar to a type of material is used for analyzing biochemical substances. A saccharimeter uses this effect to determine sugar content in solutions.

### 14.4.2 Faraday Effect

As a phenomenon, the Faraday effect is similar to optical activity. With the Faraday effect, rotation of the polarization takes place only when an external magnetic field is applied. Materials that exhibit the Faraday effect include sodium chloride (NaCl), zinc sulphide (ZnS), and yttrium iron garnet (YIG).

The relationship between the angle $\Phi$ of polarization rotation and the applied magnetic field $H$ is

$$\Phi = VH \, l \cos \gamma \tag{14.35}$$

where $V$ is the Verdet constant, $l$ is the length of the material, and $\gamma$ is the angle between the direction of the magnetic field and the direction of propagation of the light beam. The Faraday effect is one of the few magnetooptic effects that are large enough to be utilized for devices. The sense of rotation depends solely on the direction of the magnetic field $H$. Reversing the direction of $H$ reverses the sense of rotation. Reversing the direction of light propagation has no effect on the sense of rotation.

Figure 14.15 shows an example of how this characteristic of the Faraday effect can be used to make an optical isolator. The optical isolator consists of a magnetooptic material, a magnetic coil, a polarizer $P_1$ and an analyzer $P_2$. The direction of the polarizer $P_1$ is rotated from that of analyzer $P_2$ by 45 degrees.

Figure 14.15a shows how the light passes through from left to right with minimum attenuation and Fig. 14.15b shows how the light in the reverse direction is blocked. As shown in Fig. 14.15a, the polarization of the incident light is oriented in the same direction as the polarizer $P_1$ for easy passage. The direction of the polarization is rotated by $-45°$ (clockwise) in the Faraday medium. Thus the direction of the polarization becomes the same as the analyzer $P_2$ for easy passage. Next, referring to Fig. 14.15b, a wave propagating in the opposite direction (reflected wave) is considered. It can go through $P_2$, but the Faraday effect rotates it by $-45°$, the same as before, and the direction of the polarization becomes perpendicular to that of $P_1$ so that it cannot go through $P_1$. Thus, the optical isolator prevents the reflected wave from reaching the source.

It may be added that exactly the same arrangement can be used for modulating the light intensity by modulating the magnetic field. The inductance of the magnetic coil limits the maximum frequency of such a modulator.

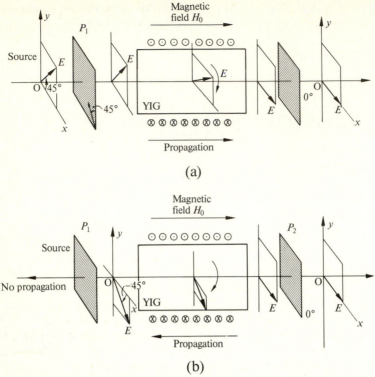

**Fig. 14.15 a, b.** Principle of the optical isolator. (**a**) Direction of transmission. (**b**) Direction of isolation

### 14.4.3 Other Magnetooptic Effects

The Cotton-Mouton effect is a magnetic effect which is observed in liquids such as nitrobenzene ($C_6H_5NO_2$). When a magnetic field is applied perpendicular to the direction of light propagation, the material behaves similarly to a uniaxial crystal whose optical axis lies along the external magnetic field. This effect is more or less a magnetic counterpart of the electrooptic Kerr effect, and the effect is proportional to the square of the magnetic field.

A similar effect observed in a vapor state like lithium gas is called the Voigt effect. The Kerr magnetooptic effect is observed when linearly polarized light is incident upon a magnetic pole-piece. The reflected wave becomes elliptically polarized.

### 14.4.4 Franz-Keldysh Effect

The Franz-Keldysh effect is an electro-absorptive effect which is caused by the shift of the absorption edge of a semiconductor due to an external elec-

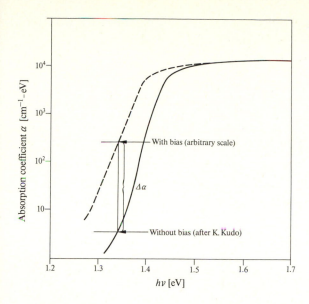

**Fig. 14.16.** Absorption edge of a semiconductor

tric field [14.10−12]. Figure 14.16 shows a plot of the absorption coefficient of a GaAs semiconductor with respect to the quanta of light. There is an abrupt increase in the absorption around 1.4 eV. This is called the absorption edge. Thus the quantity 1.4 eV is the minimum energy required to release a bound electron from its orbit thereby producing a free electron. (In terms of energy-band theory, the photon energy is used to kick an electron from the valence band into the conduction band, thereby creating a hole in the valence band and an electron in the conduction band. The absorption edge is used for determining the band structure of the semiconductor). When an external electric field is applied to the semiconductor, the absorption edge shifts, as indicated by the dotted line in the figure, and thus the absorption of light at a frequency near the absorption edge increases.

A modulator based upon the electroabsorptive effect is quite similar to a laser diode in structure, as shown in Fig. 13.13. The differences are that the composition of the layers of the modulator is designed so that the absorption edge appears at the desired wavelength, and that the modulator device is reverse biased instead of forward biased. When it is biased to almost breakdown voltage, an extinction ratio defined as the leakage power (in the off state of the switch) divided by the transmitted power (in the on state) of about 10 dB is obtainable.

## Problems

**14.1**  Consider the case where an external electric field $E_x$ is applied along the $x$ axis (perpendicular to the crystal axis) of a quartz crystal, and the direction of the incident light is along the $y$ axis. Find the allowed directions of polarization and their refractive indices.

**14.2**  An electrooptically controlled optical guide is to be fabricated using a lithium niobate substrate. The direction of the light propagation is in the $z$ direction, and the directions of polarization of the guided modes are as shown in Fig. 14.17. How should the crystal be cut in order to achieve the largest electrooptic effect? What would be the configuration of the electrodes for applying the external field corresponding to the following situations:

a)  when the guide is excited with $E$ field as shown in Fig. 14.17a?
b)  when the guide is excited with $E$ field as shown in Fig. 14.17b?

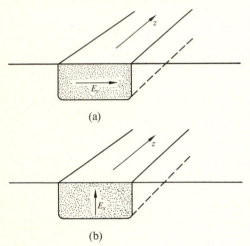

(a)

(b)

**Fig. 14.17 a, b.** Configuration of an electrooptic guide of $LiNbO_3$. **(a)** $TE_0$ mode. **(b)** $TM_0$ mode

**14.3**  A strain gauge is to be made of a lithium niobate substrate. The external electric field needed to compensate the elastooptic effect is used to determine the strain.

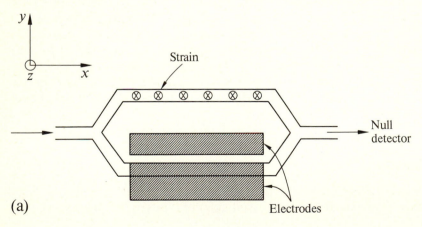

(a)

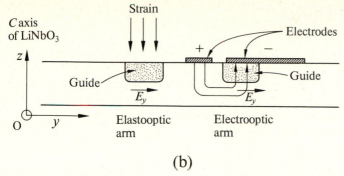

Fig. 14.18 a, b. Strain gauge. (a) Plan view. (b) Cross-sectional view

A Mach-Zehnder interferometer such as shown in Fig. 14.18 is proposed. A C-cut crystal is used and a mode polarized in the $y$ direction is excited. The strain is applied in the direction normal to the surface of the substrate.

Find a formula relating the external electric field to the applied strain.

# 15. Integrated Optics

In the case of an electronic integrated circuit (IC), e.g., for a shift register IC, one normally does not worry about how each flip-flop unit is functioning. The only concern is whether or not the input and output electrical signals are correct. The IC's are packaged in small black plastic cases, and they are literally looked upon as black boxes.

A similar objective has been the goal of integrated optics. When this goal is reached, no one will have to worry about vibration, coupling between components, temperature, dust, or moisture. Various optomodules could then presumably be integrated on a small substrate the size of a microscope deck glass for easy handling. The optomodules would contain thin-film devices such as modulators, directional couplers, switches, bistable optical devices, lenses, A/D converters and the spectrum analyzer. This chapter begins with a fundamental waveguide analysis and an introduction to coupled mode theory. Following this analysis is a description of the basic thin film devices which form the building blocks of integrated optics. A summary of the thin-film techniques that were specifically developed in connection with integrated optics is also included.

Before proceeding, a word of caution is advised concerning categorization. Quite often, a single integrated-optics device may provide a variety of functions. For example, the same basic integrated optics component may be used as a modulator, a switch, or a directional coupler. Therefore, to use the phrase "directional coupler" as the name of a subsection may be misleading. Despite this drawback, every attempt has been made to present the material in an order which is easy to comprehend.

## 15.1 Analysis of the Slab Optical Guide

The slab geometry is one of the most fundamental configurations in integrated-optics technology. Consider the propagation of a wave in a core glass with index of refraction $n_1$, when it is sandwiched by a cladding glass with index of refraction $n_2$, as shown in Fig. 15.1. When $n_2$ is smaller than $n_1$, the light propagates along the core glass in a zigzag path due to the total reflection at the upper and lower boundaries. The optical guide will be analyzed using both wave optics and geometrical optics.

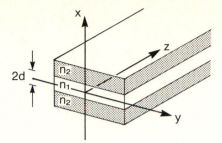

**Fig. 15.1.** Slab optical guide

### 15.1.1 Differential Equations of Wave Optics

The wave optics treatment is slightly more complicated but provides accurate information about the guide. First, the wave equations are derived from Maxwell's equation. Consider, an electromagnetic wave propagating in the $z$ direction such that

$$\mathbf{E}' = \mathbf{E}\, e^{j(\beta z - \omega t)}, \qquad \mathbf{H}' = \mathbf{H}\, e^{j(\beta z - \omega t)}. \tag{15.1}$$

By inserting these into Maxwell's equations one obtains

$$\nabla \times \mathbf{H} = -j\omega\,\varepsilon_0\,\varepsilon_r\,\mathbf{E}, \qquad \nabla \times \mathbf{E} = j\omega\,\mu\,\mathbf{H}. \tag{15.2}$$

An assumption is made that no component of the field varies in the $y$ direction, or in other words

$$\frac{\partial}{\partial y} = 0. \tag{15.3}$$

The field is decomposed into TE and TM waves. The waves that contain no electric field in the direction of propagation are TE waves, namely

$$E_z = 0. \tag{15.4}$$

Similarly, waves that contain no magnetic field in the direction of propagation are TM modes, namely

$$H_z = 0. \tag{15.5}$$

Equations (15.3–5) significantly simplify the curl operation of (15.2).
    With the assumption of $E_z = 0$, (15.2) gives the TE-mode solution as

$$E_x = 0, \qquad E_y = \frac{j}{\omega\,\varepsilon_0\,\varepsilon_r}\left(\frac{\partial H_x}{\partial z} - \frac{\partial H_z}{\partial x}\right), \qquad E_z = 0,$$

$$H_x = \frac{j}{\omega\,\mu}\frac{\partial E_y}{\partial z}, \qquad H_y = 0, \qquad H_z = \frac{-j}{\omega\,\mu}\frac{\partial E_y}{\partial x}, \tag{15.6}$$

$$\left(\frac{\partial^2}{\partial x^2} + \frac{\partial^2}{\partial z^2} + n^2 k^2\right) E_y = 0 \quad \text{where} \quad \omega^2 \mu \, \varepsilon_0 \, \varepsilon_r = n^2 k^2, \quad k = \omega \sqrt{\mu \, \varepsilon_0} \, .$$

Similarly, with the assumption of $H_z = 0$, the TM-mode solution is obtained,

$$E_x = \frac{-j}{\omega \, \varepsilon_0 \, \varepsilon_r} \frac{\partial H_y}{\partial z}, \qquad E_y = 0, \qquad E_z = \frac{j}{\omega \, \varepsilon_0 \, \varepsilon_r} \frac{\partial H_y}{\partial x},$$

$$H_x = 0, \qquad H_y = \frac{j}{\omega \mu} \left(\frac{\partial E_z}{\partial x} - \frac{\partial E_x}{\partial z}\right), \qquad H_z = 0,$$

(15.7)

$$\left(\frac{\partial^2}{\partial x^2} + \frac{\partial^2}{\partial z^2} + k^2 n^2\right) H_y = 0.$$

The expressions for the TE and TM modes contain many similarities. First, the solutions for the TE modes are sought.

### 15.1.2 General Solution for the TE Modes

As seen from (15.6), once an expression for $E_y$ is obtained, all other components of the TE wave are automatically found. A good starting point for finding an expression for $E_y$ is the differential equation at the bottom of (15.6). Making use of the fact that $z$ enters (15.1) as an exponential function, the differential equation becomes

$$\frac{1}{E_y} \frac{\partial^2 E_y}{\partial x^2} + (n^2 k^2 - \beta^2) = 0. \tag{15.8}$$

The method of separation of variables will be used to solve (15.8) for $E_y$. Since the second term of (15.8) is constant, the first term also has to be constant. As far as the values of the constant is concerned, regardless of whether a positive number or a negative number is picked, (15.8) still holds true, but the solution differs depending on which is chosen. The solution is either trigonometric or exponential. Let us first pick a negative number $- K^2$ as the value of the first term. In order that (15.8) be true, the second term has to be $K^2$ namely

$$\frac{\partial^2 E_y}{\partial x^2} + K^2 E_y = 0 \quad \text{with} \tag{15.9}$$

$$K^2 = n^2 k^2 - \beta^2. \tag{15.10}$$

The solution of (15.9) is

$$E_y = A \cos K x + B \sin K x. \tag{15.11}$$

Next, let us pick a positive number $\gamma^2$ as the value of the first term in (15.8), then,

$$\frac{\partial^2 E_y}{\partial x^2} - \gamma^2 E_y = 0 \quad \text{with} \tag{15.12}$$

$$\gamma^2 = \beta^2 - n^2 k^2. \tag{15.13}$$

The solution of (15.12) is

$$E_y = C\,e^{-\gamma x} + D\,e^{\gamma x}. \tag{15.14}$$

Both (15.11, 14) are equally valid as a solution of (15.8). In fact the sum of (15.11, 14) is also a solution. The choice of the solution depends on which one of the solutions suits the physical situation best. (One way is to use the sum of (15.11, 14) as a solution and, depending on the situation, any of $A$, $B$, $C$, or $D$ is (are) made zero. As long as $A$, $B$, $C$ or $D$ are constants, including zero, these solutions hold true.)

For the case of a slab optical guide, (15.11) is assigned as the solution in the core glass because of its oscillatory nature and (15.14) is assigned as the solution in the cladding glass because its exponential decay nature is suitable for representing the wave in the cladding glass. With this arrangement of solutions, (15.10, 13) become

$$K^2 = n_1^2\,k^2 - \beta^2, \tag{15.15}$$

$$\gamma^2 = \beta^2 - n_2^2\,k^2. \tag{15.16}$$

Note that $n_1$ has to be used in (15.15), whereas $n_2$ appears in (15.16).

For the sake of simplicity, the two terms in (15.11) are treated separately and added later. They are

$$E_y = A \cos K x, \tag{15.17}$$

$$E_y = B \sin K x. \tag{15.18}$$

The solution expressed by (15.17) is called an even mode and that by (15.18) is called an odd mode, because $\cos Kx$ is an even function with respect to $x$ and $\sin Kx$ is an odd function. In fact, the optical slab guide can be excited by either even modes alone, or odd modes alone, or a combination of the two. The ratio of $A$ and $B$ is determined by the type of incident wave to the optical guide. If the field distribution of the incident wave is symmetric with respect to the center plane of the slab, only the even modes are excited. If the distribution of the incident wave is antisymmetric, the slab will be excited only by the odd modes. If the incident field is arbitrary distribution, both modes will be present so that the incident field distribution and the excited field in the slab guide are smoothly connected.

### 15.1.3 Boundary Conditions

The propagation constant and the field distribution of each mode will be found. The treatment for even and odd modes are similar, and the even modes are considered first. The values of the constants $A$, $B$, $C$ and $D$ in (15.11, 14) are found from the boundary conditions. As shown in Fig. 15.2, there are four boundaries; two at plus and minus infinity and two others at the upper and lower core-cladding boundaries. First, the boundary at $x = +\infty$ is considered. So as to avoid (15.14) growing infinity for large $x$, $D$ has to be zero in the region $x > d$ and thus,

$$E_y = C\,e^{-\gamma x}, \quad x > d. \tag{15.19}$$

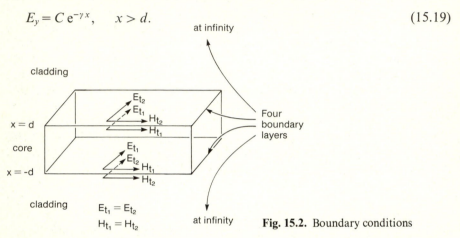

Fig. 15.2. Boundary conditions

On the other hand, in the lower cladding region, $C$ has to be set to zero for the same reason,

$$E_y = D\,e^{\gamma x}, \quad x < d. \tag{15.20}$$

The boundary condition at $x = d$ is now considered. Tangential components of both $E$ and $H$ have to be continuous across the boundary. Continuity of $E_y$ is set using (15.17, 19),

$$A \cos Kd = C\,e^{-\gamma d}. \tag{15.21}$$

Continuity of $H_z$ is set by incorporating the formula for $H_z$ in (15.6) with (15.17, 19),

$$KA \sin Kd = C\gamma\,e^{-\gamma d}. \tag{15.22}$$

The ratio of (15.21) to (15.22) gives

$$\gamma d = Kd \tan Kd. \tag{15.23}$$

Equation (15.23) is called the characteristic equation of the optical guide.

Next, the values of $\gamma d$ and $Kd$ will be determined. One more equation is necessary besides (15.23). This equation can be obtained from (15.15, 16),

$$(Kd)^2 + (\gamma d)^2 = V^2, \quad \text{where} \tag{15.24}$$

$$V = kd\sqrt{n_1^2 - n_2^2}. \tag{15.25}$$

The $V$ parameter is a function of such physical constants as the thickness of the slab optical guide, the wavelength of the light, and the refractive indices of the core and cladding glasses, and is sometimes called the normalized thickness of the guide. So as to solve the simultaneous equations of (15.23, 24), $\gamma d$ is plotted in the vertical axis, and $Kd$, in the horizontal axis in Fig. 15.3. The curve of (15.24) is a circle with radius $V$, and that of (15.23) has a shape similar to that of tan $Kd$. These two curves are shown as solid lines in Fig. 15.3. The positions of the intersection of the two curves are determined by the value of $V$. In this example, there are three intersections. Each intersection corresponds to one mode. Starting from the intersection nearest the origin they are named as even $TE_0$ mode, even $TE_1$ mode, even $TE_2$ mode and so on. In this example, the light energy propagates in three different modes with different propagation constants. Any of the three modes can be excited in this guide, depending on the configuration of excitation. The corresponding propagation constants are

$$\beta_N = \sqrt{(n_1 k)^2 - K_N^2}, \tag{15.26}$$

where $K_N$ denotes the discrete values of $K$ determined by the point of intersection in Fig. 15.3.

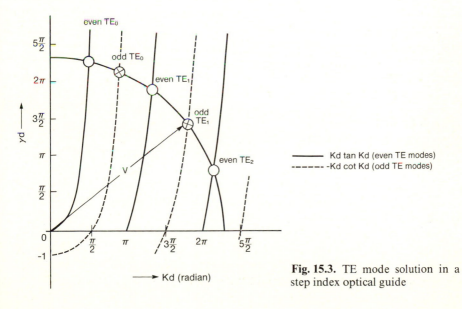

Fig. 15.3. TE mode solution in a step index optical guide

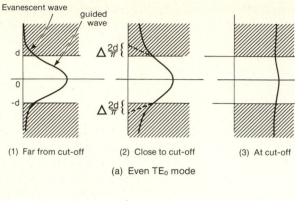

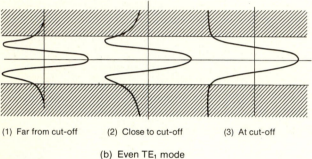

Fig. 15.4 a, b. General trend of the field $E_y$ as cutoff is approached. (a) Even $TE_0$ mode. (b) Even $TE_1$ mode

Next, the distribution of the amplitude $E_y$ is considered. The even $TE_0$ mode is considered first. Since the value of $Kd$ of the even $TE_0$ mode is slightly smaller than $\frac{\pi}{2}$, let it be represented by $(\frac{\pi}{2} - \Delta)$. From (15.17), the amplitude $E_y$ inside the core is

$$E_y = A \cos \left( \frac{\frac{\pi}{2} - \Delta}{d} \right) x.$$

The curve for this equation is shown in Fig. 15.4a. It is a cosine curve inside the core glass. Inside the cladding glass the magnitude of $E_y$ is very small, but as long as $\Delta$ is not zero, $E_y$ is not zero either, and the light penetrates into the cladding glass. Because this field decays exponentially with $x$, as seen from (15.19), this decaying field near the boundary is called an evanescent wave.

When the intensity distribution $|E_y(x)|^2$ is observed from the end of the guide, the even $TE_0$ mode displays one bright spot in the center, and the even $TE_1$ mode displays 3 bright spots, as seen from Fig. 15.4b. Generalizing, even $TE_n$ modes display $2n + 1$ bright spots.

As the value of $\gamma$ is further decreased, the penetration inside the cladding increases. When $\gamma = 0$ the light is present throughout the cladding glass. This can be interpreted as if all the light leaks into cladding glass, which corresponds to the cut-off condition of the guide. Speaking of the example in Fig. 15.3, if the value of $V$ is decreased to $2\pi$ radians, the even $TE_2$ mode goes to the cut-off condition. Similarly the cut-off condition of the even $TE_n$ mode occurs when $V = n\pi$ radians.

The cut-off of the even $TE_0$ mode is very special. This cut-off does not occur until $V = 0$, and for the region $0 < V < \frac{\pi}{2}$ only the even $TE_0$ mode is excited. The mode which is capable of being the sole excited mode is called the dominant mode. The even $TE_0$ mode is thus the dominant mode of the optical guide. The even $TE_1$ cannot be a dominant mode because, whenever the even $TE_1$ mode is excited, the even $TE_0$ mode can be excited at the same time. Hence the even $TE_1$ mode cannot be excited alone, and this is also true with all other higher-order modes.

Next, the odd TE modes are considered. The approach is quite similar to that of the even TE modes. To start with, the continuity at $x = d$ is considered. The continuity of $E_y$ is applied to (15.18, 19) and one obtains

$$A \sin Kd = C\, e^{-\gamma d}. \tag{15.27}$$

Similarly, the continuity of $H_z$, which is obtained by using (15.6), gives

$$KA \cos Kd = -C\, e^{-\gamma d}. \tag{15.28}$$

The ratio of (15.27) to (15.28) gives

$$\gamma d = -Kd \cot Kd, \tag{15.29}$$

which is the characteristic equation of the odd TE modes. This characteristic equation is plotted in Fig. 15.3 in dashed lines. These lines are almost parallel to those of the even TE modes, and interlace with them. One has to consider both even and odd modes to account for all TE modes. The spacing between the curves is approximately $\frac{\pi}{2}$.

The solutions of $E_y$ for even and odd TE modes are summarized in Table 15.1.

All other components of the TE modes in each region are obtained using (15.6).

**Table 15.1.** The $E_y$ components of even and odd TE modes

| Even TE modes | Odd TE modes | |
|---|---|---|
| $E_y = A \cos Kd\, e^{-\gamma(x-d)}$ | $E_y = A \sin Kd\, e^{-\gamma(x-d)}$ | in upper cladding |
| $E_y = A \cos Kx$ | $E_y = A \sin Kx$ | in core |
| $E_y = A \cos Kd\, e^{\gamma(x+d)}$ | $E_y = -A \sin Kd\, e^{\gamma(x+d)}$ | in lower cladding |

### 15.1.4 TM Modes

Procedures that are exactly the same as for the TE modes are applicable to the TM modes. The results are summarized here. The expression for $H_y$ is exactly the same as those in Table 15.1 with $E_y$ replaced by $H_y$. The other components of the TM modes are found using (15.7).

The characteristic equations, however, are slightly different from those of the TE modes. The characteristic equation of the even TM modes is

$$\left(\frac{n_2}{n_1}\right)^2 K d \tan K d = \gamma d, \tag{15.30}$$

and that of the odd TM modes is

$$-\left(\frac{n_2}{n_1}\right)^2 K d \cot K d = \gamma d. \tag{15.31}$$

If these characteristic equations are plotted in Fig. 15.3, together with those of the TE modes, the characteristic curves of the TM modes follow those of the TE modes but slightly below them. In practice, $n_2$ differs from $n_1$ by only a fraction of one per cent and the two sets of curves are close together.

### 15.1.5 Treatment by Geometrical Optics

In geometrical optics, the transmission of the light inside the slab optical guide is interpreted as a channelling of light by the repeated total reflections at the upper and lower boundary layers, as shown in Fig. 15.5a.

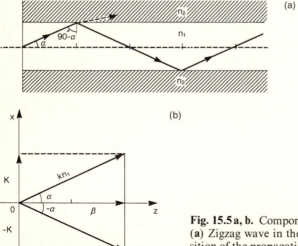

Fig. 15.5 a, b. Component waves in the optical guide. (a) Zigzag wave in the optical guide. (b) Decomposition of the propagation constant

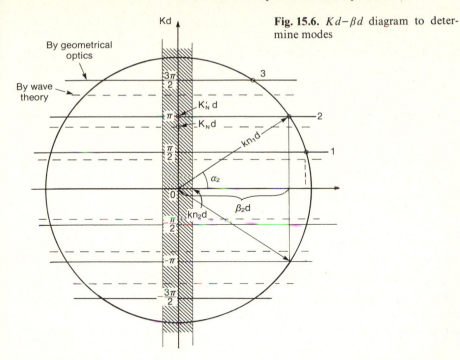

**Fig. 15.6.** $Kd-\beta d$ diagram to determine modes

If the propagation constant $n_1 k$ of the zigzag wave is decomposed, as shown in Fig. 15.5b into $K$ and $\beta$, in the $x$ and $z$ directions, respectively, then $K$ is

$$K = \sqrt{(k \, n_1)^2 - \beta^2}. \tag{15.32}$$

Equation (15.32) is rewritten as

$$(Kd)^2 + (\beta \, d)^2 = (k \, n_1 \, d)^2 \tag{15.33}$$

which represents a circle (Fig. 15.6). There are upper and lower limits that the value of $\beta d$ can take. The upper limit is obviously $k \, n_1 \, d \geq \beta \, d$ from (15.33). The lower limit is determined by the condition of total reflection. The critical condition at the boundary between the core and cladding glasses is

$$n_1 \cos \alpha > n_2 .$$

From Fig. 15.5b, $n_1 \cos \alpha = \beta/k$, hence the lower limit of $\beta d$ is

$$\beta \, d > k \, n_2 \, d,$$

and the hatched region in Fig. 15.6 should be excluded. The parameter $\beta/k$ is sometimes called the effective refractive index $n_e$.

Next, a quantization is introduced. The angle of incidence to the boundaries between the core and cladding glass is quantized in the manner described earlier in Sect. 5.8.1. The quantization condition is

$$2 K_N' d = N \pi, \quad \text{where } N = 0, 1, 2, 3, \ldots . \tag{15.34}$$

The intersections between the circle and horizontal solid lines in Fig. 15.6 are the quantized values of $K_N' d$.

The propagation constant $\beta_N'$ for the $N$th mode is given from (15.33, 34) by

$$\beta_N' = k n_1 \sqrt{1 - \left( \frac{N \pi}{2 k n_1 d} \right)^2}. \tag{15.35}$$

### 15.1.6 Comparison Between the Results by Geometrical Optics and by Wave Optics

The results obtained by wave optics will be rewritten in a form that can be easily compared with that obtained by geometrical optics. Equation (15.17) of the wave optics treatment is put into (15.1) and rewritten as

$$E_y' = \tfrac{1}{2} \{ A \exp [j(K_N x + \beta z - \omega t)] + A \exp [j(- K_N x + \beta z - \omega t)] \}, \tag{15.36}$$

where $K_N$ denotes the discrete values of $K$ obtained from Fig. 15.3. According to the vector expression (2.17) of plane wave propagation, the first term of (15.36) is a plane wave propagating in the direction expressed by

$$\boldsymbol{k} = K_N \hat{\boldsymbol{i}} + \beta \hat{\boldsymbol{k}},$$

and the second term is another plane wave in the direction

$$\boldsymbol{k} = - K_N \hat{\boldsymbol{i}} + \beta \hat{\boldsymbol{k}}.$$

The vector diagrams for these propagation constants are the same as those shown in Fig. 15.5 b, except for the slightly different value of $K$. In the wave optics approach, $K$ takes on the discrete values $K_N$ which are determined from the intersections between the characteristic equation and the normalized thickness $V$ in Fig. 15.3, whereas the values of $K_N'$ in the geometrical optics approach were determined by (15.34). The values of $K_N d$ approach $\tfrac{\pi}{2} N$ for large $V$. The solutions by geometrical optics approach those by wave optics as $V$ is increased. The values of $K_N$ are indicated by the dashed lines in Fig. 15.6. For low values of $V$, the wave optics approach is more accurate.

It should be recognized that the electric field distribution inside the core glass is nothing but a standing wave pattern of the two component waves of (15.36). The higher the mode number $N$, the larger the angle $\alpha_N$, and hence a greater number of loops and nulls of the field pattern exist with higher modes.

## 15.2 Coupled-Mode Theory

Coupled mode equations are useful for analyzing such components as directional couplers, light switches, light modulators, and wavelength filters [15.1, 2]. It is so widely used that it is worthwhile to describe it in some detail here.

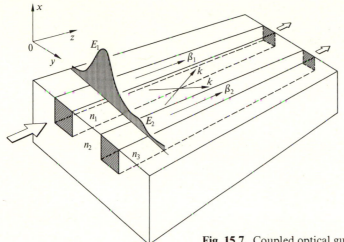

**Fig. 15.7.** Coupled optical guide

Referring to Fig. 15.7, the coupling taking place between two parallel lines is considered. Each line is located in the region of the evanescent wave of the other line, and both are loosely coupled to each other. The wave propagating along the lone guide 1 in the absence of guide 2 is expressed by

$$E_1 = A \exp\left[j\left(\beta_1^0 z - \omega t\right)\right] \tag{15.37}$$

and satisfies the differential equation

$$\frac{dE_1}{dz} = j\,\beta_1^0\,E_1. \tag{15.38}$$

When another guide 2 is introduced in the vicinity of guide 1, then two things will happen. First, the propagation constant changes because of the change in geometry. Secondly, power is exchanged between the two lines because each line is located in the region of the evanescent wave of the other. The coupled differential equations thus become

$$\frac{dE_1}{dz} = j\,\beta_1\,E_1 + k_{12}\,E_2, \qquad \frac{dE_2}{dz} = k_{21}\,E_1 + j\,\beta_2\,E_2, \quad \text{where} \tag{15.39}$$

$$\beta_1 = \beta_1^0 - k_{11}, \qquad \beta_2 = \beta_2^0 - k_{22}, \tag{15.40}$$

and where $\beta_1^0$ and $\beta_2^0$ are the propagation constants of the isolated case for guides 1 and 2, respectively; $k_{11}$ and $k_{22}$ are the changes in propagation constants of guides 1 and 2, respectively, when an adjacent guide is introduced; and $k_{12}$ and $k_{21}$ are the coefficients associated with the transfer of energy from guide 2 to guide 1 and from guide 1 to guide 2, respectively.

The basis of coupled-mode theory lies in the solution of these differential equations. Assume a solution of the form

$$E_1 = A\, e^{\gamma z}, \qquad E_2 = B\, e^{\gamma z}. \tag{15.41}$$

In order to find the value of $\gamma$, (15.41) is inserted into (15.39) to obtain

$$\begin{bmatrix} (\gamma - j\,\beta_1) & -k_{12} \\ -k_{21} & (\gamma - j\,\beta_2) \end{bmatrix} \begin{bmatrix} A \\ B \end{bmatrix} = 0. \tag{15.42}$$

The condition that non-trivial solutions of $A$ and $B$ exist is that the determinant vanishes. This condition leads to

$$\gamma_{1,2} = j\,\frac{\beta_1 + \beta_2}{2} \pm \sqrt{k_{12}\,k_{21} - \left(\frac{\beta_1 - \beta_2}{2}\right)^2}. \tag{15.43}$$

Thus, the general solutions for $E_1$ and $E_2$ are

$$E_1 = A_1\, e^{\gamma_1 z} + A_2\, e^{\gamma_2 z} \tag{15.44}$$

$$E_2 = \frac{1}{k_{12}} [A_1\,(\gamma_1 - j\,\beta_1)\, e^{\gamma_1 z} + A_2\,(\gamma_2 - j\,\beta_1)\, e^{\gamma_2 z}] \tag{15.45}$$

where (15.45) was obtained by inserting (15.44) into (15.39).

Next, the values of $k_{12}$ and $k_{21}$ are closely investigated. If one could impose the condition that the energy which has leaked out of one guide is coupled completely into the adjacent line without loss, the sum $W$ of the energy of the two guides does not alter along the line [15.2];

$$\frac{dW}{dz} = 0 \qquad \text{where} \tag{15.46}$$

$$W = (E_1\, E_1^* + E_2\, E_2^*).$$

This leads to

$$\frac{dW}{dz} = (E_1'\, E_1^* + E_1\, E_1^{*\prime} + E_2'\, E_2^* + E_2\, E_2^{*\prime}). \tag{15.47}$$

Using (15.39) and reversing the order of complex conjugation and derivative, one obtains

$$\text{Re}\{(k_{12} + k_{21}^*)\,E_1^*\,E_2\} = 0. \tag{15.48}$$

For (15.48) to be true for any values of $E_1$ and $E_2$,

$$k_{12} = -k_{21}^* \tag{15.49}$$

has to be satisfied. This is the coupling condition. When (15.43) is written using (15.49), one obtains

$$\gamma_{1,2} = j(\beta \pm \beta_b) \quad \text{where} \tag{15.50}$$

$$\beta = \frac{\beta_1 + \beta_2}{2}, \quad \Delta\beta = \frac{\beta_1 - \beta_2}{2}, \quad \Delta\beta \equiv \Delta, \tag{15.51}$$

$$\beta_1 = \beta + \Delta, \quad \beta_2 = \beta - \Delta, \quad \beta_b = \sqrt{|k_{12}|^2 + \Delta^2}.$$

Figure 15.8 indicates the relationship among $\beta_1$, $\beta_2$ and $\beta$. The symbols $\Delta\beta$ and $\Delta$ are used equivalently. Now everything except the constants $A_1$ and $A_2$ has been found. $A_1$ and $A_2$ are determined by the conditions of excitation. In the following, a few examples are given.

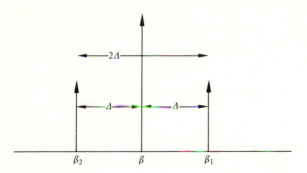

**Fig. 15.8.** Definitions of $\beta$'s

**Exercise 15.1** Find the coupled mode solutions of two identical guides when only one of the guides is excited with amplitude $U$ at the end (unbalanced excitation).

*Solution*   The following conditions hold

1) $\Delta = 0$,     $\beta_1 = \beta_2 = \beta$,

2) $E_1(0) = U$,   $E_2(0) = 0$. $\tag{15.52}$

First, the values of $A_1$ and $A_2$ are found. Inserting (15.52) into (15.44, 45) gives

$$U = A_1 + A_2, \quad 0 = A_1 - A_2. \tag{15.53}$$

The solutions for $A_1$ and $A_2$ are

$$A_1 = A_2 = \frac{U}{2}. \tag{15.54}$$

Using this result and (15.52) in (15.44, 45) gives

$$
\left.
\begin{aligned}
E_{u1}(z) &= U \left( \cos |k_{12}| z \right) e^{j\beta z}, \\
E_{u2}(z) &= - U \sqrt{k_{12}^*/k_{12}} \left( \sin |k_{12}| z \right) e^{j\beta z}.
\end{aligned}
\right\} \tag{15.55}
$$

Whereas the amplitude of $E_{u1}(z)$ is modulated by $\cos |k_{12}| z$, that of $E_{u2}(z)$ is modulated by $\sin |k_{12}| z$. This means that, when $E_{u1}(z)$ is at an extremum, $E_{u2}(z)$ is zero (and vice versa) as shown in Fig. 15.9. The energy in guide 1 is completely transferred to guide 2 after a transmission of $|k_{12}| z = \pi/2$, and the energy in guide 1 becomes zero. At the distance $|k_{12}| z = \pi$, however, the energy in guide 2 is completely transferred back to guide 1. The energy is tossed back and forth as it propagates along the guides.

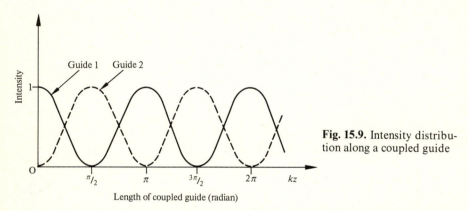

**Fig. 15.9.** Intensity distribution along a coupled guide

Such a phenomenon resembles the oscillation of two pendula connected with a weak spring as in Fig. 15.10. Let us suppose that one initiates the swing by pushing only pendulum 1. As time passes on, pendulum 2 starts gaining energy from pendulum 1 and finally reaches full amplitude when pendulum 1 is at rest. If time passes further, the situation reverses and pendulum 1 reaches full amplitude and pendulum 2 will be at rest, as illustrated at the bottom of Fig. 15.10. In this case, it is important to realize that complete energy transfer takes place only when pendulum 1 is identical to pendulum 2, and that the period of energy exchange is longer the weaker the coupling coil is.

**Exercise 15.2**   Find the solutions of the coupled guide when both guides have identical dimensions and are excited by the same input light amplitude $B$ (balanced excitation).

**Fig. 15.10.** Motion of coupled pendula

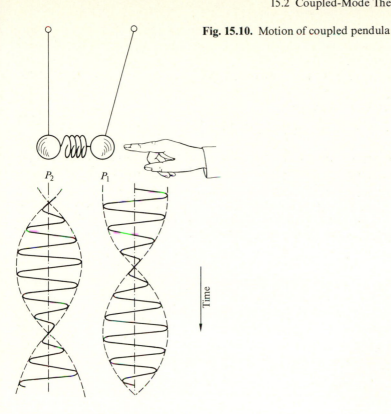

$P_2$   $P_1$

Time

*Solution*   The conditions of excitation are

1) $\Delta = 0, \quad \beta_1 = \beta_2 = \beta$

2) $E_1(0) = E_2(0) = B$. $\qquad\qquad\qquad\qquad$ (15.56)

Insertion of (15.56) into (15.44, 45) gives

$$B = A_1 + A_2,$$
$$B = j\sqrt{k_{12}^*/k_{12}}\ (A_1 - A_2), \qquad\qquad (15.57)$$

which leads to

$$A_1 = \frac{a+1}{2a}\,B, \quad A_2 = \frac{a-1}{2a}\,B, \quad a = j\sqrt{k_{12}^*/k_{12}}\ . \qquad (15.58)$$

Inserting (15.56, 58) into (15.44, 45) gives

$$E_{b1}(z) = B\left(\cos|k_{12}|\,z + \sqrt{k_{12}/k_{12}^*}\ \sin|k_{12}|\,z\right) e^{j\beta z}, \qquad (15.59)$$

$$E_{b2}(z) = B\left(\cos|k_{12}|\,z - \sqrt{k_{12}^*/k_{12}}\ \sin|k_{12}|\,z\right) e^{j\beta z}. \qquad (15.60)$$

It is interesting to note that, if an assumption is made that $k_{12}$ is a pure imaginary number, then

$$E_{b1} = B\, e^{j(\beta+|k_{12}|)z}, \tag{15.61}$$

$$E_{b2} = B\, e^{j(\beta-|k_{12}|)z}. \tag{15.62}$$

The two waves have the same amplitude but different phase velocities. Comparing the results of Exercises 15.1, 2, the amount of exchange of energy between the two guides does not simply depend upon the geometry of the guides, but it depends to a great extent upon how the guides are excited.

**Exercise 15.3**    Find the solutions of the coupled guides when both of the guides are identical, but are excited differently by $E_1(0)$ and $E_2(0)$.

*Solution*    The results of above two exercises will be used to solve this problem. The excitation can be decomposed into two modes of excitation; one is single-side feeding or unbalanced feeding, as in Exercise 15.1, and the other is identical feeding or balanced feeding as in Exercise 15.2. Using the law of superposition, the excitation can be decomposed in a convenient manner. An arbitrary excitation can be considered to consist of a combination of an unbalanced excitation $U$ and a balanced excitation $B$. Each case is solved separately, and the two solutions are summed to obtain the final result. Referring to Fig. 15.11, $E_1(0)$ and $E_2(0)$ are expressed in terms of $U$ and $B$ as

$$E_1(0) = U + B, \qquad E_2(0) = B. \tag{15.63}$$

Thus $U$ and $B$ are found to be

$$U = E_1(0) - E_2(0), \qquad B = E_2(0). \tag{15.64}$$

The solution for $E_1(z)$ and $E_2(z)$ are written in terms of the results of the last two exercises as

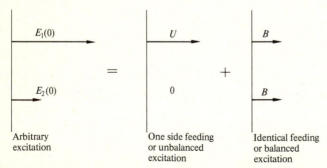

**Fig. 15.11.** Decomposition of an arbitrary excitation into an unbalanced and a balanced excitation

$$E_1(z) = E_{u1}(z) + E_{b1}(z), \qquad E_2(z) = E_{u2}(z) + E_{b2}(z). \tag{15.65}$$

Using (15.55, 59, 60, 64, 65) gives

$$\begin{pmatrix} E_1(z) \\ E_2(z) \end{pmatrix} = \begin{pmatrix} \cos |k_{12}| z & \sqrt{k_{12}/k_{12}^*} \sin |k_{12}| z \\ -\sqrt{k_{12}^*/k_{12}} \sin |k_{12}| z & \cos |k_{12}| z \end{pmatrix} \begin{pmatrix} E_1(0) \\ E_2(0) \end{pmatrix} \tag{15.66}$$

where a factor of $\exp(j\beta z)$ was suppressed. The matrix in (15.66) is called the transfer matrix.

The next exercise treats the coupling between non identical guides.

**Exercise 15.4**   Find the solutions of the coupled guides when the dimensions of the guides are not identical, and only one of the guides is excited.

*Solution*   The boundary conditions are

1) $\Delta \neq 0, \qquad \gamma_{1,2} = j(\beta \pm \beta_b),$

2) $E_1(0) = U \quad E_2(0) = 0.$
$\tag{15.67}$

First, the coefficients $A_1$ and $A_2$ are determined by using these conditions of excitation. The use of (15.44, 45, 50, 51, 67) leads to

$$U = A_1 + A_2,$$

$$0 = (\beta_b - \Delta) A_1 - (\beta_b + \Delta) A_2. \tag{15.68}$$

The solutions for $A_1$ and $A_2$ are

$$A_1 = \frac{\beta_b + \Delta}{2\beta_b} U, \quad A_2 = \frac{\beta_b - \Delta}{2\beta_b} U. \tag{15.69}$$

Inserting (15.69) in (15.44, 45), the final results are obtained after some rather straight forward manipulation;

$$E_{u1}(z) = U \left( \cos \beta_b z + j \frac{\Delta}{\beta_b} \sin \beta_b z \right) e^{j\beta z}, \tag{15.70}$$

$$E_{u2}(z) = -\frac{k_{12}^*}{\beta_b} U (\sin \beta_b z) e^{j\beta z}. \tag{15.71}$$

Let us look at the solution of Exercise 15.4 more closely. In order to see the energy transfer from this result, $E_1 E_1^*$ and $E_2 E_2^*$ are calculated,

$$E_{u1} E_{u1}^* = |U|^2 \left[ \cos^2 \beta_z z + \left( \frac{\Delta}{\beta_b} \right)^2 \sin^2 \beta_b z \right]$$

$$= |U|^2 \left( 1 - \frac{|k_{12}|^2}{|k_{12}|^2 + \Delta^2} \sin^2 \beta_b z \right) \quad \text{and} \tag{15.72}$$

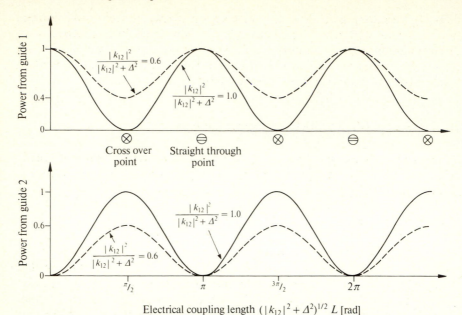

Electrical coupling length $(|k_{12}|^2 + \Delta^2)^{1/2}\, L$ [rad]

**Fig. 15.12.** Power transfer between two guides with different characteristics as a function of electrical coupling length. Input is fed to only guide 1

$$E_{u2}\, E_{u2}^* = |U|^2\, \frac{|k_{12}|^2}{|k_{12}|^2 + \Delta^2}\, \sin^2 \beta_b\, z\,. \tag{15.73}$$

As anticipated, the sum of the energy of the two guides is $|U|^2$ and is independent of the distance $z$. As shown in Fig. 15.12, the fluctuations of $E_{u1}\, E_{u1}^*$ and $E_{u2}\, E_{u2}^*$ compliment each other. It should be realized that complete transfer of energy from one guide to the other takes place when

$$\frac{|k_{12}|^2}{|k_{12}|^2 + \Delta^2} = 1\,. \tag{15.74}$$

The condition of (15.74), however, can be satisfied only when $\Delta = 0$ which means only when the two guides are identical. Another interesting fact is that an increase in $\Delta$ decreases the amount of transfer of energy. In other words, the amount of transfer can be adjusted by increasing the dissimilarity of the guides.

It is appropriate to conclude this section on coupled-mode theory by remarking on how widely used the equations are. They are a powerful tool in analyzing the interaction between two states in a conserved system. The applications include the conversion of power between two modes in a waveguide, the lower- and higher-frequency components of second-harmonic generation, and even the interaction between light and atomic systems where total energy is conserved. It is also used in connection with the population of atoms in upper and lower states, where the total number of atoms is conserved. In short, it is an extremely useful theory in dealing with coupling phenomena in a conserved system.

## 15.3 Basic Devices in Integrated Optics

Sections 15.3−6 are the climax of this and the previous chapter combined. Using the basic knowledge so far developed, various devices will be explained.

### 15.3.1 Directional Coupler Switch

Based upon coupled mode theory, an optical directional coupler that can be electrically switched has been fabricated. The structure of the switch is shown in Fig. 15.13. It consists of two closely spaced parallel guides imbedded in an electrooptic substrate like $LiNbO_3$ or GaAs. Electrodes are placed on top of each guide so that an external field can be applied to the guides. The refractive index of the guides is changed by the electrooptic effect. The applied electric fields in guide 1 and 2 are arranged to be in opposite directions so that a difference $\Delta$ in propagation constants between the two guides is created by the applied voltage. The light entering terminal (1) is channeled out to either (1′) or (2′) depending on the applied voltage.

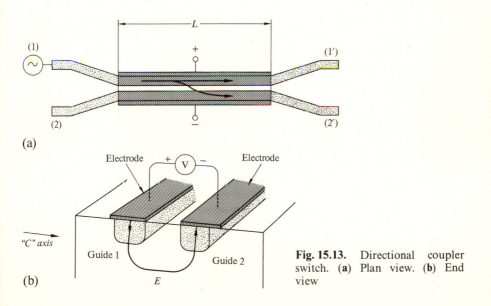

Fig. 15.13. Directional coupler switch. (a) Plan view. (b) End view

The switching operation is explained by referring back to the diagram in Fig. 15.12. The applied electric field primarily changes $\Delta$ but not $k_{12}$. As mentioned in the previous section, complete cross over takes place only when $\Delta$ is zero. When the electrical coupling length defined by $\sqrt{|k_{12}|^2 + \Delta^2}\, L$ is an odd integral multiple of $\frac{\pi}{2}$, all the energy entering (1) crosses over and comes out at (2'). Let this situation be called the cross state and be designated by $\otimes$, thus

$$\Delta = 0, \quad |k_{12}|\, L = (2\,n+1)\tfrac{\pi}{2}, \tag{15.75}$$

where $n$ is an integer. On the other hand, when all the energy entering (1) comes out at (1'), this situation is called the straight through state, or bar state, and is designated by $\ominus$. Even when $\Delta$ is non zero, as long as the electrical coupling length is an integral multiple of $\pi$, the bar state is obtained,

$$\sqrt{|k_{12}|^2 + \Delta^2}\, L = m\,\pi, \tag{15.76}$$

where $m$ is an integer. Thus, the switching action can be achieved by designing the length of the guide to satisfy exactly (15.75) when no voltage is applied. This establishes the $\otimes$ state. The $\ominus$ state is obtained by increasing the applied voltage so as to satisfy (15.76).

The graph of (15.76) in the $|k_{12}| - \Delta$ plane consists of circles while that of (15.75) contains points on the $|k_{12}|$ axis. From these two graphs, a switching diagram such as the one shown in Fig. 15.14 is made. Equations (15.75, 76) were normalized as (Note $\Delta L$ means product of $\Delta$ and $L$ and not a small displacement in $L$.):

$$\left.\begin{array}{ll} \dfrac{2\Delta L}{\pi} = 0, \quad \dfrac{L}{l} = 2\,n+1, & \otimes \text{ state} \\[3mm] \left(\dfrac{L}{l}\right)^2 + \left(\dfrac{2\Delta L}{\pi}\right)^2 = (2\,m)^2, & \ominus \text{ state} \end{array}\right\} \tag{15.77}$$

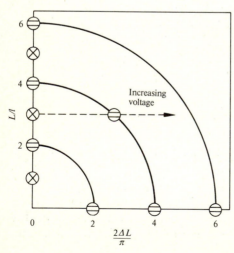

Fig. 15.14. Switching diagram for a single section [15.3]

where $l$ is the length needed for the first $\otimes$ state in Fig. 15.12 and is

$$l = \frac{\pi}{2\,|k_{12}|}. \tag{15.78}$$

The distance $l$ is sometimes called the transfer length. The cross states occur only at isolated points marked by $\otimes$ on the vertical axis in the Fig. 15.14, whereas the bar states occur at any point along the concentric circles. The length of the coupler is designed to be on one of $\otimes$ marks; and as the applied voltage is increased, i.e., $\varDelta$ is increased, the point of operation moves horizontally and meets the bar condition.

A difficulty with this coupler is that since the $\otimes$ states appear only as points in Fig. 15.14, a lossless cross state is achieved only by accurate fabrication. But a coupler with an alternating polarity of $\varDelta$ can be electrically tuned and relieves this stringent requirement. The details of this coupler will be discussed next.

### 15.3.2 Reversed $\varDelta\beta$ Directional Coupler

A directional coupler that has the capability of electrically tuning both cross and bar states is shown in Fig. 15.15. Such a capability is achieved by splitting the section into two halves and applying the voltage to each half with reversed polarity [15.3, 4].

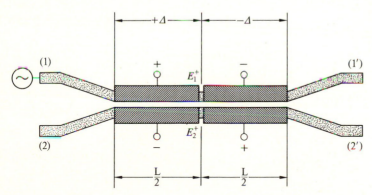

**Fig. 15.15.** Electrode configuration of a reversed $\varDelta\beta$ directional coupler

The reversed $\varDelta\beta$ is equivalent to increasing the electrical coupling length without affecting the magnitude of $\varDelta\beta$ it is as if the length $L$ of the balanced directional guides were stretched with the applied voltage.

In order to understand this electrical adjustability, the analysis developed in Sect. 15.2 will be further extended. In the left section of the reversed $\varDelta\beta$ directional coupler, only one end of the guides is driven, and the result of Exercise 15.1 is immediately applicable. In the right section, the ends of both guides are driven, and a more generalized expression is needed.

The solution with general excitation will be obtained as the sum of unbalanced and balanced excitations in a manner similar to that described in Exercise 15.3. Since the solution with unbalanced excitation has already been obtained, only the solution with balanced excitation has to be found. The boundary condition for the balanced excitation at $z = 0$ is

$$E_1(0) = E_2(0) = B. \tag{15.79}$$

Inserting (15.79) into (15.44, 45) gives

$$B = A_1 + A_2,$$

$$B = \frac{j}{k_{12}} [(\beta_b - \Delta) A_1 - (\beta_b + \Delta) A_2]. \tag{15.80}$$

The solutions for $A_1$ and $A_2$ are

$$A_1 = \frac{\beta_b + \Delta - j\,k_{12}}{2\,\beta_b}\,B, \qquad A_2 = \frac{\beta_b - \Delta + j\,k_{12}}{2\,\beta_b}\,B. \tag{15.81}$$

Insertion of (15.81) into (15.44, 45) gives

$$E_{b1}(z) = B \left( \cos \beta_b z + j\,\frac{(\Delta - j\,k_{12})}{\beta_b} \sin \beta_b z \right) e^{j\beta z}, \tag{15.82}$$

$$E_{b2}(z) = B \left( \cos \beta_b z - j\,\frac{(\Delta - j\,k_{12}^*)}{\beta_b} \sin \beta_b z \right) e^{j\beta z}. \tag{15.83}$$

The rest of the procedure is quite similar to Exercise 15.3 and the final result is

$$\begin{pmatrix} E_1(z) \\ E_2(z) \end{pmatrix} = \begin{pmatrix} A & jB \\ jB^* & A^* \end{pmatrix} \begin{pmatrix} E_1(0) \\ E_2(0) \end{pmatrix}, \tag{15.84}$$

where

$$A = \cos \beta_b z + j \left( \frac{\Delta}{\beta_b} \right) \sin \beta_b z, \qquad B = -j \left( \frac{k_{12}}{\beta_b} \right) \sin \beta_b z, \tag{15.85}$$

and a factor of $\exp(j\,\beta\,z)$ was suppressed.

This completes all that is necessary to find the transfer matrix $M$ of the reversed $\Delta\beta$ directional coupler shown in Fig. 15.15. Let $M^+$ and $M^-$ be the transfer matrices for the left and right sections of the coupler, respectively. Between the sections, $\beta_1$ and $\beta_2$ are exchanged and the sign of $\Delta$ is reversed.

The only difference between $M^+$ and $M^-$ is whether or not $A$ has a complex conjugate sign. The desired transfer matrix is the product of the two matrices, namely

$$M = M^- M^+ = \begin{pmatrix} A_r & jB_r \\ jB_r^* & A_r^* \end{pmatrix}, \quad \text{where} \tag{15.86}$$

$$A_r = |A|^2 - |B|^2, \quad B_r = 2A^* B. \tag{15.87}$$

Using the transfer matrix of (15.86), the switching condition of the reversed $\Delta\beta$ directional coupler is obtained. First, the condition for the bar state when guide 1 is excited is considered. The condition is given from (15.87), by

$$2A^* B = 0. \tag{15.88}$$

Either $A^*$ or $B$ has to be zero to satisfy this condition. Since $A^*$ is a complex number, for $A^*$ to be zero, both real and imaginary parts have to vanish,

$$\Delta = 0,$$

$$\frac{L}{2} |k_{12}| = (2n+1) \frac{\pi}{2},$$

where $L/2$ is the length of the subsection. This condition is represented by a series of isolated points at $L/l = 2, 6, 10, \dots$, with an interval of 4 along the vertical axis of the switching diagram, as shown in Fig. 15.16. If $B$ is to be zero, then

$$\frac{L}{2} \sqrt{|k_{12}|^2 + \Delta^2} = m\pi, \tag{15.89}$$

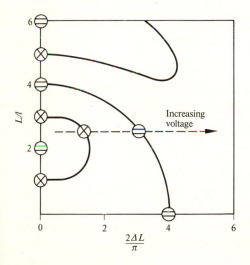

Fig. 15.16. Switching diagram for two alternating $\Delta\beta$ sections [15.3]

which can be rewritten as

$$\left(\frac{L}{l}\right)^2 + \left(\frac{2\Delta L}{\pi}\right)^2 = (4\,m)^2. \tag{15.90}$$

Equation (15.90) represents concentric circles on the switching diagram with radii 4, 8, 12, ..., at intervals of 4.

Finally, the condition for the cross state is considered. From (15.85, 87), $A_r$ is

$$A_r = 1 - 2\frac{|k_{12}|^2}{\beta_b^2}\sin^2\frac{\beta_b L}{2}. \tag{15.91}$$

The condition that $A_r = 0$ is

$$\frac{|k_{12}|^2}{|k_{12}|^2 + \Delta^2}\sin^2\frac{L}{2}\sqrt{|k_{12}|^2 + \Delta^2} = \sin^2(2\,m + 1)\frac{\pi}{4}. \tag{15.92}$$

Equation (15.92) is a transcendental equation and cannot be solved analytically, except along the vertical axis. The computer calculated graph [15.3] is shown in Fig. 15.16. The curve starts at 1, 3, 5, ..., on the vertical axis and ends on the same axis.

From the diagram in Fig. 15.16, the operation of the switch is understood. By increasing the applied voltage, the point of operation moves horizontally. Conditions of crosses and bars can be achieved with almost any length ($L/l$) of the guide, and the stringent requirement on fabrication is relieved. By increasing the number of subsections, the required voltage is reduced. A control signal of 3 V can switch the states of a directional coupler of 6 reversed $\Delta\beta$ sections formed in Ti diffused $LiNbO_3$ [15.1].

Finally, a heuristic interpretation is given for the reason why a perfect cross state does not exist with $\Delta \neq 0$ in the case of a non-reversed coupler, while a perfect cross state does exist in the case of a reversed $\Delta\beta$ coupler. In the non-reversed case, the condition for a perfect cross state is $E_1(z) = 0$ with $E_1(0) \neq 0$ and $E_2(0) = 0$, and hence $A = 0$ in (15.84). Notice that $A$ expressed by (15.85) has real and imaginary terms (in phase and quadrature components), and both terms have to be zero simultaneously.

The output from the reversed $\Delta\beta$ coupler is investigated closely. The condition of the $\otimes$ state is that the output from (1') is zero. The outputs from the left half section (at the mid-point in Fig. 15.15) are $E_1^+ = A U$ and $E_2^+ = j B^* U$. Since $A$ and $B$ are complex numbers expressed by (15.85), the phase of $E_1^+$ is *delayed* by

$$\phi_A = \tan^{-1}\left(\frac{\Delta}{\beta_b}\tan\beta_z\frac{L}{2}\right)$$

from that of $U$. The similar delay of $E_2^+$ is $\phi_B$, which depends on the value of $k_{12}$. The quantity $k_{12}$ is in general a complex number.

The right half section is excited by $E_1^+$ and $E_2^+$ and the contribution of the former to the output is $A^* E_1^+$ and that of the latter is $j B E_1^+$. This time, the phase of $E_1^+$ is *advanced* by exactly the same amount $\phi_A$. The phase of $E_2^+$ is also advanced by exactly the same amount $\phi_B$. Thus, the output (1') becomes a real quantity including an additional parameter $k_{12}$ to play with, which makes the zero condition less severe, and $A_r$ can be made zero with more varieties of combinations of $\Delta$, $L/2$, and $k_{12}$.

The directional coupler can be also used as a modulator in addition to its use as a switch but the requirements are slightly different. A modulator normally does not require a perfect cross-over state, but a switch needs it to minimize cross talk. Another difference is that a switch has to include a *dc* state. The upper limit of the modulation frequency of such a modulator can be raised by designing the electrodes as a transmission line, as was explained for the bulk modulator (Sect. 14.2.2).

### 15.3.3 Tunable Directional Coupler Filter

The tunable directional coupler filter combines the wavelength dependence of the propagation constant of the guide (dispersion of the guide) and the critical dependence of the cross-over state upon the value of $\Delta$. The pass band of the filter is tuned by a control voltage [15.5].

The following is a qualitative illustration of the operation of such a filter. A directional coupler such as that described earlier in Fig. 15.13 is modified. In order to accentuate the imbalance between the guides, the dimensions of guides 1 and 2 are purposely made different. Assume that the wavelength dependence of the propagation constants $\beta_1$ and $\beta_2$ of the guides is represented by the curves at the top of Fig. 15.17. The solid lines refer to no applied voltage, and the dashed lines refer to an applied voltage. The middle graph shows the difference $|\beta_1 - \beta_2|$ of the propagation constants with respect to wavelength, with and without applied voltage. If it is further assumed that the guides are cut at the transfer length, and that the coupling constant $k_{12}$ is essentially constant with respect to the applied voltage, then the center frequency of the maximum cross over efficiency ($\Delta = 0$) moves according to the applied voltage. The condition for good operation is rather critical, but by using the reversed $\Delta\beta$ directional coupler with independent voltage control to each electrode, the operation can be made less critical.

### 15.3.4 *Y* Junction

Figure 15.18 shows a *Y* junction of an optical guide. An optical guide is imbedded in an electrooptic substrate like lithium niobate. Guide (1) is symmetrically divided into output guides (2) and (3). External electrodes are deposited along the guides. In the absence of an external electric field, the light power is divided into equal portions, but when an electric field is applied such that the index of refraction of one guide is increased whereas the other is decreased due to electrooptic effect, the light tends to be steered toward the direction of the higher index of refraction (Sect. 5.4). The application of the electric field acts to switch one guide on and the other guide off.

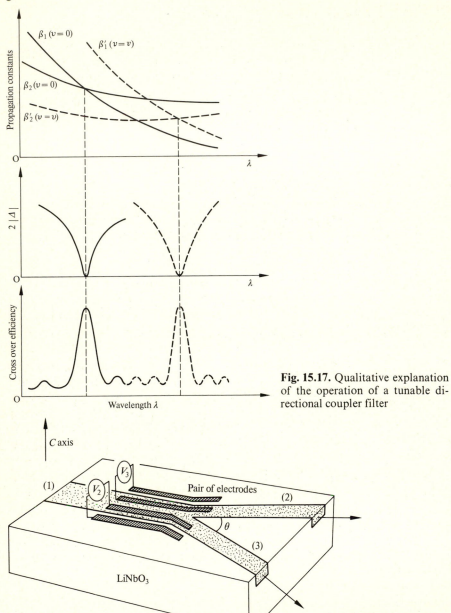

**Fig. 15.17.** Qualitative explanation of the operation of a tunable directional coupler filter

**Fig. 15.18.** $Y$ junction.

Even though the structure is simple, an on-off ratio of 17 dB is obtainable [15.6]. The output ports can be increased to more than two by increasing the number of output arms [15.7], or by staggering many $Y$ junctions in the form of a tree branch.

### 15.3.5 Mach-Zehnder Interferometric Modulator

This modulator uses the interference between two light beams which have travelled through separate paths en route to a common output port [15.8]. Figure 15.19 shows the layout of the device. An optical guide is laid out on an electrooptic material. Single-mode guides (TE$_0$ mode) are used. The input is first divided into two optical guides of equal length. One of the guides is sandwiched by gold strip electrodes.

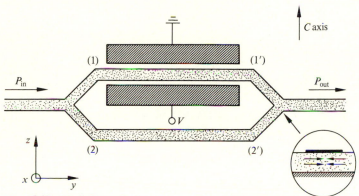

**Fig. 15.19.** Mach-Zehnder interferometric modulator

A potential applied to the electrodes changes the propagation constant of the guide between the electrodes. If the input light is divided equally between guides (1) and (2) in Fig. 15.19, and if a voltage is applied such that the phase difference between the outputs of the two guides is an odd multiple of $\pi$ radians, the outputs cancel each other. Correctly speaking, the field distribution of the output light is like the one shown in the circular insert in Fig. 15.19. This kind of field distribution belongs to a higher-order mode rather than the lowest mode excited at the input guide whose field distribution is uniform. Since the dimension of the output guide is designed such that the higher modes are cut-off, there is no output. If the output guide were widened, total cancellation would not occur.

By switching on and off the applied voltage, the device switches on and off the optical power. The optical output at an intermediate voltage is expressed by

$$I_{\text{out}} = \frac{I_{\text{in}}}{2} (1 + \cos \phi) \tag{15.93}$$

where the angle $\phi$ is the phase difference between the controlled and uncontrolled arms. The expression for $\phi$ is easily found if use is made of Exercise 14.1. Let us say that the LiNbO$_3$ crystal is $X$ cut (meaning that the cut plane is perpendicular to the $X$ axis of the crystal. A $C$ cut means that the cut plane is perpendicular to the crystal axis. $X$ is taken perpendicular to $C$). The external electric field is in the $z$ direction and the polarization of the TE$_0$ mode is parallel to the surface and also in the $z$ direction. The geometry is exactly the same as Exercise 14.1 and from (14.10)

$$\phi = \phi_0 - KLV, \quad \text{with} \quad K = \frac{\pi}{\lambda} n_e^3 r_{33} \frac{\alpha}{d},$$

where $V$ is the applied voltage, $L$ is the length of the guide, $\alpha$ is a number between 0 and 1 depending upon the geometry which influences the effectiveness of the external voltage, and $\phi_0$ absorbs all the phase shift which is independent of $V$.

Thus, if the external voltage $V$ is changed linearly with time in the form of a triangular wave, the optical output power is modulated sinusoidally according to (15.93) [15.8].

The device is simple in structure but demands high accuracy in fabrication if it is to be used as an optical on-off switch. The reason for this is that the extinction ratio depends critically upon how well the amplitudes of the output from the two guides are matched.

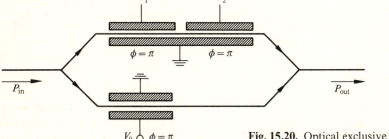

Fig. 15.20. Optical exclusive "OR" logic

The Mach-Zehnder interferometric modulator modified to be an optical exclusive "OR" gate [15.9] is shown in Fig. 15.20. The upper guide has two $\phi = \pi$ control electrodes and the lower guide has one $\phi = \pi$ control electrode. Optical power output is present only when either $V_1$ or $V_2$ is on, and the optical power output is absent for all other combinations of $V_1$ and $V_2$.

Figure 15.21 shows the Mach-Zehnder interferometric modulator when it is modified to be a digital electrical to analog optical converter. The least significant bit of the electrical signal is connected to the shortest electrical control, and the most significant bit of the electrical signal is connected to the longest. The lengths of the control electrodes are progressively increased by doubling the length of the preceding lower-order bit. The dynamic

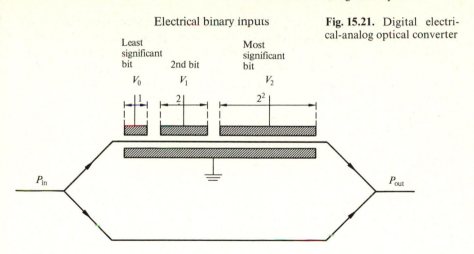

Electrical binary inputs

Least significant bit

2nd bit

Most significant bit

**Fig. 15.21.** Digital electrical-analog optical converter

range of such a converter is limited to the range in which output power can be approximated as proportional to the length of the control electrodes.

### 15.3.6 Waveguide Modulator

The modulator that is perhaps the simplest in structure is the waveguide cut-off modulator. This consists of an optical guide which operates near the cut-off region and is switched to the stop band by electrooptically changing the refractive index of the guide. A demerit of such a modulator is the rather high on state transmission loss, which cannot be avoided when operating so close to the cut-off of the optical guide.

### 15.3.7 Acoustooptic Modulator

Acoustooptic devices based upon the bulk effect have been already covered in Sect. 14.3. In this section, a similar device using the surface acoustic wave (SAW) is discussed. The SAW device is lighter and more compact, and its diffraction efficiency for a given acoustic energy is better than the corresponding bulk-effect devices [15.10].

Figure 15.22 shows various ways in which the SAW device is used. The surface acoustic wave is launched from an interdigital transducer deposited on the surface of a piezoelectric material such as lithium niobate. The interdigital transducer has a comb-like structure. The teeth of the comb establish an electric field whose polarity is spatially alternating. This creates an alternating electrically induced strain due to the piezoelectric effect which excites a surface acoustic wave. It should be noted that both a periodic variation of the refractive index and a surface deformation contribute to the diffraction effect. The variation of refractive index penetrates about one acoustic wavelength from the surface. The relative

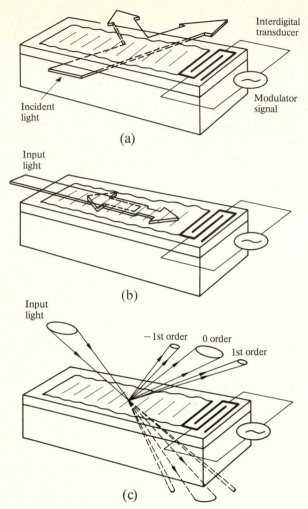

Interdigital
transducer

Incident
light

Modulator
signal

(a)

Input
light

(b)

Input
light

−1st order    0 order

1st order

(c)

**Fig. 15.22 a – c.** Various uses of a SAW device. (**a**) Light scanning by Bragg diffraction. (**b**) Filtering by selective reflection. (**c**) Probing the SAW by diffraction of the ray

orientation of the light beam with respect to the acoustic beam may vary depending on the intended use of device.

In Figure 15.22 a, the device is used for such functions as steering, modulating or switching the light beam. The light is directed so that it satisfies Bragg's condition. The direction of the Bragg diffracted wave is swept by sweeping the acoustic frequency. The intensity of the diffracted ray is controlled by the intensity of the acoustic wave.

Figure 15.22 b shows an arrangement in which a SAW device is used as a mode filter or a frequency band pass filter. It is a special case of Bragg diffraction with $\theta = 90°$ in (14.22). The modes (or frequencies) of light whose one half wavelength is exactly one acoustic wavelength are selectively reflected back, and cannot pass through the guide. The center frequency of the stop band is tunable by changing the acoustic frequency. This band-stop filter can be used as the reflector of a laser cavity to make an electrically tunable laser or to remove the side lobes of the laser diode.

Figure 15.22c shows a light beam being focused onto the surface of the SAW device in order to probe the excitation. The wavelength and degree of excitation can be determined from the diffraction beam. Both diffracted and transmitted beams can be used to gather information.

Figure 15.22d shows how the signal from a chirp radar is processed by the SAW device. Because of the slow propagation speed of the surface acoustic wave in the medium (3,485 m/s in LiNbO$_3$), a time segment of an electrical signal can be converted to a spatial distribution of the variation of the refractive index of the SAW device.

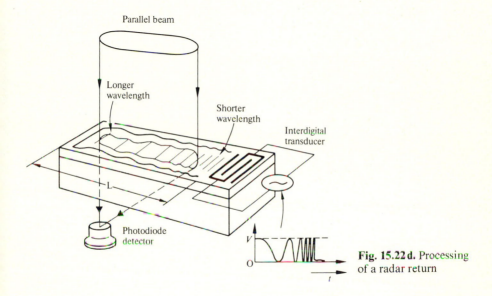

**Fig. 15.22d.** Processing of a radar return

A chirp radar uses pulses whose carrier frequency is progressively increasing with respect to time during each pulse. If the radar echo of such pulses is used to launch a surface acoustic wave on the SAW device, a surface acoustic wave whose wavelength is progressively decreasing propagates across the device toward the end and is lost at the end. The variation of the refractive index on the device resembles that of a zone plate. A parallel beam of light incident upon the device will converge to a single point. If the device is made long enough to store one entire pulse, the device displays one single point for each pulse duration. This is the basis for compressing a finite length pulse into a single point.

The chirp modulation circumvents the mutually conflicting requirements of a longer pulse for a larger energy of the returned signal (longer range of the radar) and a shorter pulse for higher target resolution.

The SAW device can also be used as a convolver. Two SAW beams $V_1$ and $V_2$ propagating in opposite directions can be generated by depositing two interdigital transducers at both ends of the substrate. The amplitude of the light diffracted from the SAW beams is proportional to $V_1 + V_2$. The output from a square law detector includes the product of $2 V_1 V_2$. The spatial integral of this product as $V_1$ and $V_2$ are sliding in opposite directions performs the convolution of $V_1$ and $V_2$.

·There are some functions such as the delay line, the band-pass filter, the pulse compression filter, and the phase coded matched filter that the SAW device can perform without using light. A brief description of these functions is given below.

A delay line can be realized by taking advantage of the low speed of propagation of the acoustic wave. Figure 15.22 e shows a SAW device on a circular shaped substrate [15.11]. The surface wave circulates around the disc and a total delay time of 400 μs can be achieved, which is about ten times that obtainable in a single pass through a rectangular chip. The delay line becomes a memory device if the output signal is amplified and fed back to the input transducer again, as shown in the figure.

A band pass filter in the VHF and UHF bands is fabricated by properly designing the length of the fingers of the interdigital transducer. Figure 15.22 f shows the design principle of the band-pass filter by the delta-function model. The transmitting transducer is a wide-band transducer and the receiver transducer has band-pass characteristics. (The device is reciprocal and the transmitter and receiver can be exchanged). The voltage

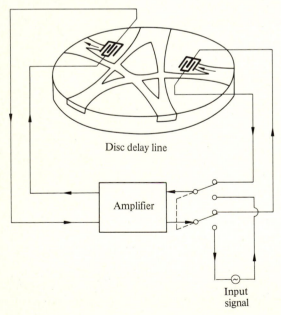

Disc delay line

Amplifier

Input
signal

**Fig. 15.22 e.** Disk delay line

induced by the piezoelectric effect in each pair of fingers is proportional to its length. The length of the fingers is modulated as sinc $z$ in the $z$ direction. When one impulse is launched from the transmitting transducer, the output voltage from the receiver transducer varies as the impulse propagates through the fingers, where $1/f_0$ is the time for the impulse to pass through only one of the pairs of the teeth. As this impulse goes through the teeth the output voltage $V_{out}$ varies as $\mathrm{sinc}\,(Bt)\cos(2\pi f_0 t)$. A filter whose impulse response is a sinc function in the time domain possesses a rectangular passband in the frequency domain. This method of design is known as the design by the delta-function model and is widely used because of its simplicity.

Figure 15.22 g shows a chirp filter which is different from that in Fig. 15.22 d. The spacing between the pairs of electrodes is made wider towards the end of the substrate. Since the transducer has maximum efficiency when the acoustic wavelength and the spacing between the pair of the electrodes coincide, the lower frequency component is excited predominantly in the outer region and the higher-frequency component, in the inner region. The shape of the receiver transducer is a mirror image of that of the transmitter. With this arrangement, the lower-frequency component

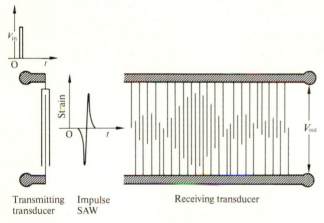

Transmitting     Impulse                    Receiving transducer
transducer       SAW

(i) Transmitting and receiving transducers

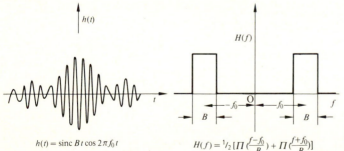

$h(t) = \mathrm{sinc}\,Bt\cos 2\pi f_0 t$

$H(f) = \tfrac{1}{2}[\Pi\,(\tfrac{f-f_0}{B}) + \Pi\,(\tfrac{f+f_0}{B})]$

**Fig. 15.22 f.** Principle of the SAW bandpass filter

(ii) Relationships between the impulse response and the transfer function

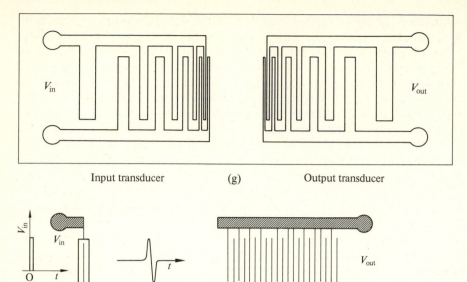

Input transducer                    (g)                    Output transducer

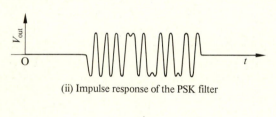

(i) Geometry of fingers of the PSK filter

(ii) Impulse response of the PSK filter

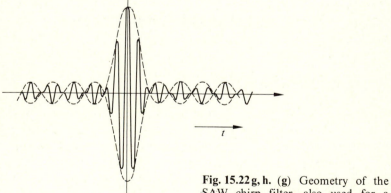

(iii) Compressed pulse
(h)

**Fig. 15.22 g, h.** (g) Geometry of the SAW chirp filter, also used for a spectrum analyzer. (h) Principle of the PSK filter

has to travel a longer distance than that of the higher-frequency component. When a chirped radar echo, such as shown at the lower right of Fig. 15.22 d, is fed to this SAW device, the lower-frequency component is launched first, but is delayed longer in the SAW device, so that the shorter-frequency component of the echo has a chance to catch up. Thus the output from the SAW device is a compressed-in-time version of the input. This device converts the frequency information into a time delay and can be used also as a spectrum analyzer.

The phase-shift keyed (PSK) filter is fabricated by taking advantage of the fact that a phase reversal of the output voltage can be easily obtained by repeating the finger belonging to the same bonding pad. Figure 15.22 h shows the geometry of the fingers of a PSK filter. The fingers of the transducer are spaced by a half wavelength, and are alternately connected to the same bonding pad, like most other interdigital transducers. However, at the locations where the phase is to be inverted, the two adjacent fingers are connected to the same bonding pad. When an impulse acoustic wave passes the particular finger pattern shown at the top right side of Figure 15.22 h, the output voltage across the terminal pad with respect to time becomes such as shown in the middle figure. Notice the sudden $\pi$ radian phase shifts.

The PSK filter is used for both coding and decoding signals. If a PSK coded signal is launched into a PSK filter, the output voltage is the convolution of the input signal and the finger pattern of the filter. As the SAW beam propagates along the fingers, at the instant in which the two patterns match up, the output signal bursts into a large signal. At all other times, the contribution of roughly half the fingers is positive and that of the other half is negative, and no voltage is seen at the output terminal, as shown at the bottom of Fig. 15.22 h.

The PSK filter is used for the same purpose as a chirp filter to compress a pulse, but the PSK filter has the advantage of discouraging electronic espionage because the signal can be decoded only by the corresponding PSK mode.

## 15.4 Bistable Optical Devices

The devices so far discussed were strictly linear, i.e., when the input light level is lowered the output light level is lowered accordingly. In this section, devices are presented which give out two different levels of output for the same level of input, depending upon the past history of the input power level. The output of these devices depends upon whether the present input level was achieved by lowering from a higher level or raising from a lower level. This phenomenon of optical hysteresis is known as optical bistability. Such a phenomenon can be used for making optical memories, optically operated switches, optical amplifiers, and optical logic gates.

### 15.4.1 Optically Switchable Directional Coupler

If the light itself is used for changing the switching state of a directional coupler, then the light can steer itself through a switch matrix. Such a switch is materialized by combining a reversed $\Delta\beta$ directional coupler with a feed-back loop (Fig. 15.23). The output from the switch is fed back to control the switching state [15.12−15].

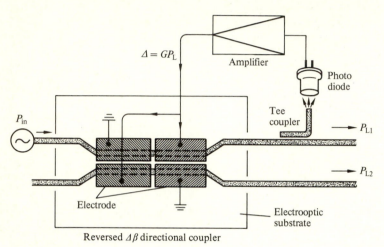

**Fig. 15.23.** Optically switched directional coupler

A more specific explanation is given as follows. The input light $P_{in}$ is incident upon guide 1. A portion of the output light $P_{L1}$ from guide 1 is taken out by a Tee coupler and detected by a photodiode. The detected signal after amplification is used to apply an electric field to induce a difference $\Delta$ in the propagation constants between the guides. If $\Delta$ is assumed proportional to $P_{L1}$, then

$$\Delta = GP_{L1}, \tag{15.94}$$

where $G$ absorbed such constants as the coupling constant of the Tee coupler, the gain of the amplifier, the electrooptic constant and dimensions of the guides. Insertion of (15.94) into (15.51) gives

$$\phi = \beta_b L = \sqrt{|k_{12}|^2 + (GP_{L1})^2}\ L. \tag{15.95}$$

The transmission $P_{L1}/P_{in}$ of guide 1 is $|A_r|^2$ and rewriting of (15.91) in terms of $\phi$ gives

$$\frac{P_{L1}}{P_{in}} = \left[1 - 2\left(\frac{|k_{12}|\,L}{\phi}\right)^2 \sin^2\frac{\phi}{2}\right]^2. \tag{15.96}$$

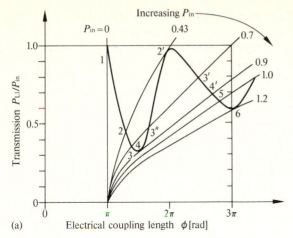

(a)

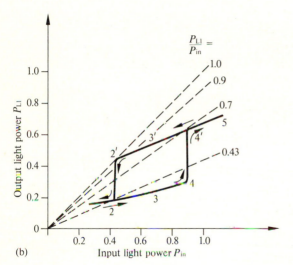

(b)

Fig. 15.24 a, b. Optical bistability. (a) Movement of operating points as the input power is varied. (b) Illustration of the graphic method for finding the output-input characteristics of optical bistability. They are with $|k_{12}| L = \pi$ and $GL = 4\pi$.

The transmission for a given input can be found from the simultaneous solution of (15.95, 96).

The solution can be found either by use of a computer or by graphical presentation, and here, the latter means will be chosen. For easier representation, (15.95) is rewritten as

$$\left(\frac{P_{L1}}{P_{in}}\right)^2 = \left(\frac{\phi}{GLP_{in}}\right)^2 - \left(\frac{|k_{12}|}{GP_{in}}\right)^2. \tag{15.97}$$

Both (15.96, 97) are plotted in the same graph in Fig. 15.24 a. The horizontal axis is the electrical coupling length $\phi$ and the vertical is the transmission $P_{L1}/P_{in}$ of guide 1. The plot of (15.96) is the curve with the sinc$^2$ characteristic. Equation (15.97) is plotted with different input powers.

The curves for (15.97) are hyperbolas, but in the region of interest they are almost linear with the slopes of $(GLP_{in})^{-1}$. The slope depends inversely on the incident light power. The movement of the point of intersection is indicated by numbers, and the corresponding points are indicated on the input-output characteristic of the coupler, shown in Fig. 15.24 b. In Fig. 15.24 a, the intersection when the input power is low is at point 1. As the input power is further increased and as the line bends beyond point 2, there are three intersections indicated by the points 3, 3′ and 3″. All three points are solutions, but point 3″ is unstable. The response to a minute fluctuation in output power in the unstable region between 4 and 2′ is considered. An increase in $P_{L1}$ leads to an increase in $\phi$, according to the sinusoidal curve in Fig. 15.24 a, and this increase in $\phi$ in this region of the transmission curve leads to a further increase in $P_{L1}$ of (15.95). On the other hand, a minute initial decrease in $P_{L1}$ leads to a decrease in $\phi$ and thus leads to a further decrease in $P_{L1}$. Regardless of what the fluctuation at point 3″ is, the fluctuation always grows bigger (positive feed back) and the system is unstable. If one makes the same argument with 3 and 3′, one soon discovers that the fluctuation is always suppressed (negative feed back) and the system is stable at these two points. Between the choice of points 3 and 3′, the system operates at point 3 if the input power is moved up from point 2. This is because, as the input power is increased, point 3 is reached first and becomes stable and the output does not grow further to reach point 3′.

All of the above discussions on stability were based upon a diagram that does not include any time dependence (static diagram). Strictly speaking, stability is inherently a time dependent phenomenon, and the preceding discussion is not a rigorous argument. The temporal behavior of such devices can be investigated by solving the differential equations with respect to time. An unstable solution of the differential equations grows indefinitely with respect to time [15.16–19].

Returning to Fig. 15.24 a, the behavior of the system beyond point 3 is considered. When the input power exceeds point 4, there is no other alternative but to jump up to 4′ and proceed onto point 5. The path of decreasing input power is different from that of increasing power. As the input power is decreased from point 5 and passes beyond point 4′, the triple state is again reached. Between the stable states of point 3 and 3′, point 3′ becomes the operating point because, in the process of decreasing the input power, point 3′ is first found stable and the operating point stays at 3′, and does not further decrease to point 3. When the input power is decreased to point 2′, the state jumps down to point 2, and it eventually reaches point 1 with a further decrease in the input power.

The arrows in Fig. 15.24 b indicate the path of change of the operating point when $P_{in}$ is cycled. The diagram takes the form of a hysteresis loop. There are higher and lower levels of the light output $P_{L1}$ for the same value of input $P_{in}$, and the choice of one of the two levels depends upon the history of $P_{in}$. Such a system is called a bistable system.

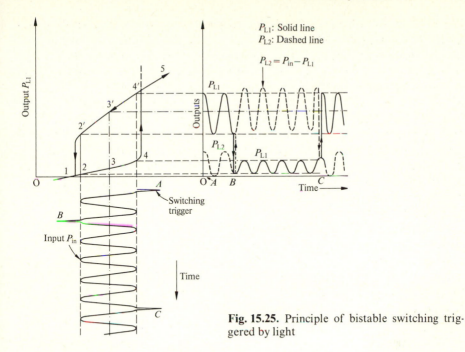

**Fig. 15.25.** Principle of bistable switching triggered by light

Figure 15.25 illustrates how such a bistable device is used. An initial triggering light pulse $A$ brings the operating point to point 5 in the upper level, and at the end of the trigger pulse, $P_{L1}$ stays in the upper level. The output $P_{L1}$ remains channelled to guide 1, as shown by the solid line. As soon as the triggering light pulse $B$ arrives, the operating point is brought down to point 1 in the lower level. After the $B$ triggering pulse, $P_{L1}$ stays in the lower level, and the light remains channelled to guide 2 until a further triggering pulse is received. The output from guide 2 is $P_{L2} = P_{in} - P_{L1}$, and is shown by the dotted line. The light signal is thus channelled through the switch according to a switch pulse contained in the incident light itself.

## 15.4.2 Optical Triode

Making use of the fact that the output $P_{L1}$ of the optically controlled directional coupler critically depends upon the input power, an optical triode, such as shown in Fig. 15.26 a, can be assembled [15.14, 17]. The steep slope of the input-output curve of a modified hysteresis is utilized. The loop of the hysteresis shown in Fig. 15.24 b is flattened by making the spacing between points 2 and 4 conform to the shape shown in Fig. 15.26 b. The spacing between points 2 and 4 in Fig. 15.24 b can be reduced by making the two curves in Fig. 15.24 a as parallel as possible between points 2' and 4 by adjusting the values of $|k_{12}|^2$, $G$, or $L$. Amplitude modulated light properly biased to the center of the hysteresis curve can be amplified as shown in Fig. 15.26 b.

It should be recognized that the light energy was not amplified. Rather, the distribution of the input energy between the two guides was changed by the input. The total energy of the output from guides 1 and 2 is constant at all times.

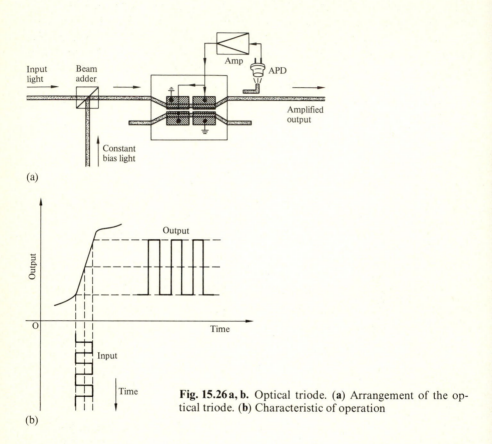

(a)

(b)

**Fig. 15.26 a, b.** Optical triode. (**a**) Arrangement of the optical triode. (**b**) Characteristic of operation

### 15.4.3 Optical AND and OR Gates

Using a modified hysteresis loop, such as shown in Fig. 15.26 b, the operation of AND and OR logic can be performed optically [15.14].

When the optical input power is set below the lower knee point of the curve, the output light is off, unless both of the incident lights are on, and the operation of AND logic is achieved.

When each input is set near the upper knee, as long as one of the inputs is on, the output light is on, and the logic of OR is performed.

### 15.4.4 Other Types of Bistable Optical Devices

A few other systems that operate on the principle of optical bistability will be covered in this section.

First, a bistable optical device using a Mach-Zehnder interferometric modulator is mentioned. The optically switched directional coupler of Sect. 15.4.1 used a reversed $\Delta\beta$ directional coupler, as shown in Fig. 15.23, but a Mach-Zehnder interferometric switch as shown in Fig. 15.19 can be used for the same purpose [15.19]. This is because the output power $P_L/P_{in}$ of the Mach-Zehnder interferometric modulator varies with the electrical length $\phi$ as $P_L/P_{in} = \cos^2(\phi/2)$ in a manner exactly the same as the reversed $\Delta\beta$ directional coupler. The difference is that the device with the Mach-Zehnder interferometric modulator has only one output port, while that of the reversed $\Delta\beta$ directional coupler has two output ports.

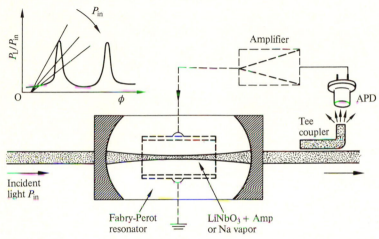

**Fig. 15.27.** Bistable optical device using a Fabry-Perot resonator

It is also possible to replace the reversed $\Delta\beta$ directional coupler by a Fabry-Perot resonator (Fig. 15.27). The cavity of the resonator contains an electrooptic material which is controlled by the output light. The transmission $P_L/P_{in}$ of the Fabry-Perot resonator as a function of $\phi$ has a much sharper peak than that of the reversed $\Delta\beta$ directional coupler, as indicated at the upper left corner of the figure, but in other respects they both have the same general shape. Thus, the Fabry-Perot device has a bistable characteristic similar to that of the reversed $\Delta\beta$ directional coupler [15.14].

As seen from the above examples, whether a reversed $\Delta\beta$ directional coupler, a Mach-Zehnder interferometric switch, or a Fabry-Perot resonator is used, what is essential is that the transmission characteristic can be controlled by a positive feedback of the optical output power.

Materials such as sodium vapor, carbon disulfide $CS_2$, and nitrobenzene ($C_6H_5NO_2$) have an intensity dependent index of refraction. When such a material is enclosed in the Fabry-Perot resonator, a change in the input power changes the electric length of the resonator, and bistable operation is achieved without the need for an external feedback loop. All one needs is a resonator filled with this intensity dependent material, and one can remove the electrooptic material and the amplifier drawn in dotted lines in Figure 15.27. The range of the bistability of the device with this built-in feedback, however, is much narrower than that with an external feedback loop.

### 15.4.5 Self-Focusing Action and Optical Bistability

Another noteworthy phenomenon connected with optical bistability is the self-focusing effect [15.20−23]. When a material is illuminated by an intense beam of light, the index of refraction of the medium will change by an amount that depends on the intensity, and consequently the propagation characteristics of the medium will also change. In this situation, doubling the input does not necessarily double the output. Such phenomena have been under intense study in the field of nonlinear optics. The dependence of the refractive index on the light intensity $I$ can be approximately expressed as

$$n = n_0 + n_2 I \qquad (15.98)$$

if higher-order terms of $I$ are ignored.

Quite often, the intensity distribution of a light beam in the transverse plane is bell shaped, e.g. a Gaussian distribution. If the intensity of such a beam is high enough to create nonlinear effects, the transverse distribution of the refractive index also becomes bell shaped (if $n_2$ is positive like that of $CS_2$) and the medium starts to behave as a converging lens. This is the basis of self-focusing. Materials such as InSb have a negative value of $n_2$ and possess properties of self defocusing. The self-focusing effect is useful not only for beam manipulation, but also for producing bistable optical devices.

Figure 15.28 shows an example of bistability achieved through the self-focusing phenomenon. A light beam with a Gaussian intensity distribution is focused onto the input face of a cell which contains a nonlinear medium. The output from the cell falls upon a convex lens which focuses the image of the input face of the cell onto a half mirror. The half mirror transmits a portion of the light through as the output, and reflects the rest into the cell again, so that both incident and reflected beams participate in the self-focusing effect.

When the input intensity is low, no self-focusing action takes place, and the output power is low (lower level). Once the input-intensity reaches a critical point sufficient to initiate self-focusing, the intensity at the center of the beam increases and the gradient of the refractive index in the cell increases.

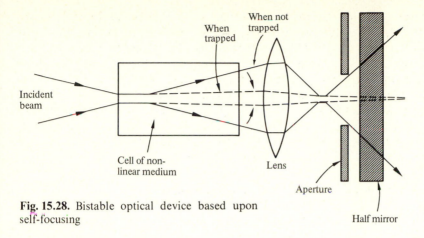

When trapped

When not trapped

Incident beam

Cell of non-linear medium

Lens

Aperture

Half mirror

**Fig. 15.28.** Bistable optical device based upon self-focusing

The increased gradient means better focusing, and better focusing means higher intensity, and higher intensity means increased gradient, and so on. The intensity of the light beam grows rapidly. Once a high intensity is established, the focusing action continues. Even after the input light power is lowered, the output power is still maintained high (higher level) by the self-focusing action, and the input-to-output curve displays a hysteresis loop characterizing the bistability. The device uses the light beam as a feedback loop but needs neither an external circuit nor a resonant cavity, and the switching speed can be as fast as a fraction of a picosecond. In contrast, the switching speed of a bistable device with external feedback is limited by the time constant $RC$, which is the product of the input resistance $R$ of the amplifier and the capacitance of the electrode. In resonator devices, the build up time of the resonant cavity normally slows down the switching time. In Fig. 15.28, the aperture adjusts the degree of the feedback to reach the condition of bistability.

## 15.5 Consideration of Polarization

So far nothing has been mentioned about the influence of the mode of excitation upon the function of the electrooptic devices. The modes in an optical guide are classified into TE (transverse electric) and TM (transverse magnetic) modes. The former does not have an $E$ component in the direction of propagation, and the latter does not have an $H$ component in the direction of propagation. If the $z$ axis is taken as the direction of propagation, the components present in each mode are

$$\underbrace{E_y \quad H_x \quad H_z}_{\text{TE mode}} \tag{15.99}$$

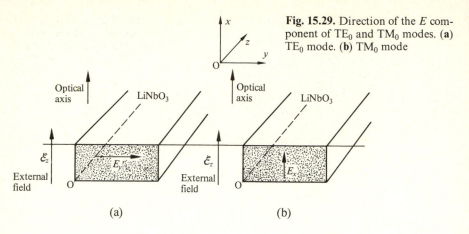

**Fig. 15.29.** Direction of the $E$ component of $TE_0$ and $TM_0$ modes. (**a**) $TE_0$ mode. (**b**) $TM_0$ mode

$$\underbrace{H_y \quad E_x \quad E_z}_{\text{TM mode .}}$$

(15.100)

The direction of the $E$ component is especially important for the operation of electrooptic devices. In the coordinate system shown in Fig. 15.29, the $E$ component of the $TE_0$ mode is in the $y$ direction, i.e., parallel to the surface of the layer whereas that of $TM_0$ mode is in the $x$ direction, i.e., in a plane perpendicular to the surface of the layer. The $TE_0$ mode has a lower cut-off frequency than the $TM_0$ mode.

The devices that use the electrooptic effect are normally sensitive to the type of mode excited because the electrooptic constant for one direction of polarization (direction of $E$) is different from that for another direction. For instance, when both the external electric field and the optical axis of $LiNbO_3$ are in the $x$ direction, as shown in Fig. 15.29, the $E$ component of the $TM_0$ mode sees the $r_{33}$ coefficient, whereas that of the $TE_0$ mode sees the $r_{13}$ coefficient. The value of $r_{13}$ is one quarter to one third of that of $r_{33}$. Some devices have to avoid this difference and others make use of this difference. The solution to eliminating the difference is either to use a mode stripper such as mentioned in connection with Fig. 15.22 b or to design the guide so as to pass only one of the modes. Otherwise, one must use a device which is designed to operate for both $TE_0$ and $TM_0$ modes.

Figure 15.30 shows a Mach-Zehnder interferometric modulator which can operate for both $TE_0$ and $TM_0$ modes [15.4]. The electrodes are divided into two sections, each having a slightly different arrangement. The lower electrode in Section I in the figure is deposited right on the optical guide. This provides an electric field $\mathscr{E}$ which is primarily vertical and parallel to the $E$ component of the $TM_0$ mode. On the other hand, the electrodes in Section II are placed symmetrically with respect to the optical guide and the $\mathscr{E}$ field is primarily horizontal and parallel to the $E$ component of the $TE_0$ mode. Thus, by adjusting the applied voltages and

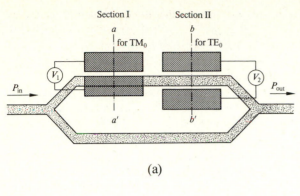

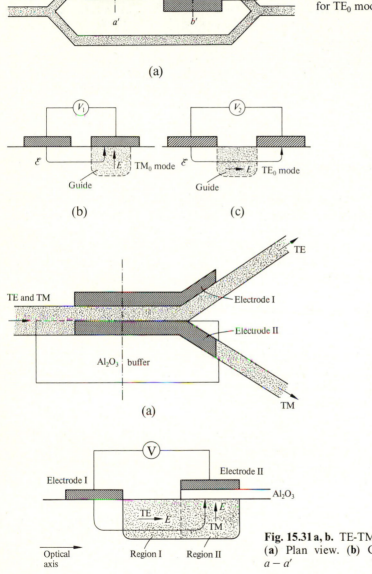

Fig. 15.30 a – c. Polarization insensitive Mach-Zehnder interferometric modulator. (a) Plan view. (b) Cross section at $a-a'$. Primarily for $TM_0$ mode. (c) Cross section at $b-b'$. Primarily for $TE_0$ mode

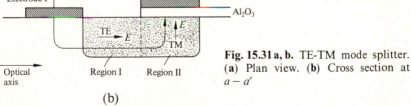

Fig. 15.31 a, b. TE-TM mode splitter. (a) Plan view. (b) Cross section at $a-a'$

length of each section, it is possible to find the condition that both $TE_0$ and $TM_0$ modes satisfy the switching condition.

Figure 15.31 shows a $TE \leftrightarrow TM$ mode splitter using a $Y$ junction [15.6]. As shown in Fig. 15.31 a, the electrodes are displaced from the line of symmetry so that the upper half of the guide in the figure has the external field $\mathscr{E}$ parallel to the surface, whereas the lower half has the $\mathscr{E}$ field perpendicular to the surface. Moreover, in the lower half of the guide, a buffer layer of $Al_2O_3$ film is deposited between the electrode and the guide to increase the effective refractive index.

When the external field is absent, both $TE_0$ and $TM_0$ modes converge into Region II in Fig. 15.31 b where the effective refractive index is higher because of the buffer layer. But once the external field $\varepsilon$ is applied, a horizontal field is established in Region I. The direction of polarization of the $TE_0$ mode is horizontal and the associated electrooptic constant $r_{33}$ is the largest (optical axis is horizontal). If the intensity of the external field is raised to such an extent that the effective refractive index for the $TE_0$ mode is larger in Region I than in Region II, the $TE_0$ mode converges to Region I.

On the other hand, the external electric field $\mathscr{E}$ in Region II is predominantly vertical and the $TM_0$ mode is even more confined in this region. With optimum adjustment of the external voltage, $TE_0$ and $TM_0$ modes exit from separate output ports.

## 15.6 Integrated Optical Lenses and the Spectrum Analyzer

Even though the operation of a lens fabricated on the substrate is restricted to one dimension, the lens is useful for such operations as imaging, focusing, spatial filtering, Fourier transforming, and convolving operations. As seen from Fig. 15.32, the physical length of the ray along the axis is the shortest and the rays get longer towards the sides. The focusing requirement is to equalize the optical length of all these rays whose physical

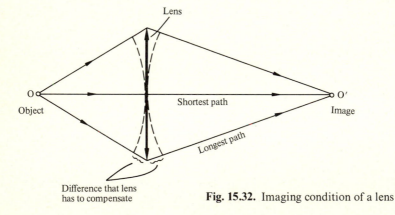

**Fig. 15.32.** Imaging condition of a lens

lengths are different. The focusing is achieved by either detouring the path or by adjusting the phase velocity of the different rays. Specific geometries will be introduced.

### 15.6.1 Mode Index Lens

A mode index lens profiles its effective refractive index by varying the thickness of the guiding layer [15.24]. Figure 15.33 illustrates how the effective index of refraction, $N = n_1 \cos \alpha = \beta/k$, of the $TM_0$ mode is increased with the thickness of the guiding layer. Here $\beta$ is the propagation constant in the direction of propagation $z$, and $k$ is that of free space. As the thickness is increased, the angle of inclination $\alpha$ decreases so that the zero lines still match the guide boundary and thus, $N$ is increased with the thickness.

In order to equalize the optical path in the center of the lens with that on the sides, the effective index of refraction is raised towards the center. For this purpose, the thickness of the guiding layer is increased by depositing the same material as the guiding layer as shown in Fig. 15.34.

A larger increase in effective refractive index is achieved if a material with a higher index of refraction $n_3$ than $n_1$ is used for depositing. The light ray goes into the deposited region and the amount of increase in effective

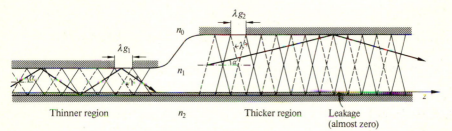

**Fig. 15.33.** Comparison of the angle $\alpha$ for the $TM_0$ mode in the thinner and thicker guiding layer. The thicker the layer is, the smaller $\alpha$ is, hence the larger the effective refractive index $N = n_2 \cos \alpha$ is

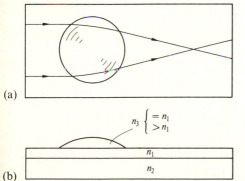

**Fig. 15.34a, b.** Luneberg lens. (a) Plan view. (b) Side view

refractive index has to be analyzed as a four-layer problem. The focusing power, however, is limited by the availability of usable transparent materials with a high refractive index.

In the case of a spherical lens, Maxwell found that a fish-eye lens whose refractive index distribution is $n(r) = 2/(1 + r^2)$ can make the image of an object on the sphere at a diametrically opposite point on the same sphere. Luneberg found that a sphere with $n(r) = \sqrt{2 - r^2}$ can converge a parallel beam into a point. Among the refractive index distribution of mode-index lenses, the one-dimensional Luneberg type distribution is most often employed.

### 15.6.2 Geodesic Lens

Figure 15.35 shows the geometry of a geodesic lens [15.25]. The thickness of the guiding layer is kept constant, but the contour length near the center is made longer by indentation. A protrusion would also give the same effect. For the lens to be aberration free, the contour of the indentation has to follow an aspheric complex profile. The difficulty in doing this is to polish the ground profile without destroying the original profile. The edge of the indentation has to be ground to prevent unwanted scattering. The ground edge should be taken into consideration in designing the profile to achieve a diffraction limited lens.

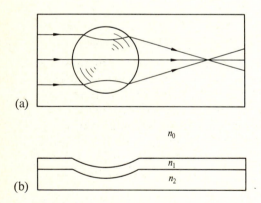

(a)

(b)

**Fig. 15.35 a, b.** Geodesic lens. (a) Plan view. (b) Side view

### 15.6.3 Fresnel Zone Lens

The principle of the Fresnel zone plate can be incorporated into an optical guide to produce a lens [15.26]. The phase distribution that a lens has to set up so as to converge a parallel beam to a point is, from (6.9),

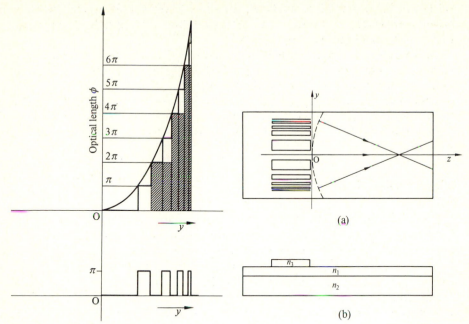

**Fig. 15.36.** Binary approximation of quadratic phase distribution (with $\phi_0 = 0$)

**Fig. 15.37 a, b.** Fresnel lens. (**a**) Plan view. (**b**) Side view

$$\phi = - k \frac{y^2}{2f} + \phi_0 , \tag{15.101}$$

where $f$ is the focal length of the lens. Figure 15.36 shows a method for approximating this quadratic expression. The curve is approximated by a staircase function. The phase $\phi$ is quantized into an integral number of $\pi$ radians in the upper graph. Since subtraction of an integral multiple of $2\pi$ radians (represented by the hatched sections) does not change anything, the required phase distribution for one half of the lens becomes like that shown in the lower graph. By depositing $\pi$ radian phase shift pads in accordance with the pattern shown in the lower graph, a Fresnel lens is fabricated. Figure 15.37 illustrates this Fresnel lens.

It is also possible to replace these phase pads by absorption pads thereby forming a zone plate of the absorption type, as mentioned in Sect. 4.5, with the sacrifice of half of the incident power. This type of lens generates background noise. The noise is due to higher-order beams which are inherent to the Fresnel zone plate lens discussed in Sect. 4.5.

A modified Fresnel lens is shown in Fig. 15.38 in which the phase pads have been tilted inwards so that the beams of specular reflection from the phase pads are directed toward the focus [15.24].

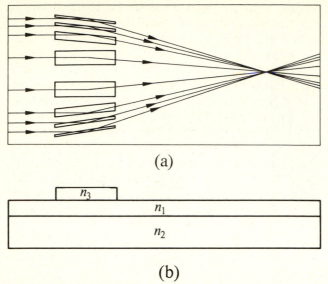

(a)

$n_3$

$n_1$

$n_2$

(b)

**Fig. 15.38a, b.** Grating lens. (**a**) Plan view. (**b**) Side view

### 15.6.4 Integrated Optical Spectrum Analyzer

Two geodesic lenses and a surface acoustic device can be combined to build an integrated optical spectrum analyzer [15.27]. Such a device can display the frequency spectrum distribution of a radio frequency electrical signal.

Figure 15.39 shows a sketch of an integrated optical spectrum analyzer. The first geodesic lens $L_1$ collimates the beam from a laser diode. The

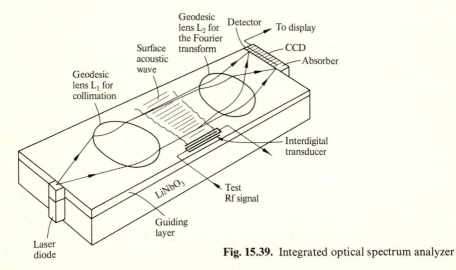

**Fig. 15.39.** Integrated optical spectrum analyzer

collimated beam intercepts the surface acoustic wave launched from the acoustic transducer. The surface acoustic wave introduces a periodic strain and spatially modulates the refractive index of the guiding layer. The light wave is diffracted in the direction that satisfies the Bragg condition, i.e. in the direction proportional to the acoustic frequency. The diffracted light is focused, or equivalently, is Fourier transformed onto an array of detectors at the end of the substrate. The light intensity distribution across the detector array represents the frequency spectral distribution of the acoustic wave. The electrical output from the detector is stored on the comb shaped electrodes of a charge coupled device (CCD) [15.28]. The CCD sequentially shifts the charge on the comb electrodes to the output, in accordance with clocked advance pulses, in a manner similar to a shift register that clocks out a parallel signal as a serial signal. The orientation of the acoustic transducer is slightly tilted so that the Bragg condition is satisfied at the center of the band of acoustic frequencies.

The required time of operation, physical dimensions and weight of the integrated optical spectrum analyzer are substantially less than the electronic counterpart of the device.

## 15.7 Methods of Fabrication

A localized increase in refractive index forms an optical guide. In this section, we describe methods of fabrication for optical guides, followed by those of forming the patterns for the guides.

### 15.7.1 Fabrication of Optical Guides

There are two major approaches to fabricating optical guides. One approach involves the deposition of a higher-index material onto the lower-index substrate. The other approach is to introduce a local increase in the refractive index of the substrate. Methods that belong to the former approach [15.29] are (1) epitaxial growth and (2) sputtering; and those in the latter are (3) ion-implantation, (4) out-diffusion, (5) in-diffusion, (6) ion migration and (7) photochromic film.

The method of epitaxial growth is a method that uses the growth of one crystalline substance on another having a similar crystal structure [15.30]. It is used to build layers of different optical characteristics, for example, the growth of $LiNbO_3$ onto $LiTaO_3$ is used to build a layer of higher refractive index. InGaAs is grown over GaAs to build a layer with a lower energy band gap for an integrated Schottky barrier photodetector. Epitaxy, however, requires careful control of the growth process before good results can be obtained.

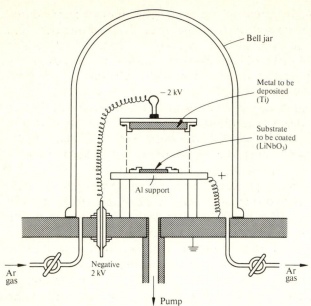

**Fig. 15.40.** Apparatus for sputtering

Bell jar

Metal to be deposited (Ti)

−2 kV

Substrate to be coated (LiNbO₃)

+

Al support

Negative 2 kV

Ar gas

Ar gas

Pump

The sputtering method [15.31] is a widely used technique. Figure 15.40 shows an apparatus for sputtering. A material (metallic or non-metallic) such as Ti is held onto the cathode. The substrate to be coated, such as LiNbO₃, is placed on the anode. An inert gas such as argon is enclosed at a constant gas pressure between $10^{-2}$ and $10^{-1}$ torr by constant pumping. A few thousand volts is applied between the cathode $(-)$ and the anode $(+)$ in order to accelerate the ions of the enclosed gas. The bombardment of accelerated Ar ions into the cathode ejects the atoms of the material (Ti). The ejected material, which is either neutral or slightly negatively charged, travels to the anode and is deposited onto the surface of the substrate (LiNbO₃) to form a film. Some of the materials used for sputtering are LiNbO₃, Ta, and Nb. Corning 7059 glass is one of the most common substrate materials used.

The next method to be mentioned is ion implantation [15.32]. When a crystalline substrate like fused silica (SiO₂) is bombarded by highly accelerated (a few hundred KeV) heavy ions like Li, Ne, Ar, Kr or Xe, the index of refraction is changed due to implantation. For instance, the implantation of Li into a fused silica substrate reduces the index of refraction by 0.8%.

Another method is out-diffusion [15.33, 34]. When a substrate is heated to a high temperature in vacuum, some of the elements in the compound literally boil out. The method of out-diffusion uses this fact. Crystallized LiNbO₃ can take the form of $(Li_2O)_v(Nb_2O_5)_{1-v}$. (This is called nonstoichiometric form. With $v = 0.5$, it becomes LiNbO₃.) The value of $v$ ranges from 0.48 to 0.5, and the refractive index of the crystal depends upon the value of $v$. A crystal with a smaller value of $v$ has a higher index of refraction. When $(Li_2O)_v(Nb_2O_5)_{1-v}$ is heated in vacuum, more $(Li_2O)$ boils out than $(Nb_2O_5)$ because Li is not as strongly bound to the lattice as Nb. The refractive index of the out-diffused part becomes higher. For example, an optical guide is made by heating a lithium niobate substrate to 1,100 °C for 21 h in a $6 \times 10^{-6}$ torr vacuum.

As a matter of fact, when $LiNbO_3$ is heated for the purpose of some other treatment to the substrate, guiding layers are formed all over the surface because of the out-diffusion. This out-diffusion, however, can be prevented by maintaining an equilibrium vapor pressure by externally providing $Li_2O$ vapor [15.34].

The method of in-diffusion [15.35] diffuses the material into the surface of the substrate to change the refractive index. In-diffusion of such transition elements as Ti, V or Ni into either a $LiNbO_3$ or $LiTaO_3$ substrate raises its refractive index. As a matter of procedure, a thin layer (200 − 800 Å) of a metal such as those mentioned above is evaporated on the surface of a $LiNbO_3$ substrate. This prepared substrate is heated at 850 ∼ 1,000 °C in a vacuum chamber containing Ar gas. After several hours of heating, a layer of higher refractive index a few microns thick is formed, due to the in-diffusion process.

Ion migration [15.36] uses the migration of metal ions into a glass substrate. When $AgNO_3$ is heated, it changes from solid form into liquid form. If a soda lime glass is immersed in the molten $AgNO_3$, a migration of Ag into the glass takes place, and Na in the glass is replaced by Ag. The refractive index of the soda lime glass is raised in the areas where Na has been replaced by Ag.

A technique which is quite different from all the others is the photochromic technique [15.37]. A normally transparent acrylic sheet coated with photochromic material changes its colour to dark blue when it is exposed to ultraviolet light. With the change in colour, the refractive index of the film is also changed. The refractive index of the colored state is higher than that of the uncolored state for wavelengths longer than the photochromic's absorption peak.

The pattern can be fabricated at room temperature. The lines, however, are subject to fading, but this can be an advantage or a disadvantage depending on whether permanence or erasability is desired.

## 15.7.2 Fabrication of Patterns

The methods of fabricating the guiding layers were the subject of the previous subsections. In order to make a useful device out of these guiding layers, the guides have to have a pattern [15.38 − 40].

The sputtering technique can be used for this purpose. If the substrate is placed on the cathode rather than on the anode, the bombardment of ions

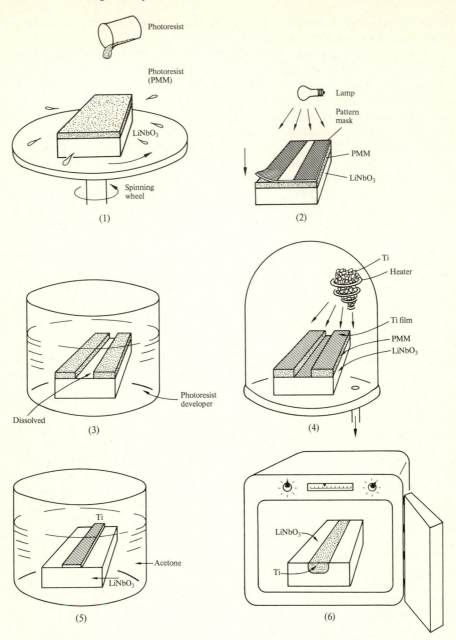

**Fig. 15.41** (*1–6*). Procedure for fabricating a strip of optical guide with photoresist. (*1*) Photoresist (PMM) is poured and the excess resist is spun off. (*2*) Pattern mask is laid over the photoresist and exposed to light. (*3*) Photoresist is developed and exposed part is dissolved (positive photoresist). (*4*) Ti is evaporated over the entire top surface. (*5*) Photoresist is dissolved by acetone. Ti on the photoresist layer is removed. (*6*) Heated at 970 °C for several hours to allow Ti film to diffuse into LiNbO$_3$

erodes the substrate material. This process can be used for pattern fabrication, and is called sputter etching.

Photoresist is one of the most commonly used techniques for pattern fabrication. Figure 15.41 illustrates the photoresist process. First (1), the substrate is coated evenly by a photoresist such as poly-methyl-methacrylate (PMM) by spinning the substrate, and then (2) it is exposed to light through a mask which has the desired pattern. The substrate is then (3) immersed into a developer bath where the portions exposed to light are dissolved. After development, a thin film of a metal like Ti is (4) deposited by evaporation. It is then put into an acetone bath where all photoresist is dissolved (5), including the photoresist under the thin metal film. Now, only the pattern of thin Ti film is left intact on the substrate. The substrate is then (6) put into an oven for several hours at 970 °C so that in-diffusion of Ti takes place. Thus, the process of fabricating the desired pattern of the optical guide is completed.

Finer resolution of the pattern can be achieved by writing in a pattern directly onto the photoresist rather than using a mask. A fine laser beam, electron beam, or even an x-ray beam is used for the writing [15.41, 42].

### 15.7.3 Summary of Strip Guides

Figure 15.42 summarizes the configurations of strip guides used in integrated optics [15.43−45]. The ridged guide shown in Fig. 15.42 a is bordered

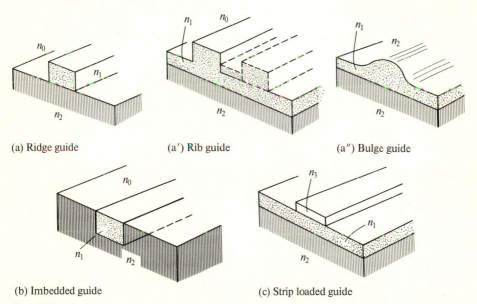

(a) Ridge guide          (a′) Rib guide          (a″) Bulge guide

(b) Imbedded guide                    (c) Strip loaded guide

**Fig. 15.42 a − c.** Various configurations of channel guides. (**a**) Ridge guide. (**a′**) Rib guide. (**a″**) Bulge guide. (**b**) Imbedded guide. (**c**) Strip loaded guide

by air on three sides. Since the dielectric discontinuity is so large, there is practically no evanescent wave, hence no cross-talk. A variation of this configuration is the rib guide shown in Fig. 15.42 a'. The raised portion uses the same material as that of the guiding layer. The light energy is still confined in the raised portion, but a larger evanescent wave is present at the bottom. By adding another rib guide, as indicated by the dotted lines, a coupler can be made. The amount of coupling is adjustable by the amount of filling between the guides. Again, because of the large dielectric discontinuity on the surface, the surface roughness of the guide critically influences the transmission loss of both the ridge and rib guides. The bulge guide in Fig. 15.42 a'' is a more realistic shape for fabrication. This type of guide is produced by removing the unmasked parts by either chemical means or sputter etching. Alternatively epitaxial growth can be used to make a bulge guide.

Figure 15.42 b shows an imbedded guide. Only one surface is bordered by air, and all other sides are bordered by a medium with only a slightly lower refractive index. The evanescent waves are present in all three sides. This is the most widely used guide. The amount of isolation as well as the coupling is easily controlled by the geometry and the refractive index. This type of guide is made by the methods of in- or out-diffusion.

Figure 15.42 c shows a strip loaded guide. A strip of dielectric of higher refractive index is placed over the guiding layer. The light energy is confined in the guiding layer under the strip, and the tolerance in fabricating the loading strip line is not so severe as for ridge or rib guides. If the same material as the guiding layer is used for the loading strip, it becomes a rib guide. Sputtering or epitaxial growth is used as the method of fabrication.

Sometimes a metal is used as the loading strip. In this case, the strip gives a means of applying an external electric field, as well as confining the light, because the metal at the light frequency acts as a lossy dielectric, even though its degree of influence is very minute [15.29]. Sometimes a $SiO_2$ or $Al_2O_3$ layer is placed between the metal strip and the guiding layer as a buffer layer to minimize the transmission loss.

### 15.7.4 Summary of Geometries of Electrodes

Figure 15.43 summarizes the configurations of the electrodes so far described. The configuration of the electrode is determined in accordance with the requirement of the direction of the applied field. In Fig. 15.43 a, b the direction of electric line of force is horizontal. Figure 15.43 a is for an imbedded guide, and Fig. 15.43 b for a rib guide. The effectiveness of the applied electric field in the latter configuration is much larger than the former, because of the smaller area of divergence of the electric line of force.

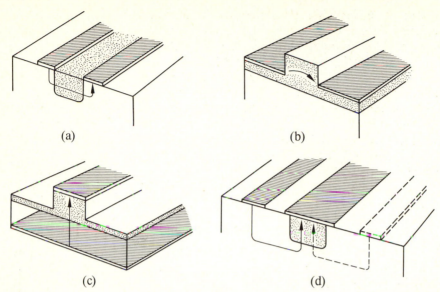

**Fig. 15.43 a – d.** Configurations for applying an external electric field in various directions. (**a**) Horizontal field in imbedded guide. (**b**) Horizontal field in rib guide. (**c**) Vertical field in rib guide. (**d**) Vertical field in imbedded guide

When the electric field has to be vertical, the configurations shown in Fig. 15.43 c, d are used. The fact that the electric line of force always terminates perpendicular to the surface of a perfect conductor is used. With the configuration in Fig. 15.43 d, the field intensity can be doubled by placing another electrode, as shown in dotted lines.

## Problems

**15.1**  Based upon the delta-function model, design an interdigital transducer of a SAW bandpass filter. The frequency of the pass band is from 80 to 120 MHz. The velocity of the surface acoustic wave in the lithium niobate substrate is 3,485 m/s.

**15.2**  Show that the loops of the $\otimes$ modes of the reversed $\Delta$ directional coupler (Fig. 15.16) are all confined in the sector bounded by the vertical axis and the 45 degree line (a line bisecting the horizontal and vertical axis).

**15.3**  An optical directional coupler such as shown in Fig. 15.44 is to be deposited on a $LiNbO_3$ substrate with the following parameters by locally raising index of refraction using titenium diffusion. The electrical coupling length of the $\otimes$ switching mode is selected as $\frac{\pi}{2}$ rad.

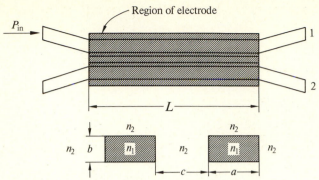

**Fig. 15.44.** Geometries of the optical directional coupler

$a = 1.0\,\mu\text{m}$,     $b = 0.5\,\mu\text{m}$,     $c = 1.5\,\mu\text{m}$

$n_1 = 1.1\,n_2$,     $\alpha = 0.3$ (Electrode efficiency)

$\lambda = 1.3\,\mu\text{m}$

Assume total length $d$ of electric line of force of the dc field is $10\,\mu\text{m}$ and the ratio of propagation constants $k_z/k_1$ is unity for the dominant mode.

a) What is the orientation of the crystal and the electrode for minimizing the control voltage?

b) What is the control voltage to achieve the $\ominus$ switching mode?

Only excitation of the dominant mode (TE mode) is desired. The condition that only the dominant mode be excited is approximately

$$a < \frac{0.8\,\lambda}{\sqrt{n_1^2 - n_2^2}}\,.$$

The coupling constant for the given dimensions and the refractive indices are found from the universal curves prepared by *Marcatili* [15.46] in Fig. 15.45.

**15.4**   Derive the expressions which would be useful for drawing the cross bar diagram of a two-bay reversed $\Delta\beta$ directional coupler (two reversed $\Delta\beta$ directional couplers in tandem), and make a rough sketch of the cross bar diagram for the two-bay reversed $\Delta\beta$ directional coupler. Also, verify that the required control voltage for the two-bay coupler is less than that of the one-bay coupler.

**15.5**   In Fig. 15.24a only the range of $0 < \phi < 3\,\pi/2$ of the curve was considered. Complete the input-output characteristic curve like the one in Fig. 14.34b taking a wider range of $3\,\pi/2\ \ \phi < 3\,\pi$ as shown in Fig. 15.46 into consideration. Assume $|k_{12}|\,L = \pi$, and $GL = 4\,\pi$.

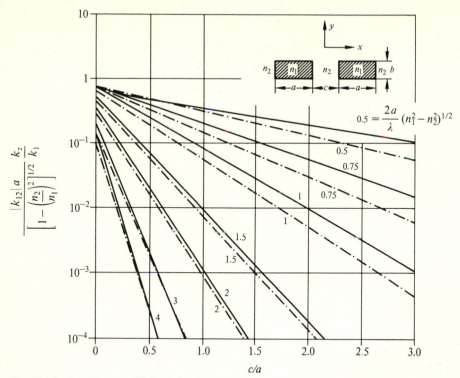

**Fig. 15.45.** Coupling coefficient for $E_{1q}^x$ modes. (——) $E_{1q}^x$ coupling for $n_1/n_2 = 1.5$; (—·—·—) $E_{1q}^x$ coupling for $n_1/n_2 = 1.1$

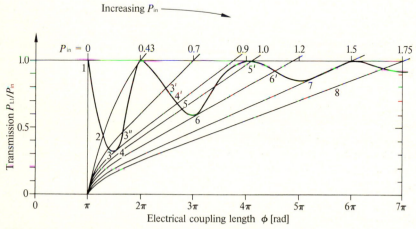

**Fig. 15.46.** Operating points of the optical bistability

# References

**Chapter 1**

1.1 R. J. Unstead: *Egypt and Mesopotamia* (Adam and Charles Black, London 1977)
1.2 S. N. Kramer: *From the Tablets of Sumer* (The Falcon's Wing Press, Indian Hills, CO 1956)
1.3 S. N. Kramer: *History Begins at Sumer* (The University of Pennsylvania Press, Philadelphia 1981)
1.4 C. G. Fraser: *Half-hours with Great Scientists; The Story of Physics* (The University of Toronto Press, Toronto 1948)
1.5 L. Gorelick, A. J. Gwinnett: "Close Work without Magnifying Lenses?", Expedition **23,** No. 2, pp. 27–34 (Winter 1981)
1.6 T. Whittle: *The World of Classical Greece* (William Heinemann, London 1971)
1.7 G. Gamow: *Biography of Physics* (Hutchinson, London 1962)
1.8 V. Ronchi: *The Nature of Light, an Historical Survey,* translated into English by V. Barocas (William Heinemann, London 1970)
1.9 K. Iizuka, H. Ogura, J. L. Yen, V. K. Nguyen, J. Weedmark: A hologram matrix radar, Proc. IEEE **64,** 1493–1504 (1976)
1.10 D. C. Lindberg: *Theories of Vision from Al-Kindi to Kepler* (The University of Chicago Press, Chicago 1976)
1.11 A. I. Sabra: *Theories of Light from Descartes to Newton* (Cambridge University Press, Cambridge 1967)
1.12 M. Born, E. Wolf: *Principles of Optics* (Pergamon, Oxford 1964)
1.13 News column, Nobel Physics Prize to Bloembergen, Schawlow and Siegbahn, Physics Today, pp. 17–20 (Dec. 1981)

**Chapter 2**

2.1 C. L. Andrews: *Optics of the Electromagnetic Spectrum* (Prentice-Hall, Englewood Cliffs, NJ 1960)
2.2 J. W. Goodman: *Introduction to Fourier Optics* (McGraw-Hill, New York 1968)
2.3 R. Bracewell: *The Fourier Transform and its Applications* (McGraw-Hill, New York 1965)
2.4 M. L. Boas: *Mathematical Methods in the Physical Sciences* (Wiley, New York 1966)
2.5 A. Papoulis: *Systems and Transforms with Applications in Optics* (McGraw-Hill, New York 1968)

**Additional Reading**

Barakat, R.: In *The Computer in Optical Research,* ed. by R. Frieden, Topics Appl. Phys., Vol. 41 (Springer, Berlin, Heidelberg 1980)
Churchill, R. V.: *Operational Mathematics* (McGraw-Hill, New York 1958)
Erdelyi, A., Magnus, W., Oberhattinger, F., Tricomi, F.: *Tables of Integral Transforms,* Vol. 1 and 2 (McGraw-Hill, New York 1954)
Lighthill, M. J.: *Introduction to Fourier Analysis and Generalized Functions* (Cambridge University Press, Cambridge, London 1958)
Sneddon, I. N.: *Fourier Transforms* (McGraw-Hill, New York 1951)

## Chapter 3

3.1  J. A. Stratton: *Electromagnetic Theory* (McGraw-Hill, New York 1941)
3.2  J. W. Goodman: *Introduction to Fourier Optics* (McGraw-Hill, New York 1968)
3.3  A. Papoulis: *Systems and Transforms with Applications in Optics* (McGraw-Hill, New York 1968)
3.4  A. Sommerfeld: *Partial Differential Equation in Physics* (Academic, New York 1949)
3.5  D. C. Stinson: *Intermediate Mathematics of Electromagnetics* (Prentice-Hall, Englewood Cliffs, NJ 1976)

### Additional Reading

Babič, V. M., Kirpičnikova, N. Y.: *The Boundary-Layer Method in Diffraction Problems,* Springer Ser. Electrophys., Vol. 3 (Springer, Berlin, Heidelberg 1979)
Born, M., Wolf, E.: *Principles of Optics* (Pergamon, New York 1964)
Economou, E. N.: *Green's Functions in Quantum Physics,* 2nd ed., Springer Ser. Solid-State Sci., Vol. 7 (Springer, Berlin, Heidelberg 1983)
Fowles, G. R.: *Introduction to Modern Optics* (Holt Rinehart and Winston, New York 1975)
Lotsch, H. K. V.: Multiple-Diffraction Formula, Optik **25,** 598–605 (1967)
Sommerfeld, A.: *Optics* (Academic, New York 1954)
Stone, J. M.: *Radiation and Optics* (McGraw-Hill, New York 1963)
Young M.: *Optics and Lasers,* 2nd ed. Springer Ser. Opt. Sci., Vol. 5 (Springer, Berlin, Heidelberg 1984)

## Chapter 4

4.1  M. Rousseau, J. P. Mathieu, J. W. Blaker: *Problems in Optics* (Pergamon, New York 1973)
4.2  A. Papoulis: *Systems and Transforms with Applications in Optics* (McGraw-Hill, New York 1968)
4.3  E. Hecht, A. Zajac: *Optics* (Addison-Wesley, Reading, MA 1974)
4.4  S. Miyaoka: Digital audio compact and rugged, IEEE Spectrum, 35–39, (March 1984)
4.5  D. J. Broer, L. Vriens: Laser-induced optical recording in thin films, Appl. Phys. A **32,** 107–123 (1983)
4.6  A. Engel, J. Steffen, G. Herziger: Laser machining with modulated zone plates, Appl. Opt. **13,** 269–273 (1974)

### Additional Reading

Bracewell, R.: *The Fourier Transform and its Applications* (McGraw-Hill, New York 1965)
Feynman, R. P., Leighton, R. B., Sands, M.: *The Feynman Lectures on Physics* Vol. I (Addison-Wesley, Reading, MA 1964)
Goodman, J. W.: *Introduction to Fourier Optics* (McGraw-Hill, New York 1968)
Nussbaum, A., Phillips, R. A.: *Contemporary Optics for Scientists and Engineers* (Prentice-Hall, Englewood Cliffs, NJ 1975)
Petit, R. (ed): *Electromagnetic Theory of Gratings,* Topics Current Phys., Vol. 22 (Springer, Berlin, Heidelberg 1980)

## Chapter 5

5.1  E. Kreyszig: *Advanced Engineering Mathematics* (Wiley, New York 1972)
5.2  A. A. Khinchin: *A Course of Mathematical Analysis* (translated from Russian) (Hindustan Publishing, Delhi 1960)

5.3  M. Kline, I. W. Kay: *Electromagnetic Theory and Geometrical Optics* (Interscience, New York 1965)
5.4  R. K. Luneburg: *Mathematical Theory of Optics* (University of California Press, Berkeley 1964)
5.5  K. Maeda, I. Kimura: *Electromagnetic Theory* (Ohm-Sha, Tokyo 1970)
5.6  A. Sommerfeld: *Optics* (Academic, New York 1954)
5.7  G. Petit Bois: *Table of Indefinite Integrals* (Dover, New York 1961)
5.8  L. Jacomme: A model for ray propagation in a multimode graded index-fiber, Optics Commun. **14,** 134–138 (1975)
5.9  S. Shimada: *Optical Fiber Communication Techniques* (Ohm-Sha, Tokyo 1980)
5.10 M. Ikeda: Propagation characteristics of multimode fibers with graded core index, IEEE J. QE-**10,** 362–371 (1974)
5.11 T. Uchida, M. Furukawa, I. Kitano, K. Koizumi, H. Matsumura: Optical characteristics of a light-focusing fiber guide and its applications, IEEE J. QE-**6,** 606–612 (1970)
5.12 S. E. Miller, E. A. J. Marcatili, T. Li: Research toward optical-fiber transmission systems, Proc. IEEE **61,** 1703–1751 (1973)
5.13 H. K. V. Lotsch: Physical-optics theory of planar dielectric waveguides, Optik **27,** 239–254 (1968)
5.14 H. K. V. Lotsch: Beam displacement at total reflection: the Goos-Hänchen effect, Optik **32,** 116–137, 189–204, 299–319 and 553–569 (1970/71)

## Additional Reading

Adams, M. J.: *An Introduction to Optical Waveguides* (Wiley, New York 1981)
Born, M., Wolf, E.: *Principles of Optics* (Pergamon, New York 1964)
Burrus, C. A., Standley, R. D.: Viewing refractive-index profiles and small-scale inhomogeneities in glass optical fibers: some techniques, Appl. Opt. **13,** 2365–2369 (1974)
Gloge, D.: Optical fibers for communication, Appl. Opt. **13,** 249–254 (1974)
Kaiser, P., Marcatili, E. A. J., Miller, S. E.: A new optical fiber, Bell Syst. Tech. J. **52,** 265–269 (1973)
Kumagai, N., Sawa, S.: Propagation of $TE_{01}$ mode in a lens-like medium, J. Inst. Electronics and Commun. Eng. Jpn. **54,** 39–44 (1971)
Marcuse, D.: *Light Transmission Optics* (Van Nostrand Reinhold, New York 1972)
Sharma, A. B., Halme, S. J., Butusov, M. M.: *Optical Fiber Systems and Their Components*, Springer Ser. Opt. Sci., Vol. 24 (Springer, Berlin, Heidelberg 1981)
Synge, J. L.: *Geometrical Optics, An Introduction to Hamilton's Method* (Cambridge University Press, London 1937)

## Chapter 6

6.1  J. W. Goodman: *Introduction to Fourier Optics* (McGraw Hill, New York 1968)
6.2  A. Nussbaum, R. A. Phillips: *Contemporary Optics for Scientists and Engineers* (Prentice-Hall, Englewood Cliffs, NJ 1976)
6.3  S. H. Lee (ed.): *Optical Information Processing, Fundamentals,* Topics Appl. Phys., Vol. 48 (Springer, Berlin, Heidelberg 1981) Chap. 1
6.4  L. P. Yaroslovsky: *Digital Picture Processing,* Springer Ser. Info. Sci., Vol. 9 (Springer, Berlin, Heidelberg, in preparation)
6.5  A. Papoulis: *Systems and Transforms with Applications in Optics* (McGraw Hill, New York 1968)

## Additional Reading

Casasent, D. (ed.): *Optical Data Processing, Applications,* Topics Appl. Phys., Vol. 23 (Springer, Berlin, Heidelberg 1978)

Frieden, B. R.: *Probability, Statistical Optics and Data Analysis,* Springer Ser. Infor. Sci., Vol. 10 (Springer, Berlin, Heidelberg 1982)
Hecht, E., Zajac, A.: *Optics* (Addison-Wesley, Reading, MA 1974)
Huang, T. S. (ed.): *Picture Processing and Digital Filtering,* 2nd ed., Topics Appl. Phys., Vol. 6 (Springer, Berlin, Heidelberg 1979)
Meyer-Arendt, J. R.: *Introduction to Classical and Modern Optics* (Prentice-Hall, Englewood Cliffs, NJ 1972)
O'Neal, E. L.: *Introduction to Statistical Optics* (Addison-Wesley, Reading, MA 1963)
Stavroudis, O. N.: *Modular Optical Design,* Springer Ser. Opt. Sci., Vol. 28 (Springer, Berlin Heidelberg 1982)

## Chapter 7

7.1   J. W. Cooley, J. W. Tukey: An algorithm for the machine calculation of complex Fourier series, Math. Comput. **19,** 297–301 (1965)
7.2   H. J. Nussbaumer: In *Two-Dimensional Digital Signal Processing* II, ed. by T. S. Huang, Topics Appl. Phys., Vol. 43 (Springer, Berlin, Heidelberg 1981) Chap. 3
7.3   H. J. Nussbaumer: *Fourier Transform and Convolution Algorithms,* 2nd ed. Springer Ser. Inform. Sci., Vol. 2 (Springer, Berlin, Heidelberg 1982)
7.4   W. T. Cochran, J. W. Cooley, D. L. Favin, H. D. Helms, R. A. Kaenel, W. W. Lang, G. C. Maling Jr., D. E. Nelson, C. M. Rader, P. D. Welch: What is the Fast Fourier Transform?, Proc. IEEE **55,** 1664–1674 (1967)

### Additional Reading

Cooley, J. W., Lewis, P. A. W., Welch, P. D.: Historical notes on the Fast Fourier Transform, Proc. IEEE **55,** 1675–1677 (1967)
Gold, B., Rader, C. M.: *Digital Processing of Signals* (McGraw Hill, New York 1969)
Bergland, G. D.: A guided tour of the Fast Fourier Transform, IEEE Spectrum 41–52 (July 1969)
Brigham, E. O.: *The Fast Fourier Transform* (Prentice Hall, Englewood Cliffs, NJ 1974)

## Chapter 8

8.1   J. W. Goodman: *Introduction to Fourier Optics* (McGraw-Hill, New York 1968)
8.2   W. T. Cathey: *Optical Information Processing and Holography* (Wiley, New York 1974)
8.3   M. Akagi, T. Kaneko, T. Ishiba: Electron micrographs of hologram cross sections, Appl. Phys. Lett. **21,** 93–95 (1972)
8.4   H. M. Smith (ed.): *Holographic Recording Materials,* Topics in Appl. Phys., Vol. 20 (Springer, Berlin, Heidelberg 1977)
8.5   R. J. Collier, C. B. Buckhardt, L. H. Lin: *Optical Holography* (Academic, New York 1971)
8.6   T. S. Huang: Digital holography, Proc. IEEE, **59,** 1335–1346 (1971)
8.7   W. H. Lee: Sampled Fourier transform hologram generated by computer, Appl. Opt. **9,** 639–643 (1970)
8.8   M. C. King, A. M. Noll, D. H. Berry: A new approach to computer-generated holography, Appl. Opt. **9,** 471–475 (1970)
8.9   W. J. Dallas: In *The Computer in Optical Research,* ed. by B. R. Frieden, Topics in Appl. Phys., Vol. 41 (Springer, Berlin, Heidelberg 1980)
8.10  S. A. Benton: In *Applications of Holography and Optical Data Processing,* ed. by E. Marom, A. A. Friesem, W. Wiener-Avenear (Pergamon, New York 1977) pp. 401–409

8.11 S. A. Benton: *Hologram reconstruction with expanded light sources,* J. Opt. Soc. Am. **59,** 1545 A (1969)

8.12 J. C. Dainty (ed.): *Laser Speckle and Related Phenomena,* 2nd ed., Topics in Appl. Phys., Vol. 9 (Springer, Berlin, Heidelberg 1984)

8.13 D. Gabor: *Light and information,* Prog. Opt., **1,** 109–153 (1961)

8.14 B. R. Frieden: *Probability, Statistical Optics, and Data Testing,* Springer Ser. Inform. Sci., Vol. 10 (Springer, Berlin, Heidelberg 1983)

8.15 W. Schumann, M. Dubas: *Holographic Interferometry,* Springer Ser. Opt. Sci., Vol. 16 (Springer, Berlin, Heidelberg 1979)

8.16 Y. I. Ostrovsky, M. M. Butusov, G. V. Ostrovskaya: *Interferometry by Holography,* Springer Ser. Opt. Sci., Vol. 10 (Springer, Berlin, Heidelberg 1980)

8.17 C. Knox, R. E. Brooks: Holographic motion picture microscopy, Proc. Roy. Soc. (London) **B174,** 115–121 (1969)

8.18 D. H. McMahon: Holographic ultrafiche, Appl. Opt. **11,** 798–806 (1972)

8.19 Y. Takeda, Y. Oshida, Y. Miyamura: Random phase shifters for Fourier transformed holograms, Appl. Opt. **11,** No. 4, 818–822 (1972)

8.20 A. Ioka, K. Kurahashi: Grating modulated phase plates for holographic storage, Appl. Opt. **14,** 2267–2273 (1975)

8.21 Y. Tsunoda, Y. Takeda: High density image-storage holograms by a random phase sampling method, Appl. Opt. **13,** 2046–2051 (1974)

8.22 L. K. Anderson: *Holographic Optical Memory for Bulk Data Storage,* Bell Laboratories Records, 319–325 (Nov. 1968)

8.23 K. Kubota, Y. Ono, M. Kondo, S. Sugama, N. Nishida, M. Sakaguchi: Holographic disk with high data transfer rate; its application to an audio response memory, Appl. Opt. **19,** 944–951 (1980)

8.24 D. J. Channin, A. Sussman: In *Display Devices,* ed. by J. I. Pankove, Topics Appl. Phys., Vol 40 (Springer, Berlin, Heidelberg 1980) Chap. 4

8.25 A. Engel, J. Steffen, G. Herziger: Laser machining with modulated zone plates, Appl. Opt. **13,** 269–273 (1974)

8.26 S. Amadesi, F. Gori, R. Grella, G. Guattari: Holographic methods for painting diagnostics, Appl. Opt. **13,** 2009–2013 (1974)

8.27 G. von Bally (ed.): *Holography in Medicine and Biology,* Springer Ser. Opt. Sci., Vol. 18 (Springer, Berlin, Heidelberg 1979)

8.28 A. J. MacGovern, J. C. Wyant: Computer generated holograms for testing optical elements, Appl. Opt. **10,** 619–624 (1971)

8.29 K. Bromley, M. A. Monahan, J. F. Bryant, B. J. Thompson: Holographic subtraction, Appl. Opt. **10,** 174–181 (1971)

8.30 C. Agren, K. A. Stetson: Measuring the wood resonance of treble viol plates by hologram interferometry, J. Acoust. Soc. Am. **46,** 120 (1969); N.-E. Molin: Private communication

8.31 W. Schmidt, A. Vogel, D. Preusser: Holographic contour mapping using a dye laser, Appl. Phys. **1,** 103–109 (1973)

8.32 L. O. Heflinger, R. F. Wuerker: Holographic contouring via multifrequency lasers, Appl. Phys. Lett. **15,** 28–30 (1969)

## Chapter 9

9.1 *Basic Photographic Sensitometry Workbook,* (Kodak Publication, No. Z-22-Ed 1968)

9.2 T. H. Jeong: *Gaertner-Jeong Holography Manual,* Gaertner Scientific Corporation (1968)

9.3 M. Lehmann: *Holography, Technique and Practice* (The Focal Press, London 1970)

9.4 G. T. Eaton: *Photographic Chemistry* (Morgan and Morgan Inc., Hastings on Hudson, NY 1957)

9.5 J. Upatnieks, C. Leonard: Diffraction efficiency of bleached photographically recorded interference patterns, Appl. Opt. **8,** 85–89 (1969)

**Additional Reading**

Chathey, T. W.: *Optical Information Processing and Holography* (Wiley, New York 1974)
Fujinami, S.: *Photography,* Kyoritsu series No. **62** (Kyoritsu Shuppan, Tokyo 1963)
Fujinami, S.: *Advanced Photography,* Kyoritsu series No. **118** (Kyoritsu Shuppan, Tokyo 1966)
Smith, H. M. (ed.): *Holographic Recording Materials,* Topics Appl. Phys., Vol. 20 (Springer, Berlin, Heidelberg 1977)


**Chapter 10**

10.1    J. W. Goodman: *Introduction to Fourier Optics* (McGraw-Hill, New York 1968)
10.2    M. Born, E. Wolf: *Principles of Optics* (Pergamon, New York 1964)
10.3    W. T. Cathey: *Optical Information Processing and Holography* (Wiley, New York 1974)
10.4    J. R. Meyer-Arendt: *Introduction to Classical and Modern Optics* (Prentice-Hall, Inglewood Cliffs, NJ 1972)


**Additional Reading**

Lee, S. H. (ed.): *Optical Information Processing, Fundamentals,* Topics Appl. Phys., Vol. 48 (Springer, Berlin, Heidelberg 1981)
Levi, L.: *Applied Optics* (Wiley, New York 1968)
Suzuki, T.: Optical signal processing, Proc. of Institute of Electronics and Communication Engineers of Japan **57,** 1278–1284 (1974)
Troup, G. J.: *Optical Coherence Theory, Recent Developments* (Methuen and Co. Ltd., London 1967)


**Chapter 11**

11.1    *Basic Photographic Sensitometry Workbook* (Kodak Publication, No. Z-22-Ed 1968)
11.2    H. M. Smith (ed.): *Holographic Recording Materials,* Topics Appl. Phys., Vol. 20 (Springer, Berlin, Heidelberg 1977)
11.3    Y. Taki, M. Aoki, K. Hiwatashi: *Picture Processing* (Corona-sha, Tokyo 1972)
11.4    S. H. Lee (ed.): *Optical Information Processing, Fundamentals,* Topics Appl. Phys., Vol. 48 (Springer, Berlin, Heidelberg 1981)
11.5    L. Gregoris, K. Iizuka: Additive and subtractive microwave holography, Appl. Opt. **12,** 2641–2648 (1973)
11.6    R. G. Eguchi, F. P. Carlson: Linear vector operations in coherent optical data processing systems, Appl. Opt. **9,** 687–694 (1970)
11.7    A. Vander Lugt: Coherent optical processing, Proc. IEEE **62,** 1300–1319 (1974)
11.8    I. McCausland: *Introduction to Optimal Control* (Wiley, New York 1979)
11.9    J. L. Horner: Optical restoration of images blurred by atmospheric turbulence using optimum filter theory, Appl. Opt. **9,** 167–171 (1970)
11.10   D. Casasent (ed.): *Optical Data Processing, Applications,* Topics Appl. Phys., Vol. 23 (Springer, Berlin, Heidelberg 1978)
11.11   J. W. Goodman: *Introduction to Fourier Optics* (McGraw-Hill, New York 1968)
11.12   F. T. S. Yu, T. H. Chao: Color signal correlation detection by matched spatial filtering, Appl. Phys. B **32,** 1–6 (1983)
11.13   D. P. Jablonowski, S. H. Lee: Restoration of degraded images by composite gratings in a coherent optical processor, Appl. Opt. **12,** 1703–1712 (1973)
11.14   A. Wouters, K. M. Simon, J. G. Hirschberg: Direct method of decoding multiple images, Appl. Opt. **12,** 1871–1873 (1973)
11.15   K. A. Frey, D. M. Wieland, L. E. Brown, W. L. Rogers, B. W. Agranoff: Development of a tomographic myelin scan, Ann. Neurology **10,** 214–221 (1981)

11.16 A. W. Lohmann, H. W. Werlich: Incoherent matched filtering with Fourier holograms, Appl. Opt. **10,** 670–672 (1971)

11.17 H. Kato, J. W. Goodman: Nonlinear filtering in coherent optical systems through halftone screen processes, Appl. Opt. **14,** 1813–1824 (1975)

11.18 G. T. Herman (ed.): *Image Reconstruction from Projections,* Topics Appl. Phys., Vol. 32 (Springer, Berlin, Heidelberg 1979)

11.19 W. J. Meredith, J. B. Massey: *Fundamental Physics of Radiology,* 2nd ed. (Wright, Bristol, UK 1972)

11.20 R. A. Brooks, G. DiChiro: Principles of computer assisted tomography (CAT) in radiographic and radioscopic imaging, Phys. Med. Biol. Vol. 21, 689–732 (1976)

## Additional Reading

Berrett, A., Brünner, S., Valvassori, G. E. (eds): *Modern Thin-Section Tomography* (Charles C. Thomas Publisher, Springfield, IL 1973)

Françon, M.: *Modern Applications of Physical Optics* (Wiley, New York 1963)

Kak, A. C.: Computerized tomography with x-ray, emission, and ultrasound sources, Proc. IEEE **67,** 1245–1272 (1979)

Yaroslavsky, L. P.: *Digital Picture Processing,* Springer Ser. Inform. Sci., Vol. 9 (Springer, Berlin, Heidelberg 1984)

## Chapter 12

12.1 R. P. Dooley: X-band holography, Proc. IEEE **53,** 1733–1735 (1965)

12.2 Y. Aoki: Microwave hologram and optical reconstruction, Appl. Opt. **6,** 1943–1946 (1967)

12.3 H. Shigesawa, K. Takiyama, T. Toyonaga, M. Nishimura: Microwave holography by spherical scanning, Trans. Inst. Electr. Commun. Engs. Jpn., **56**-B, No. 3, 99–106 (1973)

12.4 N. H. Farhat, A. H. Farhat: Double circular scanning in microwave holography, Proc. IEEE **61,** 509–510 (1973)

12.5 T. Hasegawa: A new method of observing electromagnetic fields at high frequencies by use of test paper, Bull. Yamagata Univ., Yamagata, Jpn., IV, 1 (1955)

12.6 K. Iizuka: Mapping of electromagnetic fields by photochemical reaction, Electr. Lett. **4,** 68 (1968)

12.7 C. F. Augustine, C. Deutsch, D. Fritzler, E. Marom: Microwave holography using liquid crystal area detectors, Proc. IEEE **57,** 1333–1334 (1969)

12.8 K. Iizuka: In situ microwave holography, Appl. Opt. **12,** 147–149 (1973)

12.9 L. Gregoris: Aspects of Microwave Holography, Ph.D. Thesis, University of Toronto (1975)

12.10 K. Iizuka: Microwave holography by means of optical interference holography, Appl. Phys. Lett. **17,** 99–101 (1970)

12.11 L. Gregoris, K. Iizuka: Moiré microwave holography, Appl. Opt. **16,** 418–426 (1977)

12.12 L. G Gregoris, K. Iizuka: Visualization of internal structure by microwave holography, Proc. IEEE **58,** 791–792 (1970)

12.13 N. H. Farhat, W. R. Guard: Millimeter wave holographic imaging of concealed weapons, Proc. IEEE, **59,** 1383–1384 (1971)

12.14 K. Iizuka: Microwave hologram by photoengraving, Proc. IEEE **57,** 813–814 (1969)

12.15 K. Iizuka: Use of microwave holography in the study of electromagnetic phenomena, Intern. Symp. on Antenna and Propagation, Japan 163–164 (1971)

12.16 K. Iizuka: Subtractive microwave holography and its application to plasma studies, Appl. Phys. Lett. **20,** 27–29 (1972)

12.17 P. F. Checcacci, G. Papi, V. Russo, A. M. Scheggi: Holographic VHF antennas, IEEE Trans. AP-**19,** 278–279 (1971)

12.18 K. Iizuka, M. Mizusawa, S. Urasaki, H. Ushigome: Volume type holographic antennas, IEEE Trans. AP-**23**, 807–810 (1975)

12.19 E. B. Joy, C. P. Burns, G. P. Rodrigue: Accuracy of far field patterns based on near field measurements, 1973 IEEE G-AP Symp., 57 (1973)

12.20 J. C. Bennett, A. P. Anderson, P. A. McInnes, A. J. T. Whitaker: Investigation of the characteristics of a large reflector antenna using microwave holography, 1973 IEEE G-AP Symp. 298–301 (1973)

12.21 M. W. Brown, L. J. Porcello: An introduction to synthetic aperture radar, IEEE Spectrum, 52–62 (Sept. 1969)

12.22 L. J. Cutrona, E. N. Leith, L. J. Porcello, W. E. Vivian: On the application of coherent optical processing techniques to synthetic aperture radar, Proc. IEEE **54**, 1026–1032 (1966)

12.23 E. N. Leith: In *Optical Data Processing, Applications,* ed. by D. Casasent, Topics Appl. Phys., Vol. 23 (Springer, Berlin, Heidelberg 1978) Chap. 4

12.24 K. Iizuka, H. Ogura, J. L. Yen, V. K. Nguyen, J. Weedmark: A hologram matrix radar, Proc. IEEE **64**, 1493–1504 (1976)


**Chapter 13**

13.1 A. B. Sharma, S. J. Halme, M. M. Butusov: *Optical Fiber Systems and Their Components,* Springer Ser. Opt. Sci., Vol. 24 (Springer, Berlin, Heidelberg 1981)

13.2 Special issue of J. Inst. Electr. Commun. Jpn. **63,** (Nov. 1980)

13.3 Y. Suematsu, K. Iga: *Introduction to Optical Fiber Communications* (Ohm-Sha, Toky 1976)

13.4 H. Takashima, N. Uchida, S. Takashima: Overview on techniques of optical fiber cables, Special Issue on fiber optical communication, J. Shisetsu **31**, 28–43 (1979)

13.5 S. D. Personick: *Optical Fiber Transmission Systems* (Plenum, New York 1981)

13.6 H. Takanashi, M. Kunita: Photodetectors, J. Inst. of Electr. Commun. Engs. Jpn. **63,** 1178–1182 (1980)

13.7 R. H. Kingston: *Detection of Optical and Infrared Radiation,* Springer Ser. Opt. Sci., Vol. 10 (Springer, Berlin, Heidelberg 1978)

13.8 H. Kressel (ed.): *Semiconductor Devices for Optical Communication,* 2nd ed., Topics Appl. Phys., Vol. 39 (Springer, Berlin, Heidelberg 1982)

13.9 J. Hamasaki, Y. Fuji: Photoelectric detectors, J. Inst. Electr. Commun. Engs. Jpn. **56,** 524–529 (1973)

13.10 S. D. Personick: Receiver design for optical fiber systems, Proc. IEEE **65,** 1670–1678 (1977)

13.11 J. Gowar: *Optical Communication Systems* (Prentice-Hall, Englewood Cliffs, NJ 1984)

13.12 R. J. Keyes (ed.): *Optical and Infrared Detectors,* 2nd ed., Topics Appl. Phys., Vol. 19 (Springer, Berlin, Heidelberg 1980)

13.13 Y. Suematsu: Electrooptics sources, J. Inst. Electr. Commun. Engs. Jpn. **56,** 535–540 (1973)

13.14 W. Koechner: *Solid-State Laser Engineering,* Springer Ser. in Opt. Sci., Vol. 1 (Springer, Berlin, Heidelberg 1976)

13.15 K. Seeger: *Semiconductor Physics,* 2nd ed., Springer Ser. Solid-State Sci., Vol. 40 (Springer, Berlin, Heidelberg 1982)

13.16 C. H. Gooch: *Injection Electroluminescent Devices* (Wiley, London 1973)

13.17 *Optical Fiber Communications Link Design* (ITT, Technical Note R-1, 1978)

13.18 U. Haller, W. Herold, H. Ohnsorge: Problems arising in the development of optical communication systems, Appl. Phys. **17,** 115 (1978)

13.19 J. Millman, C. C. Halkias: *Integrated Electronics: Analog and Digital Circuits and Systems* (McGraw-Hill, New York 1972)

**Additional Reading**

Bendow, B., Mitra, S. S.: *Fiber Optics* (Plenum, New York 1979)

Giallorenzi, T.: Optical communications research and technology: fiber optics, Proc. IEEE **66,** 745–780 (1978)

Howes, M. J., Morgan, D. V.: *Optical Fiber Communications* (Wiley, New York 1980)

Iizuka, K.: *Fundamentals of Engineering Optics* (Ohm-Sha, Tokyo 1980)

Midwinter, J. E.: *Optical Fibers for Transmission* (Wiley, New York 1979)

Nonaka, S.: Fiber optic components, J. Inst. Electr. Commun. Engs. Jpn. **63,** 1183–1191 (1980)

Shimada, S.: *Fiber Optical Communication Techniques* (Ohm-Sha, Tokyo 1980)

Sodha, M. S., Ghatak, A. K.: *Inhomogeneous Optical Waveguides* (Plenum, New York 1977)

Tanaka, M., Izawa, T.: Fundamental of optical fiber communications (III), J. Inst. Electr. Commun. Engs. Jpn. **60,** 783–794 (1977)

Young, M.: *Optics and Lasers,* 2nd ed., Springer Ser. Opt. Sci., Vol. 5 (Springer, Berlin, Heidelberg 1984)

## Chapter 14

14.1  I. P. Kaminow, E. H. Turner: Linear Electrooptical Materials, in *Handbook of Lasers with Selected Data on Optical Technology,* ed. R. J. Pressley (Chemical Rubber Company, Cleveland, OH 1971) pp. 447–459

14.2  T. Okada (ed.): *Crystal Optics* (Morikita Publishing Company, Tokyo 1975)

14.3  J. Koyama, H. Nishihara: Optoelectronics (Corona-Sha, Tokyo 1978) pp. 298–300

14.4  C. L. M. Ireland: A-20 ps resolution crystal streak camera, Opt. Commun. **30,** 99–103 (1979)

14.5  T. Sueta, M. Izutsu: Optical guide modulators, in: Editorial committee of the special project research on optical guided-wave electronics, Tokyo (1981) pp. 397–418

14.6  B. D. Cullity: *Elements of X-ray Diffraction* (Addison-Wesley, Reading MA 1956)

14.7  B. K. Vainshtein: *Modern Crystallography* I, Springer Ser. Solid-State Sci., Vol. 15 (Springer, Berlin, Heidelberg 1981)

14.8  A. A. Chernov (ed.): *Modern Crystallography* III, Springer Ser. Solid-State Sci. Vol. 36 (Springer, Berlin, Heidelberg 1984)

14.9  J. F. Nye: *Physical Properties of Crystals* (Oxford at the Clarendon Press, Oxford 1979)

14.10 K. Kudo: *Fundamentals of Optical Properties of Matter* (Ohm-Sha, Tokyo 1977)

14.11 L. V. Keldysh: Behavior of non-metallic crystals in strong electric fields, Sov. Phys. JETP **6 (33),** 763–770 (1958)

14.12 J. C. Campbell, J. C. Dewinter, M. A. Pollack, R. E. Nahory: Buried heterojunction electroabsorption modulator, Appl. Phys. Lett. **32,** 471–473 (1978)

## Chapter 15

15.1  R. V. Schmidt, R. C. Alferness: Directional coupler switches, modulators, and filters using alternating $\Delta\beta$ Techniques, IEEE Trans. CAS-**26,** 1099–1108 (1979)

15.2  W. H. Louisell: *Coupled Mode and Parametric Electronics* (Wiley, New York 1960)

15.3  H. Kogelnik, R. V. Schmidt: Switched directional couplers with alternating $\Delta\beta$, IEEE J. QE-**12,** 396–401 (1976)

15.4  R. C. Alferness: Guided wave devices for optical communication, IEEE J. QE-**17,** 946–959 (1981) and T. G. Giallorenzi, R. A. Steinberg: US Patent. 4291,939

15.5  R. C. Alferness, R. V. Schmidt: Tunable optical waveguide directional coupler filter, Appl. Phys. Lett. **33,** 161–163 (1978)

15.6   J. Koyama, H. Nishihara, M. Haruna, T. Seihara: Optical guide switches and their applications, in: "Editorial committee of the special project research on optical guided-wave electronics", Tokyo (1981) pp. 419–431

15.7   T. R. Ranganath, S. Wang: Ti-diffused LiNbO₃ branched-waveguide modulators; performance and design, IEEE J. QE-**13**, 290–295 (1977)

15.8   F. J. Leonberger: High-speed operation of LiNbO₃ electrooptic interferometric waveguide modulators, Opt. Lett. **5**, 312–314 (1980)

15.9   W. D. Bomberger, T. Findakly, B. Chen: Integrated optical logic devices, Proc. of SPIE, Vol. 321, Integrated Optics II, 38–46 (1982)

15.10  A. A. Oliner (ed.): *Acoustic Surface Waves,* Topics Appl. Phys., Vol. 24 (Springer, Berlin, Heidelberg 1978)

15.11  I. M. Mason, E. Papadofrangakis, J. Chambers: Acoustic-surface-wave disc delay lines, Electr. Lett. **10**, 63–65 (1974)

15.12  A. Schnapper, M. Papuchon, C. Puech: Remotely controlled integrated directional coupler switch, IEEE J. QE-**17**, 332–335 (1981)

15.13  R. S. Cross, R. V. Schmidt, R. L. Thornton, P. W. Smith: Optically controlled two channel integrated-optical switch, IEEE, J. QE-**14**, 577–580 (1978)

15.14  P. W. Smith, W. J. Tomlinson: Bistable optical devices promise subpicosecond switching, IEEE Spectrum, **18**, 26–33 (1981)

15.15  D. Vincent and G. Otis: Hybrid bistable devices at 10.6 µm, IEEE J. QE-**17**, 318–320 (1981)

15.16  H. J. Reich, J. G. Skalnik, H. L. Krauss: *Theory and Applications of Active Devices* (Van Nostrand, Princeton, N.J. 1966)

15.17  S. Tarucha, M. Minakata, J. Noda: Complementary optical bistable switching and triode operation using LiNbO₃ directional coupler, IEEE J. QE-**17**, 321–324 (1981)

15.18  M. Okada, K. Takizawa: Instability of an electrooptic bistable device with a delayed feedback, IEEE J. QE-**17**, 2135–2140 (1981)

15.19  H. Ito, Y. Ogawa, H. Inaba: Analyses and experiments on integrated optical multivibrators using electrooptically controlled bistable optical devices, IEEE J. QE-**17,** 325–331 (1981)

15.20  J. E. Bjorkholm, P. W. Smith, W. J. Tomlinson, A. E. Kaplan: Optical bistability based on self-focusing, Opt. Lett. **6**, 345–347 (1981)

15.21  A. E. Kaplan: Optical bistability that is due to mutual self-action of counter-propagating beams of light, Opt. Lett. **6**, 360–362 (1981)

15.22  S. A. Akhmanov, R. V. Khokhlov, A. P. Sukhorukov: Self-focusing, self-defocusing and self-modulation of laser beams, in *Laser Handbook,* ed. by F. T. Arecchi and E. O. Schulz-DuBois (North-Holland, Amsterdam 1982) pp. 1151–1228

15.23  J. E. Bjorkholm, P. W. Smith, W. J. Tomlinson: Optical bistability based on self-focusing; an approximate analysis, IEEE J. QE-**18**, 2016–2022 (1982)

15.24  S. Tanaka: Planar optical components, in: "Editorial committee of the special project research on optical guided-wave electronics", Tokyo (1981) pp. 449–462

15.25  B. Chen, O. G. Ramer: Diffraction-limited geodesic lens for integrated optic circuit, IEEE J. QE-**15**, 853–860 (1979)

15.26  W. C. Chang, P. R. Ashley: Fresnel lenses in optical waveguides, IEEE J. QE-**16,** 744–753 (1980)

15.27  T. R. Joseph, T. R. Ranganath, J. Y. Lee, M. Pedioff: Performance of the integrated optic spectrum analyzer, SPIE **321**, 134–140 (1982)

15.28  D. F. Barbe (ed.): *Charge-Coupled Devices,* Topics Appl. Phys., Vol. 38 (Springer, Berlin, Heidelberg 1980)

15.29  J. Koyama, H. Nishihara: *Optoelectronics* (Corona-Sha, Tokyo 1978)

15.30  K. Ploog, K. Graf: *Molecular Beam Epitaxy of III–V Compounds* (Springer, Berlin, Heidelberg 1984)

15.31  R. Behrisch (ed.): *Sputtering by Particle Bombardment* I and II, Topics Appl. Phys., Vols. 47 and 52 (Springer, Berlin, Heidelberg 1981 and 1983)

15.32 H. Ryssel, H. Glawischnig (eds.): *Ion Implantation Techniques,* Springer Ser. Electrophys., Vol. 10 (Springer, Berlin, Heidelberg 1982)

15.33 I. P. Kaminow, J. R. Carruthers: Optical waveguiding layers in LiNbO₃ and LiTaO₃, Appl. Phys. Lett. **22**, 326–328 (1973)

15.34 T. R. Ranganath, S. Wang: Suppression of Li₂O out-diffusion from Ti-diffused LiNbO₃ optical waveguides, Appl. Phys. Lett. **30**, 376–379 (1977)

15.35 R. V. Schmidt, I. P. Kaminow: Metal-diffused optical waveguides in LiNbO₃, Appl. Phys. Lett. **25**, 458–460 (1974)

15.36 G. Stewart, C. A. Millar, P. J. R. Laybourn, C. D. W. Wilkinson, R. M. DeLa Rue: Planar optical waveguides formed by silver-ion migration in glass, IEEE, J. QE-**13**, 192–200 (1977)

15.37 A. G. Hallam, I. Bennion, W. J. Stewart: Photochromic stripe waveguides for integrated optics, 1st Europ. Conf. on Integrated Optics, London, 26–28 (1981)

15.38 T. Sueta, M. Izutsu: Optical guide modulators, in: "Editorial committee of the special project research on optical guided-wave electronics", Tokyo (1981) pp. 397–418

15.39 M. K. Barnoski: *Introduction to Integrated Optics* (Plenum, New York 1973) pp. 127–165

15.40 T. Tamir (ed.): *Integrated Optics* 2nd ed., Topics Appl. Phys., Vol. 7 (Springer, Berlin, Heidelberg 1979) pp. 199–241

15.41 K. E. Wilson, C. T. Mueller, E. Garmire: Integrated optical circuit fabricated with laser-written masks, SPIE **327**, 29–36 (1982)

15.42 D. F. Barbe (ed.): *Very Large Scale Integration VLSI,* 2nd ed., Springer Ser. Electrophys., Vol. 5 (Springer, Berlin, Heidelberg 1982)

15.43 H. G. Unger: *Planar Optical Waveguides and Fibers* (Oxford, London 1977) and *Fibers and Integrated Optics,* ed. by D. B. Ostrowsky (Plenum, New York 1978) pp. 183–222

15.44 R. Hunsperger: *Integrated Optics Theory and Technology,* Springer Ser. Opt. Sci. Vol. 33, (Springer, Berlin, Heidelberg 1982)

15.45 A. B. Sharma, S. J. Halme, M. M. Bustusov: *Optical Fiber Systems and Their Components,* Springer Ser. in Opt. Sci., Vol. 24 (Springer, Berlin, Heidelberg 1981)

15.46 E. A. J. Marcatili: Dielectric rectangular waveguide and dielectric coupler for integrated optics, Bell Syst. Tech. J. **48**, 2071–2102 (1969)

# Subject Index